World Water Resources

Volume 5

Series Editor
V. P. Singh, Department of Biological and Agricultural Engineering
& Zachry Department of Civil Engineering, Texas A&M University,
College Station, TX, USA

This series aims to publish books, monographs and contributed volumes on water resources of the world, with particular focus per volume on water resources of a particular country or region. With the freshwater supplies becoming an increasingly important and scarce commodity, it is important to have under one cover up to date literature published on water resources and their management, e.g. lessons learnt or details from one river basin may be quite useful for other basins. Also, it is important that national and international river basins are managed, keeping each country's interest and environment in mind. The need for dialog is being heightened by climate change and global warming. It is hoped that the Series will make a contribution to this dialog. The volumes in the series ideally would follow a "Three Part" approach as outlined below:

In the chapters in the first Part *Sources of Freshwater* would be covered, like water resources of river basins; water resources of lake basins, including surface water and under river flow; groundwater; desalination and snow cover/ice caps. In the second Part the chapters would include topics like: *Water Use and Consumption*, e.g. irrigation, industrial, domestic, recreational etc. In the third Part in different chapters more miscellaneous items can be covered like impacts of anthropogenic effects on water resources; impact of global warning and climate change on water resources; river basin management; river compacts and treaties; lake basin management; national development and water resources management; peace and water resources; economics of water resources development; water resources and civilization; politics and water resources; water-energy-food nexus; water security and sustainability; large water resources projects; ancient water works; and challenges for the future. Authored and edited volumes are welcomed to the series. Editor or co-editors would solicit colleagues to write chapters that make up the edited book. For an edited book, it is anticipated that there would be about 12–15 chapters in a book of about 300 pages. Books in the Series could also be authored by one person or several co-authors without inviting others to prepare separate chapters. The volumes in the Series would tend to follow the "Three Part" approach as outlined above. Topics that are of current interest can be added as well.

Readership
Readers would be university researchers, governmental agencies, NGOs, research institutes, and industry. It is also envisaged that conservation groups and those interested in water resources management would find some of the books of great interest. Comments or suggestions for future volumes are welcomed.

Series Editor:
V. P. Singh, Department of Biological and Agricultural Engineering & Zachry Department of Civil Engineering, Texas A&M University, TX, USA.
Email: vsingh@tamu.edu

More information about this series at http://www.springer.com/series/15410

Giuseppe Rossi • Marcello Benedini
Editors

Water Resources of Italy

Protection, Use and Control

 Springer

Editors
Giuseppe Rossi
Department of Civil Engineering and
Architecture
University of Catania
Catania, Italy

Marcello Benedini
Italian Water Research Institute (Retired)
Rome, Italy

ISSN 2509-7385   ISSN 2509-7393   (electronic)
World Water Resources
ISBN 978-3-030-36459-5   ISBN 978-3-030-36460-1   (eBook)
https://doi.org/10.1007/978-3-030-36460-1

This Springer imprint is published by the registered company Springer Nature Switzerland AG.
The registered company address is: Gewerbestrasse 11, 6330 Cham, Switzerland

The past is our heritage
The present our responsibility
The future our challenge

Preface

Great Italian scientists who contributed to the development of hydraulics, hydrology, and water works, starting from Leonardo da Vinci (1452–1519), are frequently cited in water-related scientific texts. International handbooks on water structures often underscore the inestimable value of ancient waterworks, such as aqueducts and sewer systems, built in Rome and throughout the Roman Empire more than 2000 years ago.

Nevertheless, a recent and comprehensive overview of water resources in Italy is still lacking. In addition, papers on theoretical and experimental advances in specific water-related disciplines, published in international journals by Italian researchers, hardly provide an overview of past developments, improve current problems and perspectives on water management for Italy as a whole.

This book attempts to fulfil these needs. It is intended to serve as an introduction to the complex issues involved in water governance and management, and to offer a comprehensive view of the transformations that Italy has undergone since its unification (1861) and in the last century, specifically in water-related fields. It begins with the historical roots of the development of water structures and water services and ultimately addresses a number of changes expected for the future. We hope that it will help readers not only to deepen their understanding of specific topics concerning Italian water problems, but will also to stimulate further reflection on the pivotal role that water resources play for a sustainable and equitable future, also in other countries.

As editors, we have designed the layout of the book with the aid of several colleagues, who shared their specific expertise in chapters on new or very specific topics, together with their valued suggestions and comments.

Our professional viewpoints as editors are complementary, since one of us draws his experience from university teaching and research in civil engineering, especially hydraulics, hydrology, and water management, while the other brings with him the lessons learned at a water-oriented institute of the National Research Council, with a prevalent focus on multidisciplinary work.

Given our respective ages, we might be more likely to focus on the past than on the future, since we have been spectators of the events in the fields of water resources and soil protection in the second half of the twentieth century and the first two decades of this one. However, we hope to have avoided any nostalgia regarding the technical solutions and the policy framework for water problems in Italy in this period. Moreover, we have sought to identify the technological innovations and social trends expected for the future of water infrastructures and institutions, as well as for the transformations necessitated by the challenges of improving water governance and management and reducing environmental impact in the next decades.

We have attempted to report on the motivations behind legal reforms as objectively as possible and have critically evaluated them, identifying some weaknesses in the process. The analysis has focused on identifying discrepancies between the declared goals and the real results in terms of the implementation of the proposed innovations. We have especially investigated the shortcomings of new national regulations on water resources planning, the delayed reorganization of the municipal water services, and the structure of hydrometeorological monitoring.

With regard to water planning, the book emphasizes the introduction of the unitary approach to river basin, combining the issues of protection and utilization of water resources and soil conservation, an approach largely overlooked in subsequent years. The dramatic disasters that struck many regions in Italy attracted public and political attention to the reduction of flood risk, which became the primary concern in terms of updating regulations and funding measures, especially for emergencies.

With regard to municipal water services, particular attention is devoted to the reforms introduced by Law 36/1994, which unified supply, sewage, and wastewater treatment into the integrated water service at large scale. It also underlines the importance of defining the roles of governing and controlling, which are fulfilled by a public authority, and the executive role of public and private companies.

Unfortunately, the modernization process, which was intended to improve the organizational structure and achieve full cost recovery, encountered a host of difficulties. The chief among them was the ideological dispute between the supporters of private and those of public management. Eventually, another law did away with the multi-municipality authorities, making the regions responsible for establishing new rules for the urban water sector. In turn, this situation compelled the government to strengthen its commitment to surveillance and control of integrated urban services through the Authority for Regulating Energy, Networks and Environment, in order to better ensure customer's satisfaction.

With regard to hydrometeorological monitoring, the book evaluates the establishment of multifunctional centres at the regional level, especially in terms of civil protection for managing the impacts of hydrological extremes. It argues that while this reform has contributed to the implementation of effective emergency measures for coping with flood and drought disasters, it has also increased the fragmentation of hydrometeorological monitoring at the national level and, unexpectedly, reduced the role of structural preventive measures.

The analysis of the impacts of European Union directives on the national water policy focuses on the advantages and disadvantages of the current situation. As is commonly known, Italy prides itself on having anticipated several basic principles that the EU subsequently adopted, e.g., using the river basin as the territorial unit for water management and flood risk reduction. However, some specific features of the EU directives fail to address Italy's needs. Examples include the major emphasis that the Water Framework Directive 2000/60/EC places on water quality objectives instead of water quantity issues, which are of primary interest for Italy and all the Mediterranean European countries. Similarly, the Floods Directive 2007/60/EC only addresses flood risk while neglecting landslide risk, which is of central importance to Italian legislation. Nevertheless, despite these gaps in the European legislative framework, any objective assessment of the EU's role in water policy should recognize that the European regulations have significantly contributed to the adoption of appropriate environmental quality standards in Italian legislation.

Looking toward the future, the book rests on the assumption that the water management of the next several decades should seek to face the challenge of achieving sustainable development and human solidarity, as the United Nations has worked to promote. It is worth noting that the initiative of the "Agenda 2030 for Sustainable Development," signed by 193 countries,[1] was launched in the same year in which the encyclical *Laudato si* by Pope Francis[2] was published and the Paris Agreement on climate change[3] was signed. These significant events increased the chances of converging political, religious, and scientific objectives toward the salvation of our planet and the improvement of human life.

Unfortunately, some recent political trends in many parts of the world show that several difficulties are hindering the process of meaningfully transforming the current model of production, consumption, and societal organization, in order to avoid the collapse of the socioeconomic system and the decay of our planet while also promoting the fair development of all peoples.

In light of these developments, the book reviews the most important aspects of Italian water resources, in the hope that Italy's experiences can offer readers valuable points of departure for considering the contribution of water management to the complex challenge of fostering sustainable development and a fair society.

The content of the book is divided into four parts, the first of which includes an introductory overview of the physical and institutional conditions in Italy (Chap. 1), followed by a synthesis of the main water infrastructures and of the criteria driving water governance according to the changing paradigms in Italian society and in the European Union (Chap. 2). In turn, Chaps. 3 and 4 review the evolution of water legislation and institutional aspects.

Part II shifts the focus to conventional water resources, including surface and groundwater (Chap. 5), an assessment of unconventional resources such as treated wastewater and desalinated water (Chap. 6), and an evaluation of the water requirements in all sectors using natural or reclaimed water resources (Chap. 7).

---

[1] U.N. Transforming our world: the 2030 Agenda for Sustainable Development, U.N., 2015 http://www.un.org/ga/search/view_doc.asp?symbol=A/RES/70/1&Lang=E.

[2] Francis, Encyclical Letter "Laudato si," May 24, 2015 http://w2.vatican.va/content/francesco/en/encyclicals/documents/papa-francesco_20150524_enciclica-laudato-si.html.

[3] "Paris Agreement on climate change," 12 December 2015.

Part III describes some of the most important water problems that concern the interested people and government at the national, regional, and local level, e.g. municipal water services (Chap. 8), water pollution control (Chap. 9), and ecological flow (Chap. 10).

Part IV addresses the main challenges that the country has to face in order to mitigate hydrological extremes, such as floods (Chap. 11) and droughts (Chap. 12), as well as the best measures regarding the water-food nexus (Chap. 13) and the impacts of climate change (Chap. 14).

The book's final chapter analyzes the major strengths and weaknesses of the Italian water system and discusses essential aspects of an agenda for the future, including technical, management, and water policy priorities.

Giuseppe Rossi and Marcello Benedini

Acknowledgments

As editors, we wish to thank the contributing authors for their time and effort in writing their chapters and for having patiently borne our comments. We are also indebted to a large set of institutions, which kindly provided us with the necessary data.

We gratefully acknowledge the generous support of Alberto Campisano, Antonio Cancelliere, and Bartolomeo Rejtano of the University of Catania, Michele Mossa of the Politecnico of Bari, Antonio Massarutto of the University of Udine, Ruggiero Jappelli of the University of Rome Tor Vergata, Eugenio Lazzari of the University of Cagliari, Gianfranco Boari of the University of Basilicata, and Gaspare Viviani of the University of Palermo.

Among the many colleagues and friends who have given us precious information and valuable suggestions, we like to mention Cinzia Bianchi and Paola Salvati of the Research Institute for Hydrogeological Protection, and Andrea Buffagni, Stefania Erba, Antonio Lopez, Costantino Masciopinto, Elisabetta Preziosi, Luigi Viganò, and Michele Vurro of the Water Research Institute. Special thanks for Giuseppina Monacelli and Alessio Lotti of the Institute for Environmental Protection and Research, Raffaella Zucaro of the Council for Agricultural Research and Economics, Renato Drusiani of Utilitalia, as well as Andrea Duro of the National Department of Civil Protection; Vincenzo Nicolosi of the Ministry of Infrastructure and Transport; Silvano Pecora of the Regional Agency for Prevention, Environment and Energy of Emilia-Romagna; Giuseppe Gisotti of the Environmental Geological Association; Giovanni Ruggeri of the International Commission on Large Dams; and Salvatore Alecci and Guido Zanovello of the Italian Hydrotechnical Association.

Thanks also to Domenica Santonocito and Paola Nanni for the technical assistance in preparing the book.

Contents

Part IV   Challenges

Part V   Conclusions

Editors and Contributors

About the Editors

Marcello Benedini graduated in civil engineering at the University of Padua, Italy, in 1957. As an assistant professor to the chair of hydraulics, he was also lecturer of hydraulic measures and water resources management at the International Postgraduate School of Hydrology. After a training period at the University of Liverpool and at the British National Engineering Laboratory of East Kilbride in the United Kingdom, in 1969, he joined the Water Research Institute of the National Research Council in Rome, where he was appointed director of the water management sector. His role was particularly in research on the advanced methods for the integrated use of water and for environment protection, in scientific support to national authorities responsible of water resources management, in collaboration with worldwide scientific institutions. Such position gave him the opportunity to participate in the proposition of fundamental directives for important schemes of Italian water resources, including those of River Po. One of his main commitments was the final report of the National Water Conference, a comprehensive review of the Italian water resources problems. He was member of technical and scientific committees of international institutions devoted to water problems, such as the Organisation for Economic Co-operation and Development and the International Association for Hydro-Environment Engineering and Research. He has been co-founder of the European Water Resources Association and the Italian member at the Scientific Council of the International Institute for Applied Systems Analysis in Vienna. After his formal retirement in 1999, he still had opportunities to deal with important water problems as the president of the Italian Hydrotechnical Association. His personal experience is confirmed with the participation in several international scientific conferences, frequently as invited speaker and proceeding editor, as well as with numerous papers and technical reports, relevant to actual water management problems.

Giuseppe Rossi graduated in civil engineering in 1965, is emeritus professor of hydrology and hydraulic structures at the University of Catania (Italy). He was visiting scientist at the Colorado State University (USA) and at the Universitat Politècnica de València (Spain). He was vice president of the International Water Resources Association (2010–2012) and of the Italian Hydrotechnical Association (2012–2015) and chair of the Award Committee of IWRA (2013–2015). He has a long experience in graduate and postgraduate teaching. He has organized several international conferences on hydrologic extremes (floods and droughts) and on water resources systems. The main fields of his scientific research include stochastic hydrology, water system management, and drought analysis and management. He directed several research projects, funded by the Italian Ministry of the University and by the National Research Council, and coordinated many Euro-Mediterranean projects on drought and on water management funded by the European Commission, in cooperation with several research institutions of South Europe, North Africa, and Near East countries. Author of more than 200 scientific papers, he is also author, editor, and coeditor of ten books on hydrology and water resource such as *Operation of Complex Water Systems: Operation, Planning and Analysis of Already Developed Water Systems* (Martinus Nijhoff Publishers, Boston, 1983) with E. Guggino and D. Hendricks; *Coping with Floods* (Kluwer, Dordrecht, 1994) with V. Yevjevich and N. Harmancioglu; *Tools for Drought Mitigation in Mediterranean Regions* (Kluwer, Dordrecht, 2003) with A. Cancelliere, L. Pereira, and T. Oweis; *Drought Management and Planning for Water Resources* (Taylor & Francis, New York, 2006) with J. Andreu, F. Vagliasindi, and A. Vela; and *Methods and Tools for Drought Analysis and Management* (Springer, Dordrecht, 2007) with T. Vega and B. Bonaccorso. He serves on the editorial boards of the *Water Resources Management* (Springer) and of *L'Acqua* (Italian Hydrotechnical Association). He is member of AII, IAHR, IWRA, EWRA, ICID, and IAH.

Contributors

Salvatore Alecci Eastern Sicily Section, Italian Hydrotechnical Association, Catania, Italy

Marcello Benedini Italian Water Research Institute (Retired), Rome, Italy

Giuseppe Luigi Cirelli Department of Agriculture, Food and Environment (Di3A), University of Catania, Catania, Italy

Michele Di Natale Department of Engineering, University of Campania "L. Vanvitelli", Aversa (CE), Italy

Francesco Laio Department of Environment, Land and Infrastructure Engineering, Politecnico di Torino, Torino, Italy

David J. Peres Department of Civil Engineering and Architecture, University of Catania, Catania, Italy

Bartolomeo Rejtano Department of Civil Engineering and Architecture, University of Catania, Catania, Italy

Giuseppe Rossi Department of Civil Engineering and Architecture, University of Catania, Catania, Italy

Stefania Tamea Department of Environment, Land and Infrastructure Engineering, Politecnico di Torino, Torino, Italy

Marta Tuninetti Department of Environment, Land and Infrastructure Engineering, Politecnico di Torino, Torino, Italy

Part I
Evolution of Water Management in Italy

Chapter 1
Italy's Outline

Marcello Benedini

Abstract After the main physical characteristics of the Italian territory and a synthetic description of its geological features, the climatic aspects are recalled, to which the natural consistency of surface and groundwater is connected. Due to its geographic position, Italy denotes great meteorological variability from one region to the other, which affects the availability of natural water resources. In the northern zones close to the Alpine Chain, water is generally more abundant than in the southern parts and in the islands surrounded by the Mediterranean Sea. Such variability affects also the population distribution in the country and the relevant economic aspects. Some synthetic figures about rivers, lakes and aquifers are presented, and the Italian situation is compared with that of other Mediterranean and European countries.

1.1 Geographic Characteristics

The Republic of Italy stretches in the Mediterranean Sea across more than 10 degrees of latitude, between the Alps and the Pelagie Islands facing the African shore. The Alpine Chain, which reaches the maximum level of 4,810.90 m above sea at the Mont Blanc, is the natural border that divides the Italian territory from that of France, Switzerland, Austria and Slovenia, characterizing the European location of the country. Such territory covers an area of 302,073 km² almost completely inserted in the Mediterranean basin with the exception of some small alpine valleys totalling 565 km² that belong to the Danube catchment with the mouth in the Black Sea (Batini et al. 2000).

From north to south, the geographic structure of Italy consists of a large continental area surrounded by the Alps and of a long peninsula leaning into the sea, with some islands, the major of which are Sicily (25,707 km²) and Sardinia (24,090 km²). The total length of the coasts, including the islands, is 7456 km. In the Mediterranean, the peninsula gives rise to some well-identified basins with different maritime and

M. Benedini (✉)
Italian Water Research Institute (Retired), Rome, Italy
e-mail: benedini.m@iol.it

© Springer Nature Switzerland AG 2020
G. Rossi, M. Benedini (eds.), *Water Resources of Italy*, World Water Resources 5,
https://doi.org/10.1007/978-3-030-36460-1_1

Fig. 1.1 Physical map of Italy

coastal characteristics, namely the Adriatic Sea to the east, the Tyrrhenian Sea to the west and the Ionian Sea to the south, as shown in Fig. 1.1.

Mountains above 600 m above sea level are located in the Alps at the border of the country and in the Apennines Chain along the peninsula, covering 35.2% of the national territory. Of the remaining parts, 23.2% is covered by hills with an elevation less than 600 m above sea and 41.6% by the plains, located principally in the north. While the northern landscape is crowned by the Alps, that of the long peninsula is developed around the Apennines, surrounded with hills of little elevation. In the central parts and in Sicily, large zones denote a volcanic origin, which can be confirmed

Regions

A Piedmont
B Aosta Valley
C Lombardy
D Trentino Alto Adige
E Veneto
F Friuli Venezia Giulia
G Liguria
H Emilia Romagna
I Tuscany
J Umbria
K Marche
L Lazio
M Abruzzo
N Molise
O Campania
P Apulia
Q Basilicata
R Calabria
S Sicily
T Sardinia

Fig. 1.2 Political and administrative layout of the Republic of Italy and the towns centre of regional administration

also by some lakes that now fill the exhausted crater of volcanos that were active million years ago. Actually, some eruptive phenomena are still present in the form of hot water springs, while the big volcanos Etna, Vesuvius and Stromboli characterize the Italian landscape and give rise to specific problems due to their activity.

The coastline in the peninsula and islands is mostly made up by rocky formations, shaped by an enduring erosion activity of sea; extended sandy beaches surround the mouth of the rivers discharging into the sea.

Concerning the political and administrative aspects, Italy is a state with Rome as its national capital, where the presidency of republic, the parliament bodies and the main national institutions are hosted. The national administration is split into 21 regions, having particular local jurisdiction. Figure 1.2 shows the regional consistency with the location of the regional administration centres.

1.2 Geological Characteristics

The geology of the Italian territory is the effect of the prehistoric events that characterized the development of the Earth during a sequence of million years, and such effect can be recognized through the different aspects of the land actually visible. A

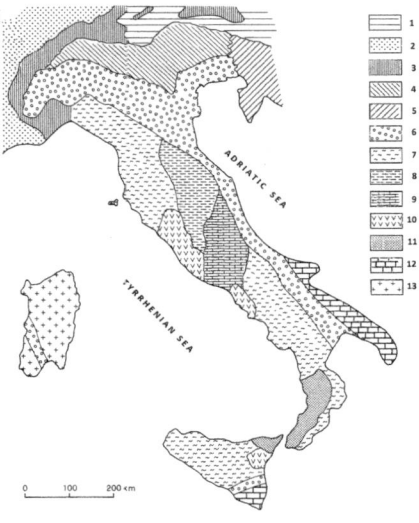

1. AUSTROALPINE UNITS (ALPS), metamorphic and sedimentary rocks
2. HELVETIC UNITS (ALPS), crystalline (metamorphic and igneous) and sedimentary rocks
3. PENNINIC UNITS (ALPS), crystalline rocks
4. SOUTH ALPINE UNITS, carbonatic , sedimentary and volcanic rocks
5. DINARIDE UNITS (ALPS), carbonatic rocks
6. PLIOCENE and PLEISTOCENE terrigenous sediments (clay, sand and gravel)
7. Terrigenous deposits (clay) and Flysch (Northern and Southern Apennines)
8. Marls formations, with limestone and clay (Umbria, Marche)
9. Limestone (Latium, Abruzzo)
10. Volcanites (volcanic complexes of Vulsini, Sabatini, Latium Vulcanoes, Roccamonfina, Vesuvius, Etna)
11. CALABRIDE-PELORITANE UNITS , crystalline and sedimentary rocks.
12. Limestone (Gargano, Murge, Salento, Iblei M.)
13. Granite (Sardinia)

Fig. 1.3 Principal geological characteristics of the Italian territory

synthetic view of the predominant geological features is in Fig 1.3, as proposed by G. Gisotti (personal communication 2018).

Massive limestone areas are in the Alpine zone and in the central parts of the Apennine Chain, beside some limited extensions of alluvial areas, surrounding the most important rivers. The actual territory configuration, with mounts and plains, as well as the coastline, has been shaped during successive geological periods.

The subsoil of some regions is affected by the evolution of deep faults, and, consequently, large part of the Italian territory is subject to the risk of frequent seismic phenomena. Also in recent years, unpredictable earthquakes have caused several conspicuous damages and casualties. Moreover, local instability of the soil, affected by the meteorological precipitation and by improper man-made interventions, still characterizes all the country territory. A recent nationwide investigation has identified numerous local potential landslides, while serious episodes of heavy rain can occur every year causing floods and inundation of extensive areas with consequent serious damage and casualties.

1.3 The Climate

The water problems of Italy are conditioned by the characteristics of its territory, on which all the surface and groundwater bodies are the effect of the local rainfall. Fundamental is therefore the identification of the climate aspects that characterize the rainfall.

Italy relies on a rich meteorological service, heritage of centuries of observations, with more than 200 thermopluviometric gauging stations spread all over its territory. Even though the collected measurement belongs to several institutions and there is still the need of a more harmonization as concerns instrumentations and an organic data proceeding, the available information allows the acquisition of acceptable climatic aspects, both at a general level of all the country and for detailed considerations regarding restricted areas.

Recent scientific investigations (Brunetti et al. 2006; Brunetti et al. 2014) have developed a complete picture of the Italian climate, focusing also some possible future outcomes. The following paragraphs describe some synthetic views, in order to outline the main aspects that characterize the water problems.

The average annual rainfall on the Italian territory is represented in Fig. 1.4, resulting from an analysis of several decades of observations (IRSA 1999). The figure puts into evidence how the precipitation varies from one region to the other.

The first reason of such variation is the latitude of the country. The northern zones are close to continental Europe, affected by the Atlantic winds that normally carry a lot of atmospheric water. The second reason is the particular orographic

Fig. 1.4 Average annual precipitation in the 1930–1990 period (Source: IRSA 1999)

configuration, as described in the preceding pages. The Alpine Chain acts as a barrier and enhances the condensation of the water accumulated in the atmosphere through the evaporation in the Mediterranean Sea. Consequently, these zones have generally high precipitation. Vice versa, the flat zones of south, which are directly exposed to the North Africa hot winds, have lower rainfall.

The highest precipitation, of the order of 2000 mm/year, is in the Alpine area, particularly in the eastern zones, where annual average values up to 2900 mm/year during the 1950–1980 period and more than 3500 mm in a particularly "wet" year have been recorded. High rainfall of more than 2000 mm/year in the same period frequently occurs in the eastern Apennine of Liguria and in some restricted valleys of Central Alps. Values between 1200 and 1800 are common in all the Apennines Chain. Restricted southern areas in the mountainous Campania and Basilicata denote a relatively increased rainfall. In the extreme areas of Calabria, the effect of elevation prevails on that of southern winds, with precipitation over 1000 mm/year. Besides the latitude, there is also a noticeable rainfall variability from the eastern Adriatic, which is frequently hit by the cold and dry winds originated in Northeast Europe, and the western Tyrrhenian coast that is more exposed to the Atlantic storms.

The lowest precipitation, with historical average between 400 and 600 mm/year in the 1950–1980 period, characterizes the plains of Apulia and southern Sicily. Exceptional low values are also frequent in narrow valleys at the foot of the Alps.

The precipitation varies noticeably during the year in the various locations of the country. In the northern regions, the annual distribution of the rainy days normally has a high peak in spring and a lower one in autumn. In the South and the largest islands, the highest rain falls in winter, while months with very low values are frequently recorded during the summer.

The above general considerations, based on the 1950–1980 period, reflect climate characteristics of the past decades that could not come true in a more recent time, which, in line with what is ascertained in all the world, shows a clear reduction of the average annual precipitation. With minor differences between the actual and the historic observations available in the scientific literature (Desiato et al. 2007; Desiato et al. 2011; Crespi et al. 2017), the overall aspects that characterize the time and space distribution are confirmed.

A recent investigation relevant to the last 10 years has been promoted (ISPRA 2015) at a regional aggregation. Figure 1.5 summarizes the results of such investigation, putting into clear evidence the scarcity of natural water in the southern regions. Exceptionally, during the last years, also the northern regions, normally considered "wet", have suffered from rainfall shortage. The scarcity is more frequent during the summer months.

The same investigations allow also some time characteristics to be evaluated, as in Fig. 1.6, in which the average annual precipitation on the entire country denotes a decreasing trend, with low values in the more recent years. In spite of the alternation of "wet" and "dry" years, the trend is in line with the progressive climatic change, as now appreciated especially in the Northern Hemisphere.

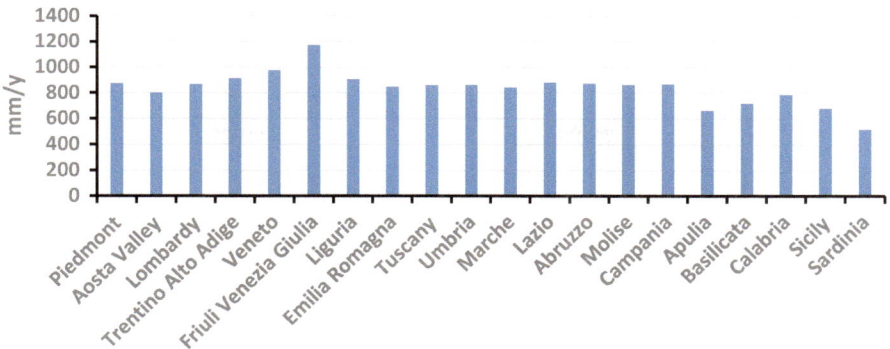

Fig. 1.5 Average annual precipitation in the Italian regions according to ISPRA (2015)

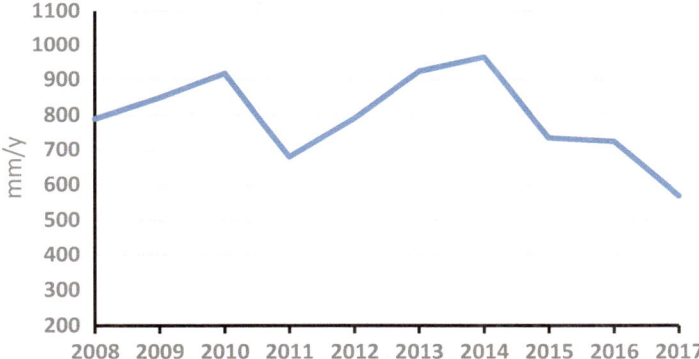

Fig. 1.6 Average precipitation in the last decade

The above considerations about the rainfall, regarding the average annual precipitation during a period of observation, do not take explicitly in due account some peculiar events occurring in short lapses of time, which are important for a more complete evaluation of the Italian climate. All the country has in fact to face intense precipitations with a short duration. Rainfall with intensity of up to 400 mm/h and more, particularly in autumn and winter (Batini and Benedini 2000), which are typical of all the Mediterranean countries, seem now more frequent and are worrisome for their unexpected effects.

A more complete description of the Italian climate requests also to consider the air temperature, an important factor tied with the natural precipitation. The space distribution of the annual average temperature is shown in Fig. 1.7, following investigations relevant to the 1961–2004 period (Toreti and Desiato 2008). A remarkable discrepancy exists between the northern continental area and the southern peninsular zones. A low annual average characterizes the Alpine mountains, with values

Fig. 1.7 Average annual temperature (Source: Toreti and Desiato 2008)

around 15 °C, while in the south, particularly along the long coastline and inside the largest islands, where values around 18 °C are recorded. The northern Valley of River Po has an almost uniform behaviour.

The already mentioned investigations relevant to the last 10 years, promoted by the Ministry of Environment, have examined also the average temperature at the regional aggregation. Figure 1.8 shows how both the maximum and minimum temperatures vary from north to south. With the exception of the Aosta Valley, where high mountains condition a restricted regional territory, the annual maximum temperature remains above 15 °C and reaches more than 20 °C in Sicily and Sardinia. Similarly, the annual minimum temperature remains below zero in the Aosta Valley and reaches more than 13 °C in Sicily.

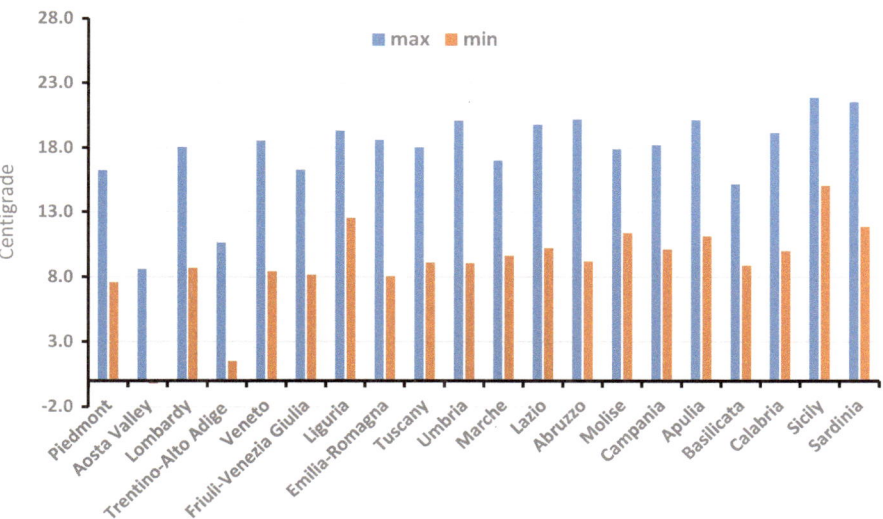

Fig. 1.8 Average annual temperature in Italian regions according to ISPRA (2015)

Table 1.1 Trend of the national average temperature during the last decades (centigrade)

Annual average temperature	1961–1990	1971–2000	1981–2010
Mean	11.8	12.2	12.6
Minimum	7.5	7.7	7.9
Maximum	16.0	16.5	16.8

Singular daily values of the minimum temperature less than -10 °C are common in the Alpine and Apennine Mountain during the winter months. High daily values above 30 °C of the maximum temperature are normal in all the regions during the summer months.

The climatic change experienced in the Northern Hemisphere (Toreti and Desiato 2009; Benedini and Giulianelli 2003), which affects the precipitation as seen in the preceding paragraphs, is confirmed also for the temperature. During the last years, daily values of more than 40 °C have been recorded. Following the investigations promoted by the Ministry of the Environment, the long series of the average annual temperature have been grouped in the three significant periods 1961–1990, 1971–2000 and 1981–2010, as shown in Table 1.1, from which the increasing trend is confirmed: in about 50 years, the maximum temperature has increased by 0.8 centigrade.

These discrepancies, existing among the various areas of Italy, show several aspects that are common to other Mediterranean countries (Portoghese and Vurro 2010; JM 2007).

1.4 The Natural Surface Water

The wide range of climatic conditions characterizes the consistency of natural resources. The northern and central regions have a relatively greater water availability, also better distributed in time and space. Surface runoff in northern and central areas at the foot of the high mountains and in the large plains is abundant.

Rivers originating from the Alps and the northern Apennines have permanent flow all over the year (Fig. 1.9).

There is a similar behaviour for the rivers in Central Italy, as shown in Fig 1.10. On the contrary, rivers in some southern areas and in the largest islands (Fig. 1.11) have frequently dry periods during the summer.

The characteristics of the principal rivers are listed in Table 1.2.

The main water body in Italy is River Po. Its catchment of 74,000 km² covers a large portion of the northern territory and includes all the largest lakes. With its numerous tributaries, it originates from the western and central parts of the Alpine Chain and from the northern Apennines, and discharges in the Adriatic Sea through a large delta. This river has a conspicuous flow that frequently, especially in early autumn, reaches high peaks of more than 10,000 m³/s, threatening the surrounding land. Several flooding events are recorded in history, hitting a territory that is of primary importance for the national economic life. River Po has traditionally called the attention of the national institutions responsible of water resources in Italy.

The second important water body is River Adige, also originating from Central Alps and reaching the Adriatic Sea after flowing through a large part of Veneto. It is also characterized by frequent high flow, which can cause dangerous floods particularly where even the bottom of the downstream reaches is above the level of the surrounding area. To mitigate this risk, a diversion tunnel can now discharge in Lake Garda part of the most dangerous flow.

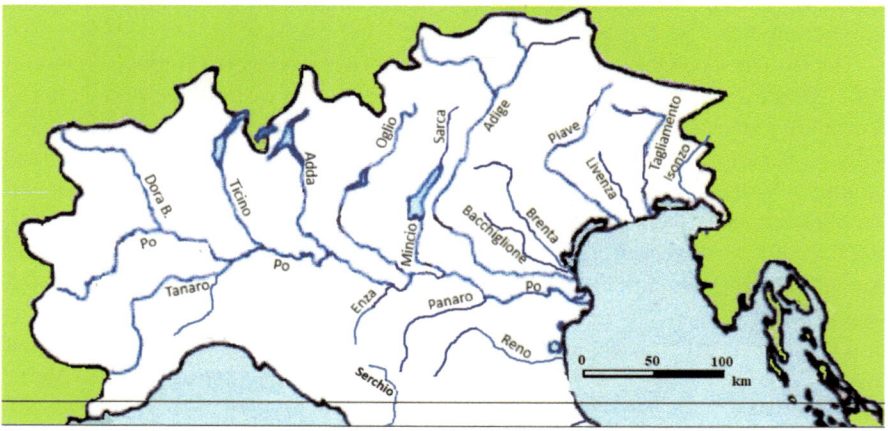

Fig. 1.9 Principal rivers in North Italy

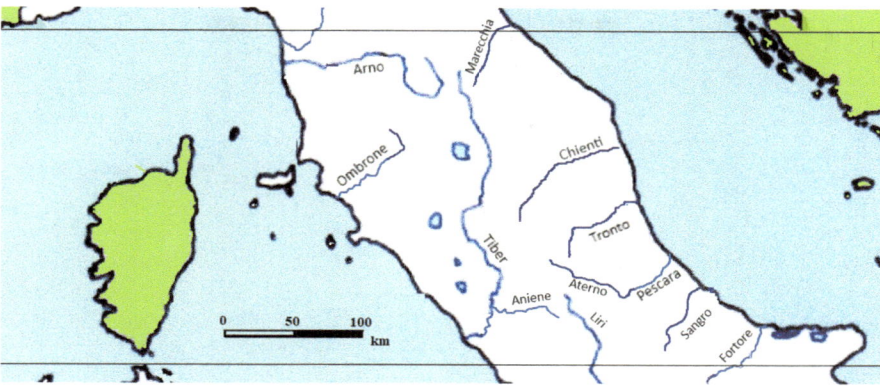

Fig. 1.10 Principal rivers in Central Italy

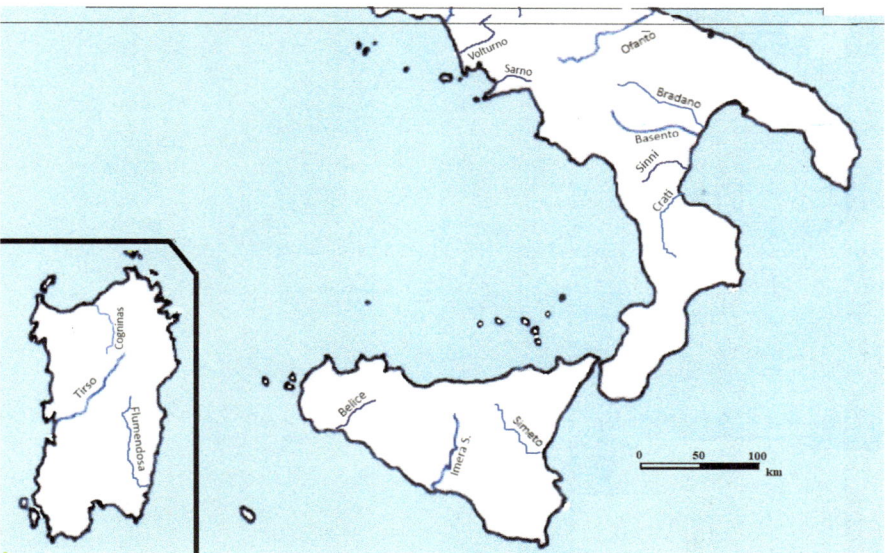

Fig. 1.11 Principal rivers in South Italy and the largest island

The third important watercourse is River Tiber, originating from Apennines and discharging in the Tyrrhenian Sea. Its catchment, ranking second after that of Po, covers large part of the peninsular territory. In the past centuries, it had several dangerous events, but now a great attention of the governmental institutions is expected to reduce the occurrence of floods. Its importance is especially tied to the presence of the urban agglomeration of Rome, sharing its main historical aspects.

River Arno, as the fourth important watercourse, also originating from Apennines and discharging in the Tyrrhenian Sea, has some records of dangerous floods, particularly in the area of Florence.

Table 1.2 Characteristics of the principal rivers of Italy

River	Region	Catchment area (km²)	Length (km)	Average flow (m³/s)	Receiver
Po	*North-west*	74,000	652	540	Adriatic Sea
Tiber	*Centre*	17,370	405	230	Tyrrhenian Sea
Adige	Trentino, Veneto	12,100	410	235	Adriatic Sea
Tànaro	Piedmont	8234	276	123	Tributary of Po
Arno	Tuscany	8228	241	110	Tyrrhenian Sea
Adda	Lombardy	7927	313	187	Tributary of Po
Ticino	*(Switzerland)* Piedmont, Lombardy	7228	248	350	Tributary of Po
Oglio	Lombardy	6649	280	130	Tributary of Po
Reno	Emilia-Romagna	5965	212	95	Adriatic Sea
Brenta	Trentino-Alto Adige, Veneto	5840	174	80	Adriatic Sea
Volturno	Molise, Campania	5550	175	82	Tyrrhenian Sea
Liri-Garigliano	Abruzzo, Lazio, Campania	5020	168	120	Tyrrhenian Sea
Simeto	Sicily	4186	113	25	Ionian Sea
Piave	Veneto	4127	220	125	Adriatic Sea
Dora Baltea	Valle d'Aosta, Piedmont	3891	168	96	Tributary of Po
Ombrone	Tuscany	3494	161	32	Tyrrhenian Sea
Isonzo	*(Slovenia)* Venezia Giulia	3460	136	**172**	Adriatic Sea
Tirso	Sardinia	3375	152	16	West Mediterranean Sea
Aterno-Pescara	Abruzzo	3190	152	57	Adriatic Sea
Tagliamento	Venezia Giulia	2916	178	69	Adriatic Sea
Ofanto	Apulia	2780	170	163	Adriatic Sea
Bradano	Basilicata	2765	120	7	Ionian Sea
Coghinas	Sardinia	2551	115	18	West Mediterranean Sea
Panaro	Emilia	2294	148	37	Tributary of Po
Crati	Calabria	2240	91	26	Ionian Sea
Livenza	Veneto, Friuli- Venezia Giulia	2221	112	85	Adriatic Sea
Imera-Salso	Sicily	2122	144	5	Mediterranean Sea
Flumendosa	Sardinia	1775	127	22	Tyrrhenian Sea
Serchio	Tuscany	1656	111	46	Tyrrhenian Sea
Fortore	Campania Molise Apulia	1650	110	15	Adriatic Sea
Sangro	Abruzzo	1545	122	8	Adriatic Sea
Basento	Basilicata	1537	149	12	Ionian Sea

(continued)

Table 1.2 (continued)

River	Region	Catchment area (km²)	Length (km)	Average flow (m³/s)	Receiver
Bacchiglione	Veneto	1400	118	30	Adriatic Sea
Chienti	Marche	1298	91	8	Adriatic Sea
Sinni	Basilicata	1292	94	15	Ionian Sea
Tronto	Marche	1192	115	17	Adriatic Sea
Belice	Sicily	964	107	4	Mediterranean Sea
Marecchia	Tuscany, Emilia	941	70	10	Adriatic Sea
Enza	Emilia	890	93	12	Tributary of Po

The numerous minor rivers listed in Table 1.2 characterize all the territory, and their particular hydrological behaviour often includes frequent unexpected high peaks but also remarkable dry periods.

The majority of Italian rivers have been harnessed for utilizing their water, especially for irrigation and electricity generation, and numerous reservoirs have been built during the twentieth century, with dams and diversion channels. Normally, the reservoir is designed in order to store large quantity of running water, also contributing to control the high flow propagation. Their capacity is designed in order to control annual and multiannual events and allow the flood routing. Now there are in the Italian territory 520 great reservoirs, which altogether correspond to a total capacity of more than 12 km³. The construction of the dams, often more than 100 m tall, has confirmed the qualified expertise of the Italian water engineering.

In a parallel way and following rooted traditions, small ponds were constructed in the hilly areas with a storage capacity of a few thousand cubic metres, with dams not more than 10 m tall. Their purpose is to store an amount of rainwater sufficient for the irrigation of nearby plots (Benedini 1995).

The construction of dams in Italy had its most flourishing period during the first half of the last century, but it slackened afterwards, and now the main interest of the responsible institutions is almost entirely on the maintenance of the existing structures, taking also in due count that many reservoirs have reduced their storage capacity as an effect of siltation (Molino et al. 2014; Di Silvio 2004).

The abundance of natural water, particularly in northern and central regions, which enhanced the development of prosperous economical activities, favoured also the realization of several artificial channels, to convey conspicuous flow mainly for irrigation and the inland navigation. A remarkable artificial network is in the River Po basin, where some canals were already in operation hundreds of years ago, like those in the area of Milan, the outline of which dates since the period of the Roman Empire. More recently new works have been done; the most important ones are in Table 1.3.

Heavy rainfalls, frequent in all seasons with an intensity higher than 100 mm/h, on ground that is now largely impermeable, cause high flow in streams and rivers. Consequently, floods and inundations are frequent in all the Italian regions.

Table 1.3 Artificial channels now in operation

Name	Location	Time of construction	Length (km)	Flow (m³/s)	Main utilization
Emiliano-Romagnolo	Emilia-Romagna	1950–1990	135	68	Irrigation
Villoresi	Piedmont, Lombardy	1920	86	48	Irrigation
Cavour	Piedmont	1880	83	85	Irrigation
Milan Navigli	Lombardy	Twelfth–sixteenth century	90	–	Navigation

According to recent estimates, about 3000 sites have been affected at least once by inundation during the twentieth century (ISPRA 2018).

Large lakes are located in Northern Italy, at the foot of the Alps, originated from ancient phenomena of glaciation. They are connected to the main rivers, which benefit for their natural regime from the great volume of water permanently stored.

Quite different is the pattern of the lakes located in Central Italy, which denote their volcanic origin, with the exception of Lake Trasimeno that is located in sandstones Pliocene deposits. These lakes, not connected to the local rivers, are independent water bodies with proper hydrologic characteristics.

The southern part of Italian peninsula is lacking of large natural lakes, as it is Sardinia, while Sicily hosts the small Lake Pergusa, originated by the sinking of ancient limestone and gypsum formations. On the contrary, small natural lakes are located in all the regions, generally connected to the minor rivers. The largest lakes have a noticeable effect on the local climate, reducing the excursion of temperature from summer and winter, and are also important for their amount of stored water, which for centuries allowed prosperous utilizations and favoured the development of high standard living conditions. The principal characteristics of the main lakes are in Table 1.4 and their location is in Fig. 1.12.

Low precipitation of recent years of drought had a strong effect on all the Italian lakes, particularly in Northern Italy, area normally considered "wet". Lake Como and Lake Maggiore, with an excursion of several metres, reached their minimum recorded level in both 1989 and 1990. Lake Garda reached its minimum level in 1990 after an excursion of 2.2 m. Similar effects have been recorded also in recent years (ISTAT 2018).

Besides the inland lakes, characterized by freshwater, the territory hosts numerous coastal lagoons of transition water, sometimes with remarkable salt concentration. The location of the largest lagoons is also in Fig. 1.12 and their main characteristics are in Table 1.5. Their presence is essential for the local environment and the maintenance of traditional living conditions, while their saline water, due to natural surging and connection with the open sea, can be used for particular purposes. The largest and most important lagoon is that of Venice, in which the tidal alternations of the connected Adriatic Sea cause frequent and worrisome variation of the water level.

Table 1.4 The principal lakes

Lake	Region	Area (km²)	Tributary	Emissary	Max. depth (m)
Garda	Lombardy, Veneto, Trentino-Alto Adige	370.0	Sarca	Mincio	346
Maggiore	Lombardy, Piedmont, Ticino	212.0	Ticino-Maggia-Toce-Tresa	Ticino	370
Como	Lombardy	145.0	Adda-Mera-Fiumelatte	Adda	410
Trasimeno	Umbria	128.0	Anguillara		6
Bolsena	Lazio	113.5			151
Iseo	Lombardy	65.3	Oglio-Borlezza	Oglio	251
Varano	Apulia	60.5	Groundwater	Connected to Adriatic Sea	5
Bracciano	Lazio	57.5	Torrente Arrone		151
Lesina	Apulia	51.4		Connected to Adriatic Sea	< 2
Lugano	*Switzerland,* Lombardy	48.7	Vedeggio-Cassarate	Tresa	288
Orta	Piedmont	18.2	Groundwater	Nigoglia	143
Varese	Lombardy	15.0	Brabbia, Tinella	Bardello	26
Vico	Lazio	13.0		Rio Vicano	48,5
Idro	Lombardy, Trentino-Alto Adige	10.9	Chiese, Re di Anfo, Caffaro	Chiese	122
Santa Croce Croce	Veneto	7.8	Canale Cellina, Tesa	Rai	44
Albano	Lazio	6.0			168
Pergusa	Sicily	1.8	Groundwater		12
Nemi	Lazio	1.7			33

1.5   The Aquifers

Large amount of water is stored in the underground of all regions, favoured by the characteristics of the subsoil and as an effect of the particular climate. Unfortunately, an assessment of the amount of water lying in the subsoil is affected by high uncertainty. Accurate surveys, based on the hydrogeological characteristics of some significant places, have given values restricted to the investigated zones, and only estimates can be proposed for the national territory (JM 2007; Civita et al. 2010; Cambi and Dragoni 2000).

Large aquifers are in the north, at various depth and geological formations, as well as in carbonate rocks permeable for fractures and karstic alterations. In the alpine zones of Veneto, Trentino and Friuli, crystalline-metamorphic rocks with limited permeability are present. The aquifers in the flat area of Adige and eastern rivers generally have an upper part with permeable soil, supplied by the same rivers,

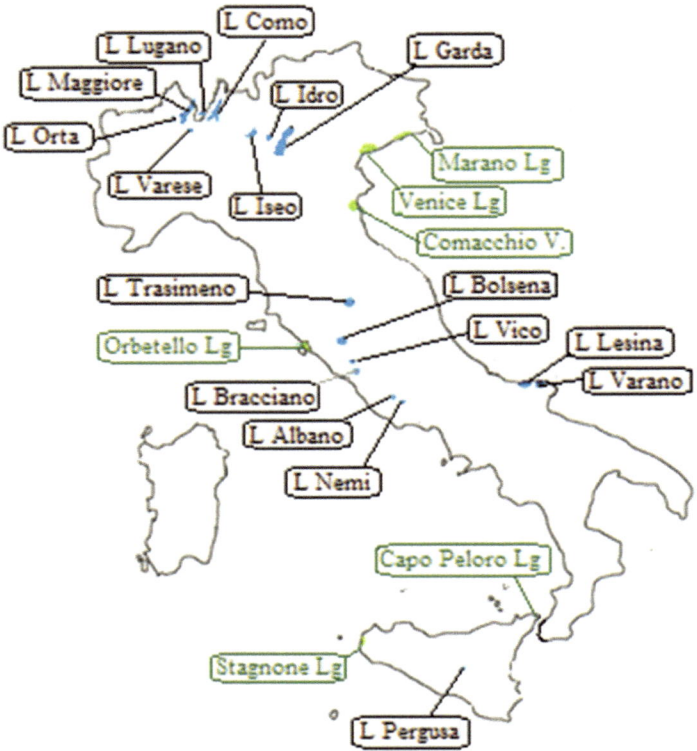

Fig. 1.12 The main lakes and transition water bodies (L = lake; Lg = lagoon)

Table 1.5 The largest lagoons in Italy

Lagoon	Region	Area (km²)	Depth (m)
Venice	Veneto	550.0	10
Marano	Veneto	93.3	1.0
Valli di Comacchio	Emilia-Romagna	110.0	1.5
Orbetello	Tuscany	26.9	1.0
Stagnone	Sicily	5.4	1.0
Capo Peloro	Sicily	0.7	1.0

and a lower part characterized by large springs. Many aquifers are already subject to high overexploitation due to an intense demand of the most productive industrial, agricultural and livestock sectors. In carbonate formations of Western Alps, large amount of groundwater is now unusable due to a high pollution level.

Along the Apennine Chain, the aquifers in the northern part belong to the carbonate and volcanic rocks of Tuscany.

In the carbonate of central Apennine, the aquifers include some of the largest springs of the Mediterranean, such as Peschiera, which has a flow rate up to 18 m³/s. In the southern part of the chain, the aquifers are present in the volcanic areas of Lazio and Campania and in the alluvial plains of Marche, Abruzzi, Campania and Basilicata.

The plains of central Italy include also the aquifers of the alluvial zones of Florence-Prato-Pistoia, which undergo dangerous overexploitation, and the coastal zones of Pisa, Lucca and Grosseto, which are threatened by sea intrusion. In addition, the small plains of Adriatic coast include several aquifers with problems of overexploitation and seawater intrusion.

The aquifers of Apulia (Tavoliere, Gargano, Murge and Salento) in carbonate rocks highly permeable for fracture and karstic alteration have high losses to sea and high risk of chemical and microbiological pollution.

Sicily hosts aquifers in the volcanic rocks of Etna, in the complexes of Iblei Mountains and in the plain of Palermo, while limited local aquifers are in the mountains along the Tyrrhenian coast and in the Trapani plain. The central part of the island does not contain remarkable aquifers due to prevalent clay formations.

Sardinia has few aquifers in the coastal areas of east and south-west, especially in carbonate rocks, while alluvial aquifers lie in the plains of Campidano, Oristano and Sulcis, with serious pollution problems.

The quantitative estimates carried out in some areas mentioned above are based on the assumption that the amount of water existing in the subsoil and naturally recharged by the rain matches that withdrawn in local pumping wells. This does not take into consideration the possibility that the quantity of exploited groundwater is sometimes greater than that naturally recharging the aquifer. This happened in several parts of the country, where intensive pumping caused a remarkable lowering of the water table (Benedini et al. 1994).

Where feasible, a more reliable evaluation of the total amount of water existing in the aquifers has been done taking into consideration the numerous natural springs in the relevant territory, in relation to the particular geological pattern of the surrounding area and to an identifiable way of natural recharge. Local investigations have been done and are still in progress promoted by the responsible authorities, also under the auspices of the European Commission (Correia 1998; Dragoni 1998; CEC 2006). Direct field observations with deep boreholes have accompanied these investigations, in a way that now the consistence of the Italian groundwater can be better appreciated.

Now, an up-to-date nationwide estimation is not yet available, but for an overall evaluation, the amounts proposed by the National Water Conference in the late 1990s can be accepted, as in Fig. 1.13 (IRSA 1999). Like the considerations about the surface water described in preceding paragraphs, also the natural availability of the Italian groundwater shows a remarkable variation in the various regions, with the predominance of the northern zones. During the last drought periods, some aquifers, already depleted for intensive potable use and irrigation, underwent an unusual lowering of the water table, which lasted long after the period of scarcity was over. In the central and southern regions, the yield of many springs diminished.

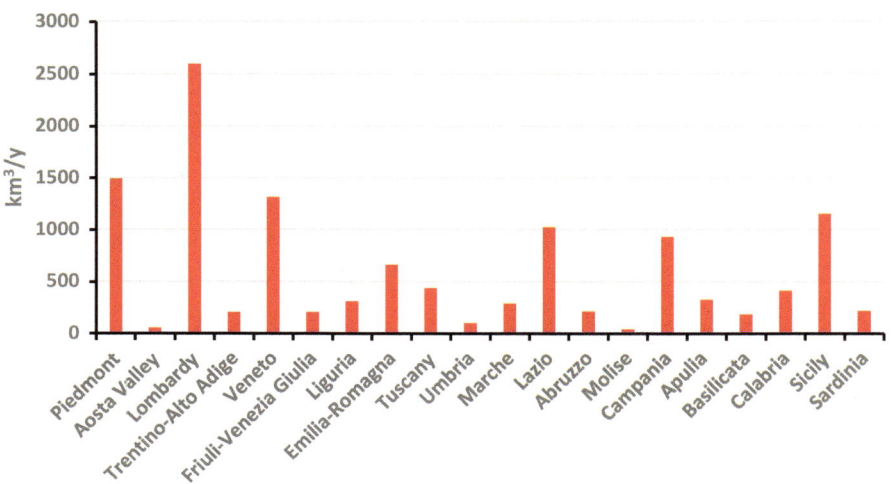

Fig. 1.13 Estimated groundwater in Italy

1.6   Population of Italy

As shown in Fig. 1.14, after the country unification (1860), a relatively high annual rate of growth characterized the Italian population, but in the last years, such rate decreased significantly and the number of original inhabitants tends now to stabilize. Nevertheless, like the other European countries facing the Mediterranean, Italy is now the final destination of a massive immigration from Asian and African countries, which probably will maintain the increase of the total number of Italian inhabitants. An internal migration still exists, which during the last centuries has seen about 15 million people moving towards the industrialized towns of the northwest, mostly leaving the countryside and particularly the southern regions and the islands (ISTAT 2018).

During the last centuries, Italy suffered an intensive emigration, particularly from the southern regions, and a remarkable number of people moved towards American and northern European countries.

Today this emigration decreased, and Italy has to face the reverse problem of inserting the immigrants from Africa and Asia into its own labour force that is already hit by a worrisome unemployment. Following a national official census, the situation of Italian population is summarized in Table 1.6.

A particular kind of emigration still exists and affects now the young generation of highly qualified people, attracted by better promising working conditions in other countries.

The table shows also the difference of the population density among the various regions that can be an effect of the internal migration. Northern regions, in particular Lombardy, denote high inhabitant concentration, which tends to grow further,

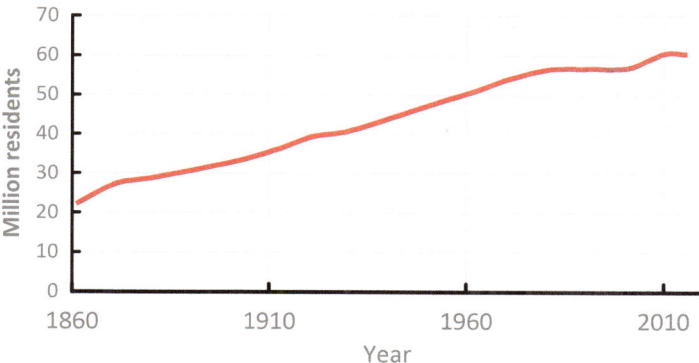

Fig. 1.14 Historic evolution of the Italian population

increasing a gap already existing with the other regions and with an expected impact on all the country economy, affecting the actual and future water management problems.

1.7 Economic Aspects

During the last decades and particularly after the Second World War, Italy has undergone substantial economic transformations, enhancing the industrial sector to the detriment of agriculture and favouring the development of large urban agglomerations. The rate of employment of the agricultural sector remarkably decreased while the industrial sector increased. Figure 1.15 shows the comparison of the 1931 situation, assumed as a significant original value, with that of 2007 (Aiello and Pupo 2012).

Since the country unification, an economic divide has grown between the northern and southern regions. Even at the present time, such regional disparity persists, and the situation of Southern Italy still calls for the attention of the governing institutions.

During the 1950s, a special organization was set up, namely, the Southern Italy Development Fund, "Cassa per il Mezzogiorno", that mastered several initiatives with the public financial support aiming at increasing the economy of the southern regions. In spite of remarkable realizations, among which huge works relevant to the local water, several objectives are still missing, and the intervention of the central government is always necessary (Leonardi 1995; Milio 2007; Barone et al. 2016).

Now, the primary productive sector devoted to exploit the natural resources, like agriculture, livestock and fishery and mineral abstraction, concerns only 4% of the total productive activity. The last national censuses show that agriculture, particularly in the southern regions, has undergone a noticeable reduction of the cultivable land, estimated at -32% in the 2000–2010 period.

Table 1.6 Actual demographic situation of Italy

Region	Area (km²)	Population (inh.)	Density (inh./km²)	Admin. centre
Abruzzo	10,832	1,322,247	122.1	L'Aquila
Aosta Valley	3261	126,883	38.9	Aosta
Apulia	19,541	4,063,888	208.0	Bari
Basilicata	10,073	570,365	56.6	Potenza
Calabria	15,222	1,965,128	129.1	Catanzaro
Campania	13,671	5,839,084	427.1	Naples
Emilia Romagna	22,453	4,448,841	198.1	Bologna
Friuli Venezia Giulia	7862	1,217,872	154.9	Trieste
Lazio	17,232	5,898,124	342.3	Rome
Liguria	5416	1,565,307	289.0	Genoa
Lombardy	23,865	10,019,166	419.8	Milan
Marche	9401	1,538,055	163.6	Ancona
Molise	4461	310,449	69.6	Campobasso
Piedmont	25,387	4,392,526	173.0	Turin
Sardinia	24,100	1,653,135	68.6	Cagliari
Sicily	25,832	5,056,641	195.8	Palermo
Tuscany	22,987	3,742,437	162.8	Florence
Trentino Alto Adige	13,606	1,062,860	78.1	Trento
Umbria	8464	888,908	105.0	Perugia
Veneto	18,407	4,907,529	266.6	Venice
ITALY	**302,073**	**60,589,445**	**183.45**	

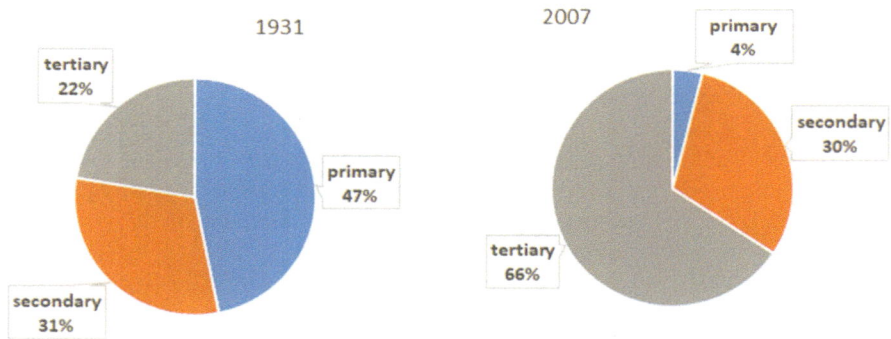

Fig. 1.15 The economic situation

The secondary sector, which includes industry, energy production and construction, numbers now few powerful multinational corporations, but also a large number of small and medium-sized enterprises particularly in the northern regions.

The tertiary sector, made up of public and private services and including information, research and planning activities, is also in continuous expansion.

The economic disparity among the regions is confirmed by the annual gross domestic product, which in 2015 was of the order of 1600 million euros for the entire country, with a per capita of 27,000 euros, but some northern regions denote higher values up to 35,000 euros.

The economic aspects mentioned above foster an encouraging challenge for the political institutions, where the national water problems, now and in view of a future development, share a primary position.

1.8   Italy and Europe

Not only its geographic position but also the new ways that characterize the national and international policies enforce the ties of Italy with the rest of Europe. In particular, Italy, being a founder and an active member of the European Union, is deeply involved in the formulation and application of directives, which, with the purpose of harmonizing the Union's life, impose specific rules to the national political issues.

Water problems play an important role and are the object of frequent interventions of the water responsible institutions. The scientific community, which is also fostering an international cooperation, is called to provide the most efficient tools (European Parliament 2000; European Union 2000; Suzenet 1997; EEA 2003; Viganò et al. 1997).

With a total area of more than 4.4 million square kilometres and a population close to 500 million inhabitants, the European Union is a very important aggregation of countries and is the heritage of centuries of historical living patterns that characterize the entire world for culture, political structures and economic development. Its importance can be expressed by the average gross domestic product, which, even though variable from one country to the other, is now assessed around 20 trillion euros. Table 1.7 summarizes the main geographic characteristics of the member states, including an estimate of the annual average precipitation, assumed as an indicator of the relevant water problems.

The table contains also Switzerland, which is not a member state but shares with the other countries many aspects of its water policies.

The great variety emphasized in the table underlines the specific nature of the water problems in each member state. Anyhow, there are some situations that call for common approach, among which are principally the use, protection and control of the natural water. On behalf of its governmental and scientific institutions, Italy is deeply involved in a continuous cooperation with the correspondent European structures. The actual Italian water policy is complying as much as possible with the Directives of the European Union, which are the fruit of a common work of expert people coming from the member states.

Table 1.7 Main characteristics of the European member states

Country	Precipitation (km³/y)	Population (Million)	Area (1000 km²)
Austria	98.00	9	84
Belgium	28.89	11	31
Bulgaria	72.58	7	111
Croatia	62.33	4	57
Cyprus	3.03	1	9
Czech Republic	54.65	11	79
Denmark	38.49	6	43
Estonia	29.02	1	45
Finland	22.20	6	338
France	500.77	67	641
Germany	278.00	83	357
Greece	115.00	11	132
Hungary	55.71	10	93
Ireland	87.63	5	70
Italy	241.10	61	302
Latvia	42.70	2	65
Lithuania	44.89	3	1
Luxembourg	2.03	1	3
Malta	0.18	0	0
Netherlands	31.62	17	42
Poland	193.96	38	313
Portugal	82.16	11	92
Romania	150.94	20	238
Slovakia	37.35	5	49
Slovenia	31.75	2	20
Spain	346.53	46	506
Sweden	342.16	10	450
United Kingdom	287.61	66	242
Switzerland	*61.21*	*8*	*41*

1.9   Italy and the Mediterranean

High variety of natural conditions characterizes the Mediterranean basin, accompanied by a variety of economic and social situations. Several countries of Europe, Africa and Asia face the Mediterranean Sea, with total or part of their relevant national territory, as described in Table 1.8. They give rise to a large catchment of more than 1,345,000 km² with a population of more than 500 million. The extension of such catchment and its geographic belonging to three continents, characterized by very different environmental conditions, underline a particular climatic aspect extremely variable from one place to the other. Consequently, the availability of natural water is extremely variable and gives rise to specific problems (Margat and

Table 1.8 The Mediterranean basin

Country	Area (1000 km²)			Country	Area (1000 km²)		
	A Total country	B Mediterranean catchm.	ratioB/A (%)		A Total country	B Mediterranean catchm.	ratioB/A (%)
Albania	28.75	28.75	100.0	Macedonia	25.71	25.70	99.9
Algeria	2381.74	133.00	5.6	Malta	0.32	0.32	100.0
Bosnia and Herzegovina	51.13	16.30	31.9	Monaco	0.00	0.00	100.0
Croatia	56.59	37.21	65.7	Montenegro	13.81	13.00	94.1
Cyprus	9.25	9.25	100.0	Morocco	710.85	8.00	1.1
Egypt	1010.00	0.20	0.0	Serbia	88.36	6.32	7.2
France	543.97	130.10	23.9	Slovenia	20.27	4.84	23.8
Gaza Strip	0.37	0.37	100.0	Spain	505.99	185.60	36.7
Greece	131.96	131.94	100.0	Syria	185.18	22.00	11.9
Israel	22.07	10.50	47.6	Tunisia	163.61	90.00	55.0
Italy	302.07	301.27	99.7	Turkey	783.36	19.00	2.4
Lebanon	10.45	9.80	93.8	West Bank	5.90	2.42	41.0
Libya	1759.54	158.86	9.0				

Vallée 1999; JM 2007; Shiklomanov 2000, Boldrin and Canova 2001). Against limited natural resources, a very high demand for the main uses results in a massive abstraction of water from the natural bodies, some of which are already close to the limit threshold of the natural renewable water. The demand is destined to increase in the future, further worsening the imbalance between withdrawal and availability. This situation stresses the vital importance of water in the entire basin.

The situation of Italy in the Mediterranean context appears very peculiar, since the country belongs almost entirely to the Mediterranean reality, while several large countries can benefit from their relevant portion that shares the territory of other catchments. This underlines the importance of an accurate evaluation of all the climatic and environmental aspects of the basin. Facing the water problems in the area entails primarily a detailed knowledge of the natural precipitation, starting from a thorough investigation of climate, geology and other geographic conditions, in order to clearly ascertain the peculiarity in time and space.

An efficient hydrological survey system should be recommended, working with instruments and procedures able to provide homogeneous data, comparable all over the various countries. The acquired information should call for the attention of the scientific community, in order to improve the knowledge of the natural phenomena governing the surface and underground natural water. In a parallel way, an accurate survey of the economic, social and environmental aspects is essential, in a close contact with the political and administrative responsible institutions. The knowledge of these aspects is necessary to assess the water demand, presently and in the future.

The short descriptions in the preceding paragraphs are given with the purpose of introducing the peculiarity of the water problems in Italy, which are strictly conditioned by the local aspects of the various territorial entities, as it will be explained in the next chapters of the book.

References

Aiello F, Pupo V (2012) Structural funds and the economic divide in Italy. J Policy Model 34(3):403–418

Barone G, David F, De Blasio G (2016) Boulevard of broken dreams. The end of EU funding: Abruzzi, Italy. Reg Sci Urban Econ 60:31–38

Batini G, Benedini M (2000) Facing a changeable hydrological regime in a Mediterranean country. In: Proceedings of international symposium the extremes of extremes. IAHS Publication N. 271. Reykjavik, Iceland, 3–8

Batini G, Benedini M, Munari M (2000) The contribution of Italy, 20th conference of Danubian countries. Check Republic, Bratislava

Benedini M (1995) Improving the practice of small ponds in the Italian hilly land as a tool to mitigate the effect of droughts. Water Research Institute, Rome

Benedini M, Giulianelli M (2003) Urban water and climate change. In: Proceedings of international N.A.T.O. research workshop enhancing urban environment by environmental upgrading and restoration. Rome, pp 311–320

Benedini M, Giuliano G, Pagnotta R, Passino R (1994) Contribution of the Italian water research institute. In: Proceedings of the European Network of Fresh Water Research Organisations Euraqua: land use change and water resources, First technical review, Wallingford (UK), pp 125–143

Boldrin M, Canova F (2001) Inequality and convergence in Europe's regions: reconsidering European regional policies. Econ Policy 16(32):206–253

Brunetti M, Maugeri M, Monti F, Nanni T (2006) Temperature and precipitation variability in Italy in the last two centuries from homogenized instrumental time series. Int J Climatol 26:345–381

Brunetti M, Maugeri M, Nanni T, Simolo C, Spinoni J (2014) High-resolution temperature climatology for Italy. Interpolation method intercomparison. Int J Climatol 34:1278–1296

Cambi C, Dragoni W (2000) Groundwater yield, recharge variability and climatic changes: considerations arising from the modelling of a spring in the Umbria-Marche Apennines. Hydrogeology, vol 4, ed. BRGM, pp 11–25

CEC (2006) Commission of the European Communities. Communication from the Commission to the Council and the European Parliament. Establishing an environment strategy for the mediterranean" – COM (2006) 475 final. Brussels, Belgium

Civita MV, Massarutto A, Seminara G (2010) Groundwater in Italy: a review. In EASAC (European Academies Science Advisory Council) Groundwater in the Southern Member States of the European Union: an assessment of current knowledge and future prospects. Accademia dei Lincei, Rome

Correia FN (ed) (1998) Water resources management in Europe. In: Volume I: Institutions for water resources management in Europe. Balkema, Rotterdam

Crespi A, Brunetti M, Lentini G, Maugeri M (2017) 1961–1990 high-resolution monthly precipitation climatologies for Italy. Int J Climatol. Published online in Wiley Online Library (wileyonlinelibrary.com). https://doi.org/10.1002/joc.5217

Desiato F, Lena F, Toreti A (2007) SCIA: A system for a better knowledge of the Italian climate. Boll Geofis Teor Appl 48(3):351–358

Desiato F, Fioravanti G, Fraschetti P, Perconti W, Toreti A (2011) Climate indicators for Italy: calculation and dissemination. Adv Sci Res 6:147–150

Di Silvio G (2004) Modelling long-term reservoir sedimentation for optimal management strategies. In: 6th conference on hydro-science and engineering. Brisbane

Dragoni W (1998) Some considerations on climatic changes, water resources and water needs in the Italian region south of the 43°N. In: Issar A, Brown N (eds) Water, environment and society in times of climatic change. Kluwer, pp 241–271

EEA (2003) European environment agency. Climate Qualità Index, Copenhagen, p 2003

European Parliament (2000) Directive 2000/60/CE of 23/10/2000, Concerning Action in Water. E.U. Official Journal of 22/12/2000, series l327

European Union (2000) Towards a sustainable/strategic management of water resources: evaluation of present policies and orientation for the future. European Commission, Study 31, ISBN 92, Luxembourg. http://onlinelibrary.wiley.com/doi/10.1002/joc.5217/epdf

IRSA (1999) Water Research Institute, Un futuro per l'acqua in Italia. [A Future for Water in Italy] (in Italian), Quaderni IRSA n.109, Roma, p 235

ISPRA (2015) Temperature and precipitation climatic normals over Italy. Climate change. http://www.isprambiente.gov.it/en/publications/state-of-the-environment

ISPRA (2018) Landslides and floods in Italy: hazard and risk indicators – 2018 Edition ISPRA Reports, 287/2018 ISBN 978-88-448-0901

ISTAT (2018) National institute of statistics annual report. The state of the Nation. https://www.istat.it/en/archivio/217955

JM (2007) Joint Mediterranean EUWI/WFD Process, Mediterranean Groundwater Report, Mediterranean Groundwater Working Group (MED-EUWI WG on groundwater. MAP. Mediterranean Action Plan. "Water demand management, progress and policies". In: Proceedings of the 3rd regional workshop on water and sustainable development in the mediterranean, Zaragoza, Spain

Leonardi R (1995) Regional development in Italy: social capital and the Mezzogiorno. Oxford Review of Economic Policy, vol 11, No. 2, Regional Policy 95, pp 165–179

Margat J, Vallée D (1999) Mediterranean vision on water, population and the environment for the XXI century. MEDTAC Plan Bleu

Milio S (2007) Can administrative capacity explain differences in regional performances? Evidence from structural funds implementation in Southern Italy. Reg Stud 41(4):429–422. https://doi.org/10.1080/00343400601120213

Molino B, De Vincenzo A, Ferone C, Messina F, Colangelo F, Cioffi R (2014) Recycling of clay sediments for geopolymer binder production. A new perspective for reservoir management in the framework of Italian legislation: the Occhito reservoir case study. Materials 7(8):5603–5616. https://doi.org/10.3390/ma7085603

Portoghese I, Vurro M (2010) The challenge of water balances and run-off projections in the Mediterranean Hydrology. Impact of climate change on water resources- 200 Years hydrology in Europe. Euraqua Symposium, Koblenz, Germany

Shiklomanov IA (2000) Appraisal and assessment of world water resources. Water Int 25(1):11–32

Suzenet G (1997) R & D policies: case study: water, present status in Greece, Italy, Portugal and Spain. Technical Report, EUR 17726 EN

Toreti A, Desiato F (2008) Temperature trend over Italy from 1961 to 2004. Theor Appl Climatol 91:51–58

Toreti A, Desiato F (2009) Changes in temperature extremes over Italy in the last 44 years. Int J Climatol 28:733–745

Viganò L, Benedini M, Badino G, Barbiero G, Buffagni A, Pagnotta R, Spaggiari R (1997) Quantity and quality aspects in the protection of the aquatic environment. In: Van de Kraats JA (ed) Let the fish speak: the quality of aquatic ecosystems as an indicator for sustainable water management, Fourth Euraqua Technical Review, Koblentz, pp 149–171

Chapter 2
Paradigm Change in Water Resources Development in Italy

Giuseppe Rossi

Abstract The analysis of water resources development of Italy requires to describe the evolution of the infrastructures and services that provide water resources utilization, water quality protection and soil defense, within the legislative and institutional framework. This evolution can be expressed by the succession of three paradigms: (i) water resources exploitation to meet human needs and to achieve economic growth, (ii) focus on water quality protection and on water-related disaster mitigation and (iii) comprehensive approach towards an integrated, sustainable and equitable water management. For each phase of development, some examples of water works, water services and government organizations can prove how the paradigms have changed due to the improvement of the general cultural and social principles and to the evolution of water legislation at national and European level. An assessment of specific features of the Italian way of facing the challenges of water-related governance is presented.

2.1 Introduction

The solution of water problems has been progressively recognized as a key issue for the definition and implementation of national and international policies aiming at the socio-economic development and environmental protection. For example, improvement of the access to safe drinking water and sanitation (recognized as human right by the Assembly of the United Nations in 2010) has been included, as sixth goal, in the United Nations Agenda 2030 for Sustainable Development (United Nations 2015). Also the recommendations of the last World Water Forums underline that stability and prosperity of all geographical regions depend on the correct water resources management.

On the other side, the development of the water sector, as a component of the global system, has been influenced, in a variety of ways, by changes in non-water sectors, particularly by legislative, institutional and political events as well as by

G. Rossi (✉)
Department of Civil Engineering and Architecture, University of Catania, Catania, Italy
e-mail: grossi@dica.unict.it

© Springer Nature Switzerland AG 2020
G. Rossi, M. Benedini (eds.), *Water Resources of Italy*, World Water Resources 5,
https://doi.org/10.1007/978-3-030-36460-1_2

general sociocultural principles. Furthermore, water management in the future is expected to be affected by more complex links with other processes. In fact, the technological advancements, the evolution of information and communication as well as the governance features, including civil society organizations, requirements of stakeholders and public participation, will have in the next decades a more remarkable effect than that in the past.

Consequently, the current conditions of water resources in Italy and the perspectives of the future cannot be analysed correctly without trying to identify the main drivers of the historic water development, connected to the evolution of the more general paradigms that have affected the Italian society in the recent past.

Therefore, the objective of this chapter is to identify the stages of the water resources development and to review the most important steps of the construction of water infrastructures and of the evolution of the water services which, in turn, depend from the legislative and organizational framework in the country and from the dominant cultural and social paradigms in the various stages.

Section 2.2 of this chapter presents the basic concepts of water system and water services. Section 2.3 identifies main trends and paradigm changes in water development. Sections 2.4, 2.5 and 2.6 describe significant examples drawn from the historic development of water infrastructures, of water service organizations and of planning priorities and decision-making features to prove the dominance of the identified paradigms in each phase of development. Section 2.7 presents few reflections about the specific features of the Italian way of facing the challenges of water governance.

2.2   Basic Concepts of Water Systems and Water Services

The management of water resources, in its most general meaning, is conceived as the comprehensive process aiming at achieving water resources utilization, water quality protection and defense from water excess in the best possible way (Fig. 2.1).

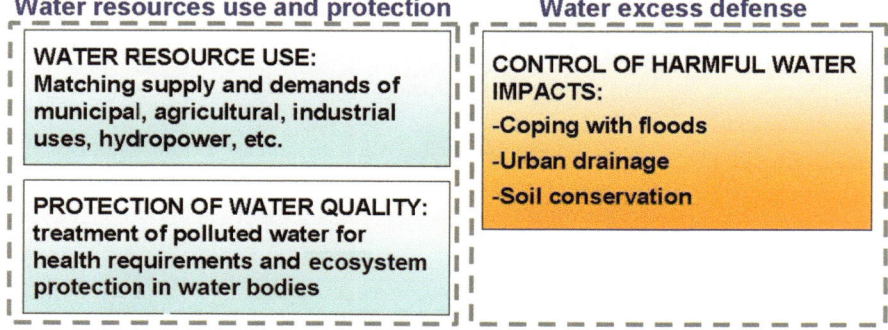

Fig. 2.1 Management of water resources

It includes the necessary activities for the development of water infrastructures (planning, design, construction, operation, maintenance and control of structural facilities), as well as many nonstructural instruments, such as the legislation framework and the institutional setting in the decision-making process, including the participation of stakeholders and of all citizens.

Despite the decreasing emphasis on system approach as a key methodology to solve the complex water problems, as it was believed during the 1970s (Hall and Dracup 1970), it seems still appropriate to adopt the concept of system to perform a thorough analysis of the various components of a water problem. In fact, the system is conceived as the combination of several elements that have a common objective and are linked mutually and strongly by several inputs and outputs of various nature, including some form of feedback, with only "weak" connections with the external environment.

In particular, the main components of a water supply system are listed in Table 2.1 under two categories: water sources and water uses. An example of a complex water system that supplies the municipality of Palermo in Sicily together with other municipalities and also agricultural and industrial areas is presented in Fig. 2.2.

In general terms, a water system includes two types of structural facilities (A and B) and two types of nonstructural elements (C and D), as follows:

A. **Facilities to match available resources and water needs (off-stream uses):**

- In quantity → withdrawal from surface water and groundwater sources
- In space → water transfer from sources to user's location (conveyance)
- In time → regulation through reservoir storage
- In quality → water treatment according to required standard of the use

Table 2.1 Components of a water supply system

Water sources	Water uses
Surface bodies:	*Off-stream uses:*
Diversion	*Water withdrawn to supply:*
Reservoir	Urban use (domestic, collective)
Groundwater bodies (aquifers):	Agricultural use (irrigation, livestock)
Spring	Industrial use and energy production
Wells	*In-stream uses:*
Infiltration gallery	*Water used in the same watercourse for:*
Nonconventional resources:	Navigation
Wastewater treatment plants	Hydroelectric power generation
Desalination plants	Protection and improvement of water bodies for recreational activities, fish propagation and river ecosystems safeguard

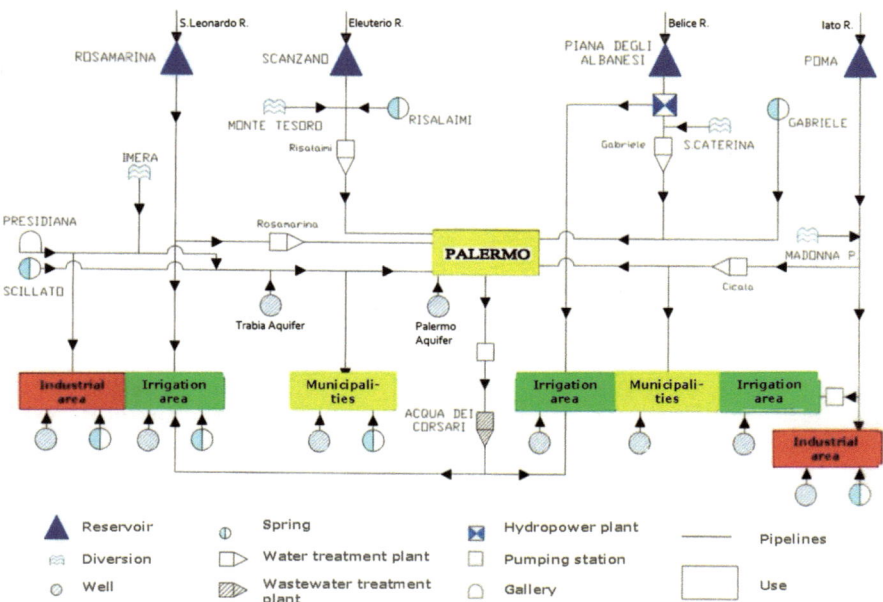

Fig. 2.2 Example of a complex water supply system (Palermo, Sicily). (Source: Rossi 2011)

B. **Facilities (and regulation) to allow in-stream uses:** navigation, recreation, aquatic ecosystem protection and landscape conservation

C. **Regulation of water rights (permit systems):**

- Ownership of water resources
- Water rights regime for different uses
- Priority among uses (first municipal use, particularly in water scarcity conditions)
- Constraints on withdrawal from watercourses to assure minimum stream-flows to preserve the river ecosystem (ecological flow)

D. **Institutions and management organizations**

- Water governance bodies (policy-making and planning) at international, national, subnational and local levels
- Monitoring and advising roles for water governance
- Water system management bodies in different water sectors (including public territorial authorities providing direction and control) and service companies (for all operational activities)
- Surveillance on operation bodies and citizen participation

Although the concept of **water systems** is not restricted to the physical components, the focus is on the structural facilities, requiring planning, engineering design, financial and economic feasibility assessment and Environmental Impact Assessment.

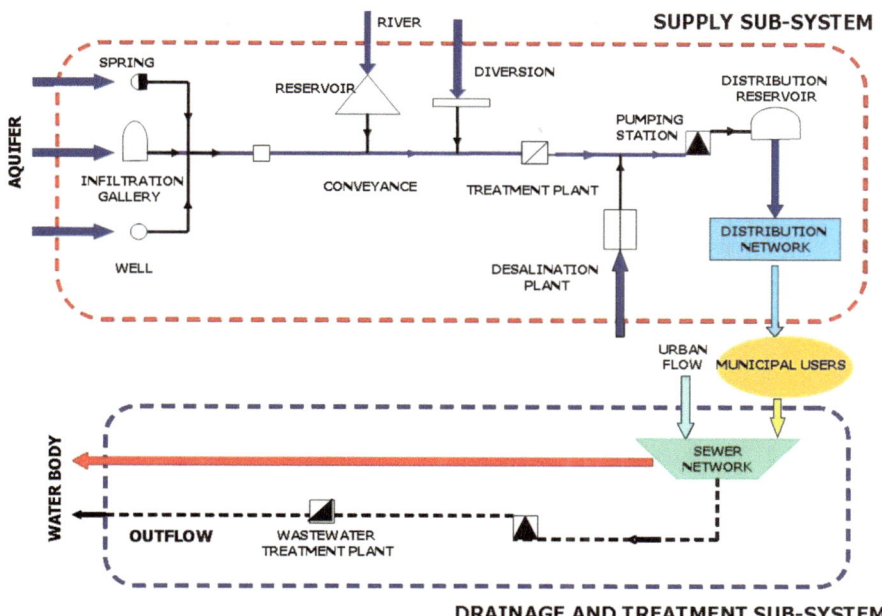

Fig. 2.3 Example of an urban water system. (Source: Rossi 2011)

A more comprehensive view considers the water management in terms of *water services*, where the focus is on the institutional organizations, generally including a political authority and one or more management companies and a few organizations in charge of supervising the service performance and tariff in order to guarantee the customers.

Figures 2.3 and 2.4 show the differences between the two concepts for the urban case. Figure 2.3 describes a general urban water system, including a supply subsystem and a combined sewer subsystem (for sanitary waste and storm runoff) and a treatment subsystem. Figure 2.4 describes a general urban water service with the interactions among the institutional bodies (which operate facilities and regulate service) and all elements of the environmental, economic and legislative framework, as well as the users and other stakeholders.

2.3   Evolution of Paradigms at Various Stages of Water Development

In spite of the differences between the water resources development features including both the development and operation of water systems and legislative and institutional framework, several common historical steps can be identified in different countries and large regions in the world.

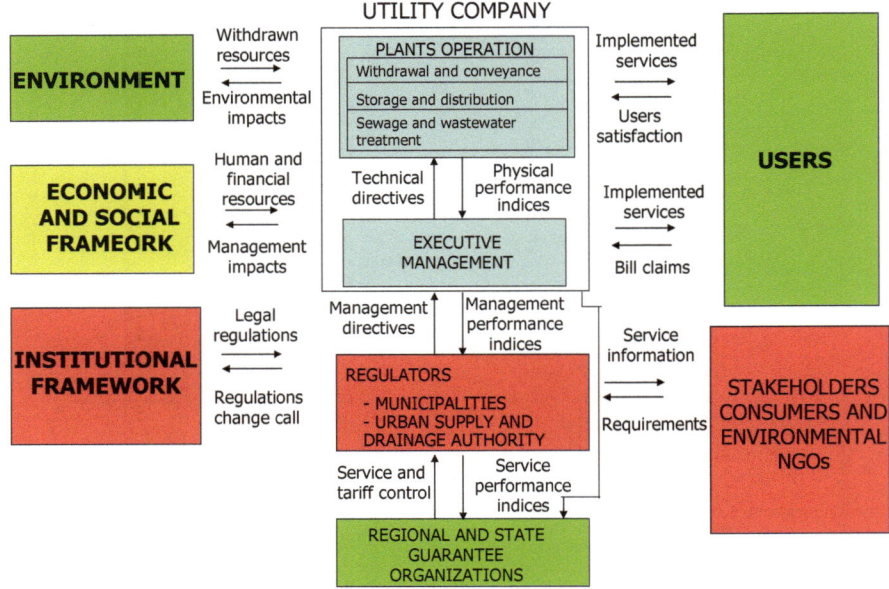

Fig. 2.4 Complexity of an urban water service. (Source: Rossi 2011)

Main trends in water infrastructures development can be described in terms of increasing complexity:

1. From single-source to multisource (withdrawal from surface and groundwater)
2. From single-purpose to multipurpose (e.g. consumptive uses, flood routing, recreation)
3. From single-structure to multistructure
4. From simple structural measures to combined structural/nonstructural measures
5. From separated quantity and quality aspects to a joint analysis
6. From separated planning for water use, pollution control and excess water defense to a comprehensive water resources and soil conservation planning
7. From the design of a single-purpose facility to the planning of a water system at large territorial scale
8. From "centralized" to "distributed multistage responsibilities" according to the principle of subsidiarity
9. From a reactive approach for coping with risks to a proactive approach

Such a complexity affects all features of water resources management in terms of:

1. **Spatial extent of water system:** the system is extended to include several river basins, physically linked (e.g. by water transfers) or linked by common users (Yevjevich 1995).

2. **Time horizon of planning:** it depends not only on physical and economic life of plants but also on obsolescence of systems and time horizon of social, economic and environmental needs as well of possible climatic changes.
3. **Controversies among users**: they refer not only to the water requirements of different categories of users but also to conflicting interests of off-stream water users versus the claims of green movements for protection of aquatic ecosystems.
4. **Institutional aspects of water management**: they include the distinction of the regulatory function (public bodies) from the service provision (carried out by a public or private company or by a public-private partnership).
5. **Growing role of the transparency in information and of the public partici-pation in the decision-making processes**.
6. **Objectives of water resources planning:** a shift occurs from a single economic objective to various socio-political and environmental objectives.

In particular, with reference to the last feature, it is necessary to underline that the enlargement of objectives does not always bear a definitive evolution of water resources management. For example, in the USA, the principles and standards published by the Water Resources Council (USWRC 1973) prescribed four objectives for evaluating alternatives: (i) national economic development, (ii) environmental quality, (iii) regional economic development and (iv) other social effects. However, more recent guidelines in water policy pay less attention to environmental and equity objectives, within a general trend which limits the mission of the government to regulation and coordination, with the focus on private initiative (e.g. water market), thus shifting the planning process from comprehensive plan to specific sub-plans such as for coping with disaster risk (flooding and drought).

Indeed, specific trends identified in the water field can be ascribed to more general changes of paradigm in the society and in the cultural framework.

Different interpretations have been provided to the change of paradigms affecting the water problem solution in the last centuries. For example, a simplified analysis distinguishes between a *water supply* model, aiming at increasing the usable resources, and a *water demand management* model aiming at reducing the water needs and/or developing recycle and reuse of water in a circular approach. As an example, the water policy in Spain has been described in terms of shift from the focus on water works construction in order to supply water to fields and control rivers and increase hydropower capacity towards the priority given to restoring the ecological quality of water resources and ecosystems and controlling groundwater overexploitation. This shift was evoked by the new government after the elections of 2004 as a motivation to justify the stop to the Ebro water transfer planned by the National Water Plan and the choice of focusing on the reuse of treated wastewater and new desalination plants (Garrido and Llamas 2010).

Dooge (1999) identified some phases in the design and implementation of hydro-projects. He stated that, up until the 1950s, engineers were the major decision-makers; during the 1960s, economists began to play a significant role, while in the

1970s environmentalists, and in the 1980s, the stakeholders started to participate, while in the following years, also the NGO took part in the decision process.

In more general terms, Allan (2003) identified the following paradigms:

1. **Pre-modernity**, connected to limited technical and organizational capacity.
2. **Industrial modernity**, characterized by engineering development in hydraulic works and large investment by either public or private. In mid-twentieth century, this "hydraulic mission" involved both western economies (Europe, the USA) and planned economies (communist countries) and was transferred to the developing countries.
3. **Reflexive modernity,** including:

   3.a. **Environmental awareness**, inspired by the environmentalist movement, in order to allocate or to leave more water to the environment.
   3.b. **Economic approach**, based on the recognition of the economic value of water as a scarce resource and oriented to foster water pricing, privatization and water market.
   3.c. **WR management as an integrated political attention** to engineering, economic and environmental aspects with the need to mediate the conflicting interests of users and management agencies. The focus is on institutions, public participation, social movements and NGOs.

Within this general interpretation of paradigm change, the evolution of water resources management in most recent decades in Italy is described shortly in Table 2.2, where three major stages are identified.

In the next sections, these three major paradigms will be detailed by analysing the Italian experience in water resources development during the last decades. The focus will be on few main water infrastructures and organization of water services at each stage. The legislative framework which has accompanied the whole process of development will be discussed in detail in Chap. 3. However, acts which have played a key role at each stage will be discussed here in order to better describe the nature of the conceptual transformation.

2.4   Water Resources Exploitation for Human Needs and Economic Growth

This first stage of water resources development in Italy is characterized by the exploitation of water resources in order to satisfy human needs and enhance the economic growth of the country. The engineering and economic approach can be considered the prevalent driver of the infrastructures built in the country and of the legislative actions and institutional adjustments.

As the water works are concerned, although the review is limited to the period after the unification of Italy (1861), i.e. the last decades of the nineteenth century and the first half of the twentieth century, it must be mentioned that the hydraulic

Table 2.2 Stages of evolution of water resources development in Italy

Stage		View	Planning focus	Decision-making features
I	WR exploitation for human needs and economic growth (prevailing engineering and economic approach)	River basin as a hydrological and geomorphological system	Structural measures to maximize yield to human needs and economic development of the country	Large funding program by public. Top-down approach
II	Focus on WR quality improvement and defense from water-related disasters	River basin as an ecosystem	Actions to reduce pollution and deterioration of natural rivers (with adversion against large structures such as dams) and to mitigate flooding risk	Concern on water quality and ecological requirements. Concern on flood disasters in river and urban areas Bottom-up approach
III	Towards an integrated sustainable and equitable WR management	River basin as a complex system (including social aspects)	Conjunct management of water and land by improving existing supply systems to maximize welfare in a social equitable way without compromising sustainability of ecosystems	Effort to shift from fragmentation to coordination with stakeholders' participation. Meeting top-down and bottom-up approaches

works in the Italian peninsula date back to at least the third-century B.C. (as demonstrated by ancient ruins), when Roman engineers built outstanding long channels for municipal supply, sewers for wastewater removal and field drainage for land reclamation.

Several water works, aiming to meet municipal, agricultural and industrial demands and to develop hydropower, were constructed starting from the country unification to the period following the Second World War (1950s). Here, only the infrastructures that can be considered symbols of the political strategies of the Italian State and the technical competence of Italian engineering will be taken into account.

Also, it could be worthy to mention that the expertise of Italian contractor firms in that period was recognized also in foreign countries, where the Italian advanced engineering technology was applied to build several hydraulic works, particularly dams, hydropower plants, irrigation networks and treatment plants.

The legislative framework of this period is characterized by a slow transformation of the modalities of water resources management and soil conservation, moving from a prevailing private initiative towards an increasing role of public administration, both in planning and construction of facilities. In fact, the Royal Decree 1775/1933 provided a regulation of the private withdrawals for various uses by means of a "licensing rule", which fostered for many years the growth of hydro-electric power and the extension of irrigation, but, at the same time, a few acts indicated the will of central government to increase the role of the state by means a more incisive planning activities. In particular, Law 184/1952 established the first river regulation plan, and Law 129/1963 identified the criteria for a drinking water master plan, developed in all regions by the peripheral structures of the Ministry of Public Works, since the regions, though established in the Constitution Chart, had not begun their activities.

2.4.1 Irrigation in the Po Valley

The major projects for the irrigation in the Po Valley, developed after the unification, follow an ancient tradition, with roots in the thirteenth century, like the construction of Naviglio Grande, mentioned in Chap. 1, which connected the Ticino river with the surroundings of Milan for irrigation and internal navigation, and Naviglio Piccolo from Adda river built in the fifteenth century. The new hydraulic works consisted of several canals to allow withdrawing of water from river Po (Cavour canal, started in 1863), from Ticino river (Villoresi canal, started in 1877) and from Adda river (Vallecchi canal, started in 1887). These canals allowed irrigation of the Po Valley in Piedmont and Lombardy with the advantage of withdrawing from rivers, which benefit of the regulation in the Pre-Alpine lakes, such as Maggiore on Ticino, Como on Adda, Iseo on Oglio and Garda on Mincio, beside the storage effect of snow and glacier. During the last decades, the network of irrigation canals was extended to the Emilia-Romagna region, thus covering almost the whole Po river plain. Most of the canals derive from the northern tributaries or directly from river Po to irrigate more than 30 irrigation or land reclamation districts in the plain at right (Fig. 2.5).

2.4.2 Reclamation of Marshes

The request of more land for increasing the agricultural production, and the need of improving living conditions of local people, particularly hit by endemic malaria, forced the government to develop large reclamation works in Veneto and in marshy zones of the Central Peninsula (especially in Lazio) and Sardinia. In the flat lands of Lombardy, the traditional meadow practice ("marcite") was extended to large areas, thus contributing also to the disposal and treatment of urban wastewater.

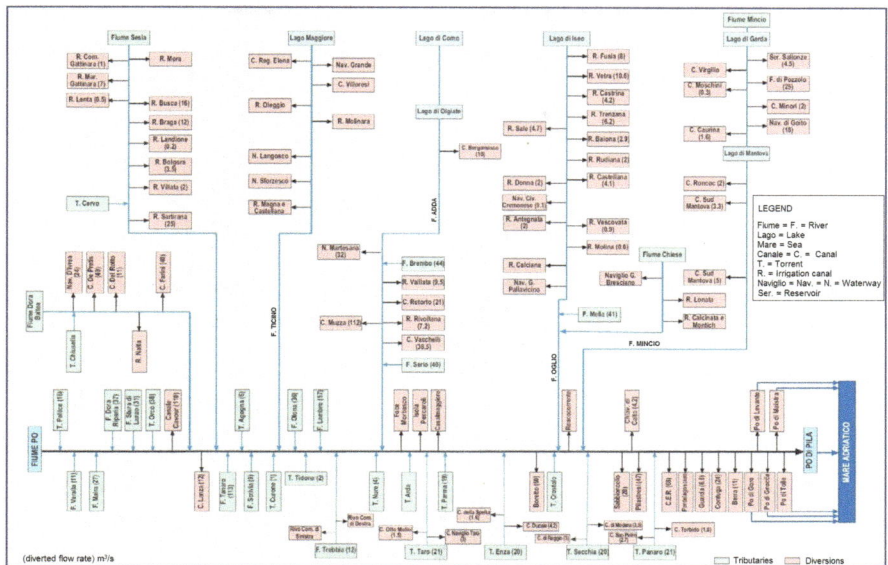

Fig. 2.5 Irrigation canals in the river Po valley. (Source: AdBP 2010)

2.4.3 Apulian Aqueduct

Among the several infrastructures built for municipal water supply, the Apulian aqueduct has an important place, as it was considered the largest municipal aqueduct of the world in the first decades of the twentieth century. Apulia is an arid region in the south of Italy without permanent rivers and with limited groundwater resources. Since the last decades of the nineteenth century (and in particular when a cholera epidemics struck the provinces of Bari and Brindisi), politicians of the region felt that local resources were insufficient to meet water needs for drinking and irrigation. Standing the failure of local authorities to solve the problem, the central government decided to build an aqueduct to convey water from the springs of Caposele in Campania, located hundreds kilometres far. A private firm was initially charged for the construction (1906), acting under the "concession role", but in 1919, after the First World War, a state body was established to complete the work and to provide operation and maintenance. The main canal from the Caposele springs to the village of S. Maria di Leuca was designed for a discharge of 6.34 m³/s, with 99 tunnels, 91 bridges and 6 siphons. The whole aqueduct with the canals and pipelines, built till 1937 to supply 301 urban centres in Apulia and 13 in Basilicata, Campania and Molise, had a length of 2290 km, more than 1300 km of urban networks, 37 pumping stations and 179 reservoirs (Fig. 2.6).

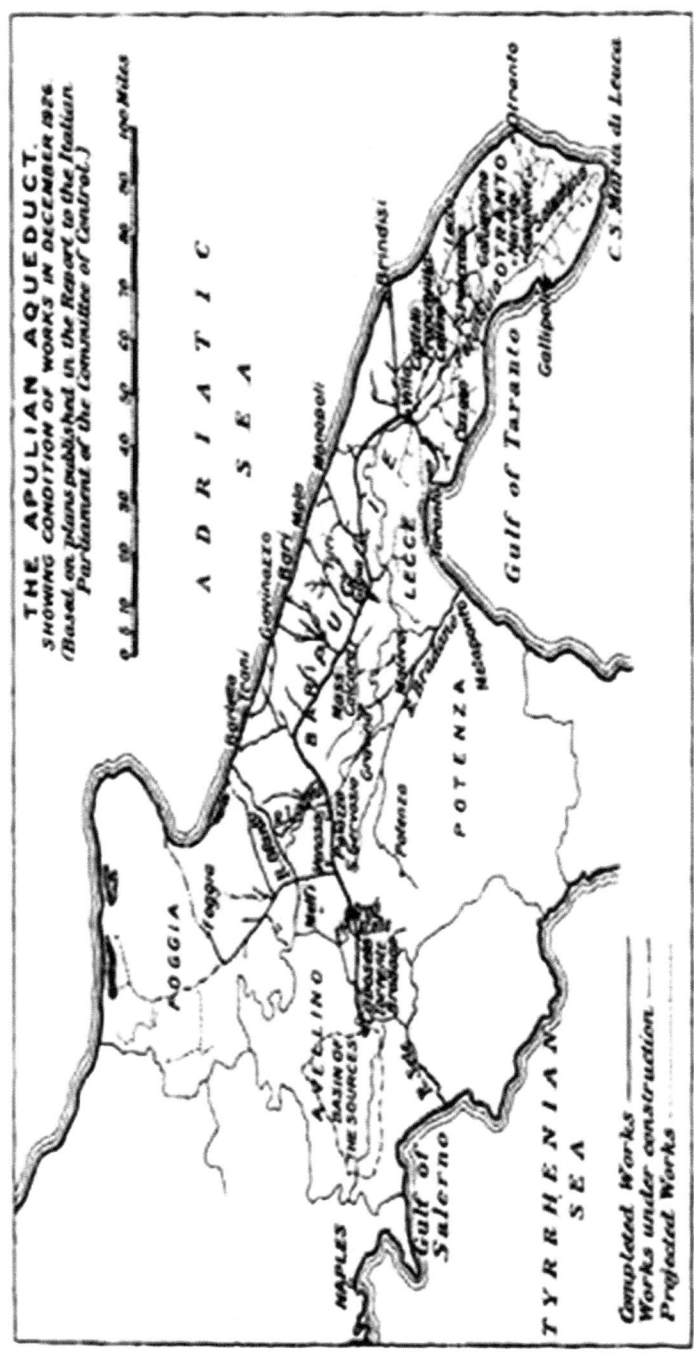

Fig. 2.6 Map of Apulian aqueduct. (Source: Viggiani 2002)

2.4.4 Improving Drinking Water Supply Schemes

Beside the Apulian aqueduct, numerous other water supply schemes were constructed in several parts of the country during the first half of the twentieth century with the purpose of improving the drinking water supply of the large urban agglomerations and of the population living in the countryside, which, in many cases, were supplied by poor local sources. Among the most important works of the period, the aqueduct designed for the peninsula of Istria (1930–1943) should be mentioned. Istria at that time belonged to the Italian territory and nowadays is part of nearby countries. Other long aqueducts were constructed in regions with limited groundwater resources, such as that in the west part of Sicily built by Ente Acquedotti Siciliani (established in 1942). After the Second World War, the need of providing water supply to islands in the Naples Gulf drove to identify innovative solutions such as setup of undersea aqueducts for Procida and Ischia islands (1954–1956).

During the 1930s, large works were completed for Rome, using some far groundwater sources in the Apennines, thus bringing the city's per capita water availability to the level of the best equipped cities. The need of increasing water availability also in places with less groundwater sources oriented the responsible authorities to rely on the surface water to be extracted from rivers and treated in advanced drinking water plants, such as those of the river Po for Turin and the river Arno for Florence. These circumstances, among others, are at the base for the development of complex water supply systems.

2.4.5 Hydroelectric Power

The use of water resources in the sector of electric power generation played a key role since the 1920s. After the First World War, Italy had the necessity to transform the war industry towards other productive fields, but it was lacking of the coal, which was the main source for the industrial development of other European countries, such as Germany, France and Great Britain (Angelini and Dolcetta Capuzzo 1978). Therefore, the "white cool" became the most adequate alternative to the fossil fuels. Since 1890 until the First World War, the first run of hydropower plants were designed to use directly the flow of the river, through an intake, an open channel, followed by a head tank, a pressurized conduit and a power station. In the period between the two wars (1918–1939), most of the new plants included a reservoir, in order to ensure a better utilization of water resources by means of inter-seasonal regulation. Under the strain of a growing demand of energy, more complex schemes were developed, considering in a unique system several plants with seasonal or multiyear storage reservoirs in the upper part of the catchment and plants with daily regulation downstream.

Up to 1960s, the hydroelectric use of water resources gave a significant contribution to the "Italian economic miracle". Several plants were constructed with high

Table 2.3 Number of dams built till 1959 and their distribution among the different uses

Use	Until 1889	1890–1899	1900–1909	1910–1919	1920–1929	1930–1939	1940–1949	1950–1959	Total 1920–1959 No.	%
Hydroelectric	0	0	8	24	79	47	35	96	257	81
Joint hydroelectric and other uses	2	1	0	2	5	1	1	14	21	7
Other uses only	8	1	0	1	2	10	1	25	38	12
Total	10	2	8	27	86	58	37	135	316	100

dams and large reservoirs, in order to supply electricity to the urban communities and the industrial settlements, nationwide. The new plants were designed to achieve integrated use of the withdrawn water, thus fostering the possible links with neighbouring catchments with multipurpose and/or sequential use of the available resources.

Among the various examples, the following systems can be mentioned: Valdossola on Toce river (Lombardy), Tirso in Sardinia, Sila system (including Ampollino, Arvo and Cecita reservoirs) in Calabria and Velino and Turano in Umbria (which included a reservoir for flood control).

In this period, the important role of the hydroelectric sector in the construction of large dams can be appreciated from Table 2.3, which lists the number of dams built in the decades from 1890 to 1959 for hydroelectric use only, for joint hydroelectric and other uses or for other uses except the hydroelectric. The hydroelectric use in the period 1920–1959 concerns 81% and the joint hydroelectric and other uses the 88% of all dams.

The most important hydroelectric schemes which were initiated in these years (although in a few cases, they were completed after 1960) include the systems of Valtellina (Lombardy), river Piave (Veneto), Sarca-Molveno (Trentino), Val d'Ultimo (Alto Adige), Sangro (Abruzzo and Molise), Platani and Simeto (Sicily), Taloro and Flumendosa (Sardinia). Figure 2.7 shows the map of the Piave hydropower system.

After 1960, the importance of the hydroelectric plants began to decrease, due to the prevailing growth of the thermoelectric power plants. Decisive was also the strong opposition of people to the new hydraulic facilities felt as a potential damage to the fluvial ecosystems, while the construction of high dams viewed as a possible threat to the safety of downstream areas.

During year 1962, the electricity sector, private until then, was nationalized, in the wake of examples of other European countries. A state-controlled entity (the Italian National Electricity Board "ENEL") was established, in order to guarantee an efficient and reliable electricity supply. ENEL incorporated all the previous private companies in the field of production, conveyance and local distribution of electric energy. The nationalization of electricity allowed to develop an integrated management over the whole national territory. At that time, the technical and eco-

Fig. 2.7 Map of the Piave hydropower system

nomic hydroelectric potential was exploited already almost entirely. So, the prevailing growth of production facilities turned to the hermoelectric plants supplied by fossil fuel. Nuclear power plants were also considered and constructed, but a nationwide referendum, motivated by the 1987 Chernobyl disaster, stopped definitively the nuclear sector.

The hydroelectric production assumed mainly the key role of meeting the peak needs of energy, due to their fast adaptation capability. According to this view, some new generation-pumping-storage plants were built as energy accumulators for peak demand. Some conventional hydroelectric facilities were also provided with pumping equipment. The main plants of this type include Gesso and Delio (Piedmont), Lete-Sava (Campania) and Anapo (Sicily).

Today, the generation-pumping-storage plants continue to play an important role in a system characterized by a growing development of renewable resources of wind and sun, in all Italian regions.

Cooling needs of thermal plants required hydraulic works to provide considerable flow to be withdrawn from rivers, lakes or lagoons, while the plants located in the coastal areas could benefit of sea water. Recently, the development of cooling towers reduced consistently the demand of freshwater for this purpose.

Since the 1980s the belief that the management of the hydroelectric sector could be more efficient in the hands of a public monopolistic company progressively reversed. Then, in 1992, following the liberalization of the electric market in Italy according to the European Directive, ENEL was transformed into a private company. The Italian State, through the Ministry of Economy and Finance, is still the main shareholder.

2.4.6   Complex Water Systems in Southern Italy

The action of the Southern Italy Development Fund for water infrastructures contributed to the improvement of water availability for human needs and economic growth. This action, started in 1950, and continued also after closing the Development Fund in 1984, in order to complete some still unfinished constructions.

The works included a large set of hydraulic schemes aiming to overcome the lack in water supply for different uses, in sanitation and in soil conservation. The most important works were planned in order to extend the irrigation to Southern Italy, where the scarcity of water resources, more than the adverse orographic, climatic and soil characteristics, had delayed the development of large irrigation districts. Several of these schemes became multipurpose water systems, aiming at meeting the increased municipal and industrial needs and improving the hydropower use. They were accompanied by reclamation works oriented to reduce the risk of flooding and to allow a good drainage of flat lands.

Such an idea had been proposed in the first years of the twentieth century by technicians, such as Angelo Omodeo (1876–1941), and politicians such as Francesco Saverio Nitti (1868–1953), in order to overcome the underdevelopment of the southern regions of Italy by means of a comprehensive planning of river basin, including both hydropower and irrigation uses, driver regulation and soil conservation. It was based on the alliance of private financial groups interested to hydropower and farmers interested to develop agriculture (Barone 1986). It was pursued by as initiative of the public commitment of Southern Italy Development Fund. Among the several schemes built in Campania, Basilicata, Apulia, Calabria, Sicily and Sardinia, the Sinni-Agri-Basento-Bradano (Basilicata and Apulia), the Campidano of Cagliari (Sardinia) and the Simeto river system (Sicily) are the largest multipurpose systems.

The current water infrastructures present very large difference in the north and south of the Italian peninsula, due to different climatic and hydrographic features of the two areas and also to the action of the above-mentioned Southern Italy Development Fund (ITAL-ICID and Italian Hydrotechnical Association 1987). In the northern regions, two separated types of systems exist: the aqueducts for municipal drinking use, typically supplied by wells and/or springs, and complex water resources systems for hydropower and irrigation uses. Instead, in the southern regions, most of water resources schemes include multipurpose regulating surface waters for municipal, agricultural, industrial and hydropower uses.

2.4.7 The Italian Patrimony of Dams

The most significant result of the engineering phase of water development is the construction of dams for hydropower generation and, mostly in more arid regions, for inter-seasonal regulation of water for multipurpose use. Today, Italy has about 520 large dams higher than 15 m with a total storage capacity larger than 12 billion cubic metres. Italy is the European country with the maximum number of dams after Spain. The distribution of the reservoirs, grouped in three classes of storage capacity in different Italian regions according to the data of the Ministry of Infrastructures and Transports, Direction for Dams and Water Infrastructures, is reported in Fig. 2.8.

2.5 Focus on Water Resources Quality Improvement and Defence from Water-Related Disasters

In a second stage of water development, the attention was focused on the water quality and pollution control. This was done referring to water resources issue and on the concern of saving lives and reducing the damages concerning water-related disasters.

This phase was fostered by the new ecological awareness, emphasized by several opinion movements (the so-called green movements), and driven by the directives of the European Commission concerning measures to reduce pollution and improve

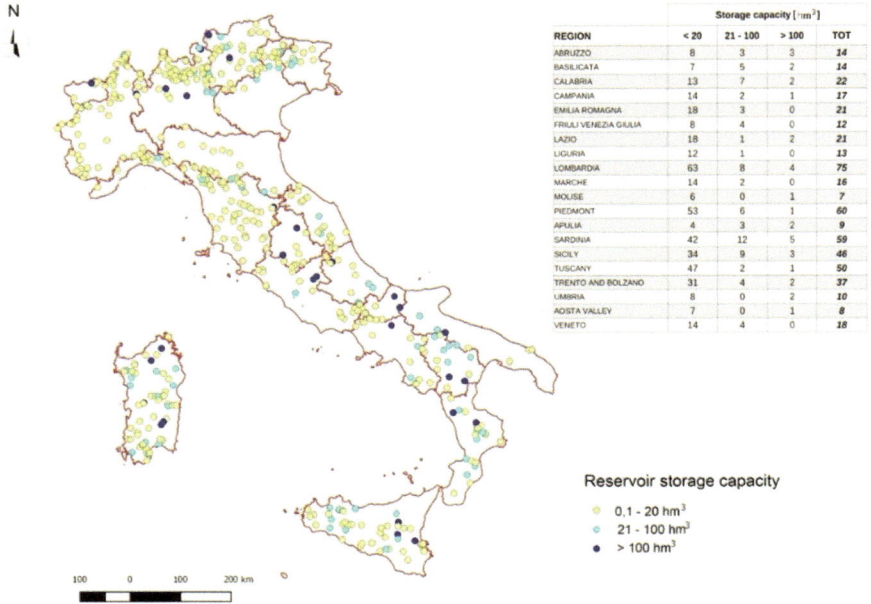

REGION	Storage capacity [hm³]			
	< 20	21 - 100	> 100	TOT
ABRUZZO	8	3	3	14
BASILICATA	7	5	2	14
CALABRIA	13	7	2	22
CAMPANIA	14	2	1	17
EMILIA ROMAGNA	18	3	0	21
FRIULI VENEZIA GIULIA	8	4	0	12
LAZIO	18	1	2	21
LIGURIA	12	1	0	13
LOMBARDIA	63	8	4	75
MARCHE	14	2	0	16
MOLISE	6	0	1	7
PIEDMONT	53	6	1	60
APULIA	4	3	2	9
SARDINIA	42	12	5	59
SICILY	34	9	3	46
TUSCANY	47	2	1	50
TRENTO AND BOLZANO	31	4	2	37
UMBRIA	8	0	2	10
AOSTA VALLEY	7	0	1	8
VENETO	14	4	0	18

Reservoir storage capacity

○   0.1 - 20 hm³
◉   21 - 100 hm³
●   > 100 hm³

100   0   100   200 km

Fig. 2.8 Distribution of dams in Italian regions

the quality of water bodies. The evolution of law played a very important role. The Law 319/1976 ("Merli Act") regulated the water pollution control, introducing standard quality levels for effluents to be discharged into water bodies, with the development of sewerage systems. A fiscal charge was also provided for the wastewater treatment, in accordance to the "polluter pays" principle, with penal liability for the polluters. Following the transfer to the regions of several duties in the water field, the new Law entrusted the regions of drafting a Water Restoration Plan, which had to provide a survey of polluted discharges into watercourses, fostering at the same time the design and construction of municipal sewerages and treatment plants. According to the goals of the European policy on water, several decrees acknowledged the EEC Directives on water quality and protection of water bodies.

The establishment of the Ministry for Environment, in 1986, can be considered as a consequence of the new European principles on environment. This new Ministry was entrusted of many duties which formerly were covered by the Ministry of Public Works. Example of a new awareness of the protection of fluvial ecosystems are the constraints on the withdrawal from rivers and streams with the obligation of maintaining a "minimum acceptable flow" (MAF), or a "minimum in-stream flow" (MIF), identified by several institutions, like the River Basin Authorities and the local administrations.

The establishment of the Law 319/1976 fostered the expansion programs of several sewerages and the planning and construction of new wastewater treatment plants. Different criteria were adopted for the design of collection works and for treatment schemes. While in a first phase the prevailing solution was a combined system, including wastewater from households and rain water runoff with outfall in some receiving bodies, afterwards a separate system was preferred, with two sewer networks, one for wastewaters and one for rain water drainage. The choice of wastewater treatment plants, serving one single municipality, was replaced by the design of larger plants able to serve several municipalities, according to the guidelines of the regional plans. A priority effort was devoted to expand the sewerage systems and to construct the wastewater treatment plants in the main urban areas of the country.

As an example, the Municipality of Rome, which at the establishment of Merli Act had a sewer network limited to the historic centre, promoted the construction of two wastewater treatment plants, respectively Rome-Ostia and Rome-east. In operation since 1974, these plants enlarged the sewer system very significantly. The company for management of municipal water services in Rome extended the sewerage network and built the new plants of Rome-north (1981) and Rome-south (1985), beside the expansion of the two existing plants. Thus the population served by the sewerage in 1976 (about 300,000 inhabitants) increased to 2.63 million inhabitants at 1999, increasing the total sewers length from 191 km to 390 km.

Only recently, the city of Milan has solved the problem of an effective treatment of urban wastewaters. In fact, in spite of the presence of ancient sewer systems since the Middle age, the disposal of urban wastewater till the first decade after the Second World War was the spreading in nearby meadows ("marcite"). This solution implied a sort of natural treatment, thus resulting in a primordial case of wastewater reuse for irrigation. The plants of Noseda and S. Rocco, completed in 2015, treat 90% of the wastewaters (about 2.3 million of equivalent inhabitants) and through historic canals (e.g. the Roggia Vettabbia) provide the water for the irrigation of almost 100 km^2 of agricultural areas.

In the metropolitan area of Naples (with about 4.5 million inhabitants), several sewerage systems and wastewater treatment plants have been built, which however did not avoid a sentence by the European Court of Justice for missing observance of water quality standards. The circumstance shows that construction does not imply service automatically! An update of the treatment processes started recently (2015) for the main wastewater treatment plants (Cuma, Napoli North, Casertana area, Acerra, Foce Regi Lagni).

In the 1980s the city of Palermo (about 670,000 inhabitants) planned works to complete the sewer system and built two treatment plants for the whole urban area, thus avoiding the outfall in the industrial harbour. Unfortunately, only one treatment plant (Acqua dei Corsari) is active; therefore, it is estimated that about half only of the produced wastewater is treated currently. Indeed, the underwater conduit of the other treatment plant (Fondo Verde) couldn't be built, due to the constraints of a protected area. Now it is planned to transfer wastewater to the first plant.

This phase of development is characterized also by several initiatives against the water-related disasters, particularly protection from floods and landslides and mitigation of drought risk. The evolution of legislation on these issues, beside the general presentation of Chap. 3, is discussed in Chap. 11 for flooding and Chap. 12 for drought and water shortage.

After the efforts of commission for flood defense and soil conservation problems (Commissione Interministeriale 1970), aiming at detecting priorities of works to reduce flood damages throughout the main Italian rivers, a more detailed list of water works for the prevention of flood risk was proposed in the river basin plans developed for several rivers by the authorities of the national, interregional and regional basins established by Law 183/1989.

The research activities carried out by the National Group for Defense from Hydrogeological Disasters (GNDCI), of the National Research Council, activated a large number of Italian universities and research institutions in contributing to studies about the problems of flood and landslide risk mitigation, developing methodologies for estimating the design floods.

Unfortunately, in spite of improved knowledge acquired by these researches and of the high number of public works included in the planning tools for hydraulic defense of several river basins, the inadequate financial resources dedicated to flood prevention did not produce a substantial reduction of flooding risk in many Italian regions.

Most of investments for flood defense were funded by the civil protection to support emergency actions for rescue, assistance, recovery, temporary or permanent relocation and restoration of infrastructures and buildings, with positive effects in mitigating the impacts on affected population but with limited improvement of the resilience of the flood-prone territory.

2.6   Towards an Integrated Sustainable and Equitable Water Resources Management

According to the increasing awareness on the potential threats of environmental pollution on the quality of life and the efforts for improving efficiency and effectiveness of public services, as well as for achieving a better solidarity among the citizens, a new phase of water development arose in recent years, and a more comprehensive approach to water management has been accepted progressively. It includes the new concept of sustainability (Brundtland 1987) and the ethical issues (Selborne 2000; Rossi 2008; Rossi 2015). In particular, the public opinion has recognized progressively the idea that a development is sustainable if it is compatible with the needs of future generations and does not damage the systems supporting the life in our planet (air, water, soil and biological systems). Thus it will be able to maintain a flux of goods and services deriving from self-regenerating natural resources and to guarantee that the social systems assure a fair distribution of benefits at familiar, local, national and international levels. Besides, the concept of hydro-solidarity (Falkenmark and Folke 2002), as well as the goal of equitable distribution of scarce resources, is progressively becoming an international reference requirement during the last World Water Forums. In Italy the principle of "water access as human right", recognized by the Assembly of United Nations in 2010, contributed likely to the success of the referendum on June 2011 against the privatization of municipal services. Moreover, an enhanced role of citizens and stakeholders in all decision-making processes has been invoked, starting from transparency in public administration acts till the public participation procedures.

Several positive aspects of this phase of water development can be identified, such as:

- Technological innovations that facilitate water resources planning (e.g. GIS, remote sensing, etc.) and operation of water systems (e.g. DSS, remote control devices)
- Requirements of European Directives that obligate national and regional governments to cope with actual management problems and ecological needs
- Improving governance of water problems at different levels, by means a wider information and public participation
- Growing awareness of considering the connection of specific problems of water management with external processes affecting water resources as global changes (e.g. uncontrolled urbanization, soil use modification, climate change, changes in world economy, etc.)
- Trend to face the water-energy-food nexus by linking their development strategies, in particular by reducing energy footprint of water facilities and water footprint of power generation systems

The legislative framework of this phase of development includes several very important acts, which characterize the current rules of water management. But it pursues some contradictory objectives. Also, it meets difficulties in the actual implementation of the reforms which it introduced in order to improve the management of the water services.

The Law 36/1994 ("Galli Act") recognized the public nature of all water resources, including groundwater, to be used according to the criteria of solidarity and environmental safeguard. It reformed the organization of municipal water services by means of major innovations such as (i) the management as a single service of all the components of the urban water system (supply, sewage and wastewater treatment); (ii) the separation of the function of providing the service, to be entrusted to an utility (public, private or public-private), from the role of government and performance control by a public authority on behalf of all individual municipalities; (iii) the establishment of organizations for users guarantee, including a National Committee to control the service quality and the tariff, with yearly reports to the parliament; and (iv) the adoption of full cost recovery pricing criteria including investments, operation and replacement costs.

After severe water-related disasters, Law 267/1998 (Sarno Act) and Law 365/2000 (Soverato Act) tried to improve the measures for coping with flooding and other water-related calamities, by means of plans, to be considered as part of river basin plans introduced by Law 183/1989. They aimed at fostering the identification of areas under flooding hazard and risk, at providing a step-by-step implementation of protection measures and at developing procedures of warning system for hydraulic risk. The civil protections (established by Law 225/1992) had the commitment of drafting the emergency plans.

Referring to pollution control, new directives were issued by the Legislative Decree 152/1999, (issued by government under commitment by the parliament) which modified the previous rules of the Law 319/1976. It introduced specific levels of water quality with reference to each use of water body (e.g. drinking water production, bathing, etc.), and established requirements for vulnerable zones and for areas sensible to eutrophication. This same Decree entrusted the regions to draft the plan for water protection, to be considered as part of the river basin plans, introduced by Law 183/1989.

Finally, the Legislative Decree 152/2006 has confirmed the previous acts, with a few changes and innovations according to recent European Directives, within a comprehensive Environment Code, dealing also with the Environmental Impact Assessments, the solid waste disposal, the remediation of polluted sites, the atmospheric emissions and other topics.

Besides, national guidelines to estimate "ecological flow" were issued by the Ministry for Environment Land and Sea in 2004 and updated in 2017 in order to protect the fluvial ecosystems. Chapter 10 will deal in detail with the topic of "minimum in-stream flow" and "ecological flow" in Italian rivers.

A broad access to information on the environmental effects of water works is allowed by the procedures for Environmental Impact Assessment (EIA). They were introduced by European Directive 85/337/EC and acknowledged by Italian legislation, as mandatory for the most relevant water works (Decree of Prime Minister 337/1988) and entrusted to the regions for minor works (Decree of the Republic President 12/4/1996).

The aim of guaranteeing efficiency, effectiveness and cost coverage in municipal water service has been enlarged by Law 290/1999 in terms of customers' satisfaction, which issued a service chart for water supply customers. The chart has been updated by a 2015 deliberation of the Authority for Electric Energy Gas and Water System (AEEGSI), which replaced the Committee for Surveillance on Use of Water Resources, and lately took the name of Authority for Regulation Energy Networks and Environment (ARERA). Finally, a specific fund was established by Law 221/2015 in order to assure drinking water access to low-income users.

Figure 2.9 shows that the acceptance of the sustainable development principles and of the enhanced role of citizens and stakeholders participation, as innovated by water laws, has contributed to improve some important issues in the water field: solving conflicts between withdrawal for off-stream uses and ecological uses, broader access to information on environmental impacts of water works, better quality and equity of urban water services (Rossi and Ancarani 2002). These significant progresses refer to (i) guidelines for minimum in-stream flow (now ecological flow) assessment, (ii) EIA procedure for water works, (iii) water supply service chart and (iv) guarantee of drinking water access to low-income users.

However this phase of water development presents some weakness points, such as:

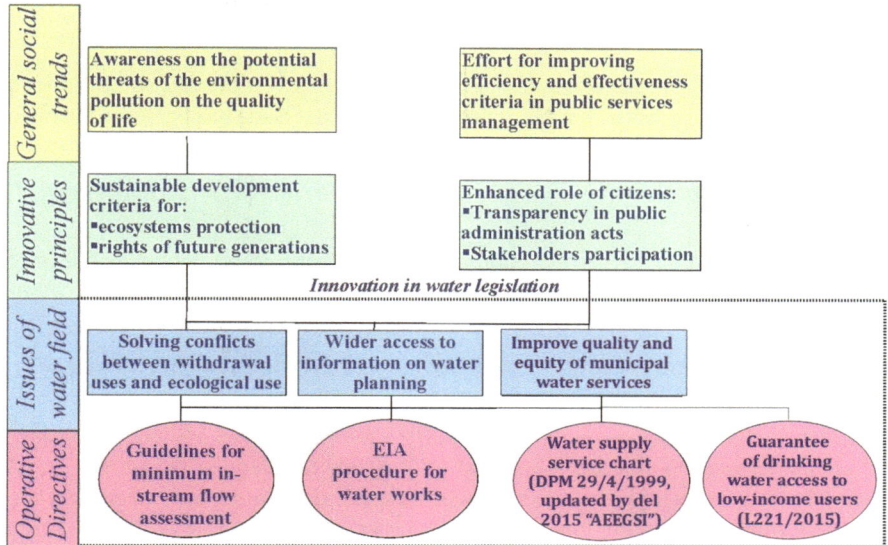

Fig. 2.9 Social trends and principles affecting water issues and related legislation

- Very limited financial public resources for all hydraulic works (supply, pollution control, mitigation of floods and droughts).
- Contradictory decisions about the reform of water legislation and improvement of governance, particularly for municipal water services.
- Dramatic delays in the construction of sewerage networks and wastewater treatment plants, particularly in southern regions with the consequence of trials and sentencing stages for breaking the rules of the European Directives 91/271/EC, to which some deliberation of the Inter-ministerial Committee for the Economic Planning (CIPE) are trying to react by using the European funds as funding source for the lacking infrastructures.
- Scarce results of the river planning approach, particularly because of delays in drafting and in the updating of the plans. In turn, this is due to the delay in the operational start of the new districts, which had to lump the large number of River Basin Authorities of the country. The most severe gaps refer to the water quality status of water bodies and to the knowledge of amount of current water deliveries to users and related economic data.
- The reform of municipal water services must be considered unsatisfactory in many regions, since the fragmentation of decision centres has been overcome only partially and the choice of financing new works by the tariff has led to a dramatic reduction of new investments.

However, the general judgment of this phase of water development can be considered positive, since some general environmental and social paradigms have found progressive acceptance in the public opinion and in legislative innovations.

2.7   Latest Controversial Trends

In the international debate which led to the vote of the Assembly of United Nations recognizing water access as human right, a controversial issue has been the intrinsic value of water. From a side, the principle approved by the Dublin Conference (1992):

> water has an economic value in all its competing uses and should be recognized as an economic good

was called to support the privatization of the water services indicated as the only possible solution to avoid water losses and inefficient management. From the other side, the conclusion of the Rio de Janeiro meeting (1992) that states:

> Integrated Water Resource Management is based on the perception of water as an integral part of the ecosystem, a natural resource, and a social and economic good

was indicated as the basic principle to support a fair solution which would consider the social component as a prevailing tool for possible coverage of the service costs. It should be based on general taxes or consumers' mechanism able to protect the right to water access by the low-income users.

These different positions have been the central topics of the political debate developed for the Italian referendum of year 2011 that abrogated the preference to private companies and cancelled the part of the tariff allocated to cover the capital costs, in order to reduce the convenience of the private subjects to invest in water supply and sanitation services. The controversy is still open, because a draft of law presented to the parliament and oriented to limit the assignment of the water municipal service to private companies conflicts with the European orientation of fostering competition among private, public or public-private partnership. Unfortunately, the other problems of water resources protection and management and of soil defence from water-related disasters, more significant and urgent of this ideological controversy, loose the attention by public opinion, except when disasters due to pollution or flooding or drought catch the interest of newspapers and media. Among the above-mentioned weaknesses, the scarce financial investments in the water fields, the complexity of the legislation and of the decision-making process with overlapping of institutions as well as the delays in planning process and in timely construction of water works represent the key issues. Furthermore, the decrease of the role of the experts from university and research institutions to help with the politician's decisions, as occurred in the very last decades, has an adverse effect on the quality of new rules and on the effectiveness of the governance of water resources for the mitigation of water related disaster risks.

References

AdBP (2010) Autorità del Bacino del Po. Piano di gestione del distretto idrografico del fiume Po [River basin management Plan] (in Italian), Parma, p 137

Allan JA (2003) Virtual water: the water, food and trade nexus. Water Int 28(1):4–11

Angelini A, Dolcetta Capuzzo M (1978) Gli impianti idroelettrici nell'ultimo cinquantennio [Hydroelectric plants in the last fifty years]. In: A.I.I (Associazione Idrotecnica Italiana), Cinquanta anni di ingegneria dell'acqua [50 years of water engineering] (in Italian). Japadre, L'Aquila, pp 279–327

Barone G (1986) Mezzogiorno e modernizzazione [Southern Italy and modernization] (in Italian). Einaudi, Torino

Brundtland GH (1987) Our common future. Oxford University Press, Oxford

Commissione Interministeriale (1970) Atti della Commissione Interministeriale per lo studio della sistemazione idraulica e della difesa del suolo. [Commission for flood defense and soil conservation problems] (in Italian). Istituto Poligrafico dello Stato, Roma, p 900

Dooge JCI (1999) Hydrological science and social problems. Arbor 646:191–202

Falkenmark M, Folke C (2002) The ethics of socio-ecohydrological catchment management towards hydrosolidarity. Hydrol Earth Syst Sci 6(1):1–9

Garrido A, Llamas MR (eds) (2010) Water policy in Spain. CRC Press/Balkema, Leiden, p 234

Hall W, Dracup JA (1970) Water resources systems engineering. McGraw-Hill, New York

ITAL-ICID and Italian Hydrotechnical Association (1987) Irrigation and land reclamation in Italy. XIII International Congress of Irrigation and Drainage, Casablanca, September 1987

Rossi G (2008) Prospettive etiche nell'uso delle risorse idriche [Ethical perspectives in water resources use] (in Italian), Bioetica e Cultura, XVI 2:49–75

Rossi G (2011) Water resources development. Towards an integrated, sustainable and equitable water management. In: Proceedings of Indo-Italian workshop on advances in Fluvial hydraulics and water resources development and management, Pune (September 15-16 20119, pp 279–295

Rossi G (2015) Achieving ethical responsibilities in water management: a challenge. Agric Water Manag 147:96–102

Rossi G, Ancarani A (2002) Innovation in water legislation in Italy: ecosystem protection, and stakeholders participation. Int J Water 2(1):17–34

Selborne L (2000) The ethics of freshwater use: a survey. World Commission on the ethics of scientific knowledge and technology. UNESCO, Paris

United Nations (2015) Transforming our world: the 2030agenda for sustainable development. http://www.un.org/ga/search/view_doc.asp?symbol=A/RES/70/1&Lang=E

US WRC (Water Resources Council) (1973) Water and related land resources: establishment of principles and standards for planning, Federal Register, 38 (174): 24778-24869

Viggiani C (ed) (2002) The Apulian aqueduct. Engineering 1928, Hevelius, Benevento, p 96

Yevjevich V (1995) Effects of area and time horizons in comprehensive and integrated water resources management. Water Sci Technol 31(8):19–25

Chapter 3
Evolution of Water Legislation

Giuseppe Rossi

Abstract The chapter refers to the development of Italian water legislation during the last century in order to identify the main innovative principles which were introduced in water management and soil defense fields, recently also under the pressure of the European Directives. The evolution of Italian water legislation has been affected by the ongoing acceptance of a few general principles, namely, (i) the growing role of the state in regulating private access to water uses and in planning the allocation of water resources among various sectors; (ii) the need to face pollution of water bodies, in order to guarantee human health and environment protection: (iii) the choice of a comprehensive approach in the river basin planning, including water resources use, water quality improvement and flood defense; (iv) the effort to assure good quality levels of municipal water services by an efficient management; and (v) the shift from structural to nonstructural measures. Some of these innovations can be considered as a consequence of the increasing awareness of the need to protect ecosystems and to preserve the rights of future generations, as well as to improve access to information and increase participation of citizens and stakeholders in the decision-making process. When dealing with defense from water-related disasters, there is a rising need of a more coordinated effort to face the risk of extreme hydrological events. Finally, the reasons for a poor implementation of plans required by European Directives are discussed, and some necessary changes in water legislation to adapt water management to the expected changes of society and climate are identified.

3.1 Introduction

During the last decades, the Italian water policy has been innovated deeply by several legislative acts based upon a few general political orientations, such as the environmental protection, the distribution of roles between state and regions, and the increasing role of private companies in the management of public services. The

G. Rossi (✉)
Department of Civil Engineering and Architecture, University of Catania, Catania, Italy
e-mail: grossi@dica.unict.it

necessity of adapting Italian legislation to the European Directives has been a key factor to foster the reform of the legislative framework. An understanding of the development of the Italian water legislation in the last century requires to identify the main innovative principles introduced into the regulation of water and soil planning, water services management, water pollution control and flood risk mitigation. Particular attention has to be paid to the relationship between Italian law and European Directives, which have progressively affected the national legislations of all states that have participated from the beginning in the European Communities while being partners of the European Union at present.

3.2 Early Development of Water Legislation in Italy

3.2.1 Laws Till 1989

The year 1989 was an important divide in Italian water legislation which initiated an era of major planning innovations. A simplified outline of the development of the water legislation in Italy until 1989 is presented in Table 3.1, which lists the main laws and decrees issued on the three main sectors of water management, namely, water resources uses, water pollution control and soil conservation (especially flood risk mitigation). The list of the main European Directives is also shown, together with the Italian legislative acts, which provide transposition of each directive.

For many decades, the Italian legal framework for the utilization of water resources was based on Royal Decree (RD) 1775/1933. It was issued with the aim of increasing hydroelectric use and of providing the regulation of private withdrawals (particularly for irrigation and hydropower) on the basis of a licensing system, although the activity of water resources planning by the state was lacking (Greco 1983). In the same year, another fundamental law (RD 215/1933) regulated the responsibilities of the organizations for land reclamation (Consorzi di bonifica), which were entrusted of developing an integrated plan in the fields of soil conservation, irrigation, rural roads and rural electricity supply.

A previous Act RD 523/1904 had introduced a classification of the hydraulic works along the watercourses. Five categories were identified according to the importance of the watercourse and duties of the state, of the local institutions (provinces and municipalities) or consortia of private owners in order to provide for their construction and maintenance. Probably, today the value of this decree is negligible due to significant changes in the responsibilities of the different organizations of the Italian Republic, but the decree has been recently mentioned within the new rules for sharing the duties of the state and regions concerning hydraulic works on watercourses, through the Legislative Decree (DLgs) 112/1998 (Art.89).

The first tool for state-level planning against flood risk can be recognized in Law 184/1952. It was issued after the catastrophic flooding of the Po River (1951), in order to provide a national plan for river regulation. The plan, published in 1954,

Table 3.1 Main Italian acts on water resources and soil conservation and related European Directives issued before Law 183/1989

Italian acts on water resources and soil conservation issues			European Directives (Italian acknowledgement acts)
Water resources uses	Water pollution control	Soil conservation and flood risk mitigation	
		RD 523/1904 Hydraulic works on rivers	
RD 1775/1933 Water licenses and hydropower development		**RD 215/1933** Integrated land reclamation	
		L184/1952 River regulation plan	
DPR 1363/1959 Regulations on dams			
L 1643/1962 Establishment of ENEL (Ente Nazionale Energia Elettrica)			
L129/1963 Drinking water supply master plan			
DPR 8/1972 Transfer of functions from state to regions on aqueducts and public works			
L382/1975 Transfer of functions from state to regions			**Dir 75/440 EEC** Surface water quality for drinking use abstraction (DPR 515/1982)
	L319/1976 Water pollution control Revised by **L650/1979**		**Dir 76/160 EEC** Water quality for bathing (DPR 470/1982 & DM Health 29/1/1992) **Dir 76/464 EEC** Measures for water pollution control, enlarged by **Dir 86/280** (DL 130/1992)
Del I.M.C. 4/2/1977 Criteria for rational water use	**Del I.M.C. 4/2/1977** Guidelines for sewage, wastewater treatment and discharge		

(continued)

Table 3.1 (continued)

Italian acts on water resources and soil conservation issues			
Water resources uses	Water pollution control	Soil conservation and flood risk mitigation	European Directives (Italian acknowledgement acts)
			Dir 78/659EEC Water quality for marine life (DL132/1992) **Dir 78/659EEC** Sampling methods of surface water for drinking use
			Dir 79/923EEC Water quality for shellfish
			Dir 80/68EEC Measures for protecting groundwater from pollution (DL 133/1992) **Dir 80/777EEC** Trade of mineral waters D. Lgs 105/1992 **Dir 80/778EEC** Water quality control for drinking use (DPR 236/1982)
	L 153/1981 Charges for sewage and wastewater treatment		
			Dir 85/337EEC Environmental Impact Assessment (EIA) for water works
	L 349/1986 Establishment of Environment Ministry and EIA		**Dir 86/2788EEC** Characteristics of sewage sludge (DL 99/1992)
DPR 236/1988 Quality of water for municipal supply			
L 183/1989 Soil Conservation and Water Planning Act (modified by L 253/1990 and L 493/1993, with criteria for implementation in DPCM 233/1990			

DL decree law, *DLgs* legislative decree, *DPR* Decree of the President of Republic, *Dell.M.C.* Deliberation of Inter-ministry Committee, *Dir* European Directive, *L* law, *RD* royal decree, *EEC* European Economic Community

provided the investments of 1550 billion liras in 30 years. Half of this sum was devoted to the construction of works along rivers (commitment of Ministry of Public Works), while the other half was devoted to forestal and agricultural actions (commitment of Ministry of Agriculture and Forestry). The Inter-ministerial Commission for hydraulic regulation and soil conservation (1970) expressed a very severe evaluation of the implementation of this plan, as it stated that "the public expenditure was very limited and not sufficient to provide even the ordinary maintenance".

The planning principle for drinking water supply was introduced by Law 129/1963, which established the drafting of the aqueduct master plan to be developed by the local offices of Ministry for Public Works. The aim was to impose constraints on water sources in order to satisfy urban water requirements as estimated for the year 2015.

During the 1970s the focus of water legislation was on water quality and environmental impacts of wastewater. As the principle of decentralization, established by the Italian Constitution (1947), had been implemented by means of the establishment of ordinary regions (1971), many responsibilities in water field have been transferred to the regions by DPR 8/1972 and by Law 382/1975. Thus, Law 319/1976 (Merli Act) entrusted the regions with the preparation of water restoration plans, including sewerage and wastewater treatment plants. The Merli Act regulated the pollution control through the introduction of standard quality levels on the effluent to be discharged in water bodies, and established a service fee for sewage system and wastewater treatment, in accordance with the *polluter pays* principle, while establishing a penal liability for polluters.

At a later time, several decrees defined technical guidelines for water pollution control, thus acknowledging the European Directives on water quality and environmental protection, adopted by the European Economic Communities (created by the Treaty of Paris, 1951 and the Treaty of Roma, 1957) in the 1970s and 1980s with focus on protecting waters of anthropogenic interest (drinking, bathing, fishing). Only during the second wave of European water legislation, i.e. in the 1990s, the focus was enlarged to the agricultural sector.

After the institution of the European Union (by the Treaty of Maastricht, 1992), a shift occurred towards a new more comprehensive legal and institutional mechanism to improve water management, in particular, by means of the 2000 Water Framework Directive (Canelas de Castro 2008). However, the acknowledgement of the European Directives on water quality in Italian legislation occurred in general many years after the directives had been issued. Furthermore, the rules of the Merli Act were revised by Law 650/1979, particularly for wastewater discharges into water bodies, and by Law 172/1995, which gave the regions the power to set up their own emission standards. The influence of the new European Directives on environment protection and sustainable development had already led to the establishment, by Law 349/1986, of the Ministry for Environment, with responsibilities for many issues which previously were covered by the Ministry of Public Works.

3.2.2   Law 183/1989

The most innovative aspects of the recent water legislation were introduced by Law 183/1989. It was the conclusion of a long process which started soon after the disrupting flooding of Florence and Veneto region (November 1966), under the drive of two technical-scientific works: (i) the De Marchi Committee for hydraulic defense and soil conservation problems (Commissione Interministeriale 1970) and (ii) the National Water Conference (Conferenza Nazionale Acque) for the problems of water resources uses, pollution control and institutional reform (CNA 1972). The new law aimed at adopting an integrated approach to water and soil conservation problems within the river basin boundaries, regardless of the administrative boundaries. It established the River Basin Authorities with the task of coordinating all the activities of water-related planning, water works construction and control.

The key concept of the act is the river basin plan, conceived as a tool to collect relevant information and to identify the needed actions for (1) flood defense and soil conservation, (2) water supply for different uses and (3) pollution control of water bodies (rivers, lakes and aquifers).

With reference to the first aim (soil conservation), it is often emphasized that this Italian law had envisaged a more complete and valid approach than the following European Directive 2007/60 EC, since Italian law considered both the flooding and the landslide risk according to the geological features of the territory and not just flooding risk as in the European Directive.

The law identified three levels of river basins, based on their importance, size and location:

- The first level considered basins of national interest under the direct responsibility of the state; it included six River Basin Authorities in the largest Italian river basins (Po, Adige, Piave, Arno, Tevere, Liri-Garigliano-Volturno).
- The second level considered basins of interregional interest including rivers whose territory belongs to 2 or more contiguous administrative regions, though not so important to be considered of national level; 18 interregional River Basin Authorities were established to be run jointly by the governments of the affected regions.
- The third level included basins of regional interest, belonging entirely to one single region. The law entrusted the regional governments with the setup of regional river authorities.

For the basins of national interest, each authority is composed of a technical committee, a General Secretary (appointed directly by the government) and an institutional committee (formed by the different ministers and the presidents of the governments of the regions located within the river basins).

Furthermore, the law established a National Committee for Soil Defense aimed at coordinating the actions, which however operated only for a short period. It also reformed the National Technical Services, which were transferred under the jurisdiction of the Prime Minister, with the task of organizing and managing the information system on hydro-meteorological data. Another important innovation was the introduction of concept and practice of the "minimum river flow", i.e. an ecological in-stream flow to be guaranteed downstream of diversion authorized by a withdrawal license.

In 2006, Law 183/1989 was abrogated formally, but all its fundamentals and principles were maintained, with only minor changes, in the context of the broader Environmental Code of Decree 152/2006 (Maglia and Galotto 2009).

3.3 Development of Water Legislation in Italy After 1989

3.3.1 Legislation Following Law 183/1989

Main laws issued after 1989 are listed in Table 3.2, according to the three main sectors of water management. The table includes also the main European Directives as well as the Italian transposition acts. The comments on the evolution of legislation for each sector are presented in the following sections.

3.3.2 Urban Water Service Management

A sweeping reform of the municipal water services was introduced by Law 36/1994 which modified the previous management structure based upon the prevailing responsibility of the municipalities in their territory or of organizations supplying water (e.g. Apulian Aqueduct) or treating wastewater effluents for a number of municipalities in some parts of Italy. First of all, Law 36/1994 introduced important innovations in the general principles of water resources government. It stated that all surface and groundwater resources must be considered public and should be used according to criteria of solidarity and sustainability. It stated also that drinking use has priority over all other uses, followed by agricultural use under water scarcity conditions.

The management of municipal water services has been reformed significantly by considering it as a unitary service. The law put the three elements of the urban water cycle (water supply, sewerage, and wastewater treatment) under a single responsibility in order to simplify the management structure, to account for large-scale interconnected systems, to obtain scale economies and to improve the protection of water sources from pollution. The reform included territorial, functional and financial features:

- Territorial scale was established in terms of "optimal territorial areas" (OTAs), in Italian ATO, (i.e. Ambiti Territoriali Ottimali) to be defined by regional governments, thus including several municipalities, in order to overcome the fragmentation of water management at municipal level (more than 5000 management bodies) and to develop organizations with increased size in terms of both served population and supplied volume.

Table 3.2 Main Italian acts and European Directives on water resources and soil conservation after 1989

Italian acts on water resources and soil conservation issues			European Directives (Italian acknowledged acts)
Water resources uses	Water pollution control	Soil conservation and flood risk mitigation	
L 142/1990 Regulation of local service bodies (municipalities and provinces)			
DRP 85/1991 National Technical Services reform **Decree 13/12/1991** Ministry of Health on aqueducts and control of drinking water			**Dir 91/271 EEC** Municipal wastewater treatment **Dir 91/676 EEC** Water protection from nitrate pollution
L 498/1992 Urgent action on public finance (Art.12 on change of mixed companies) **DPR 7/1/1992** Criteria for information acquisition for river basin plan		**L 225/1992** Establishment of Civil Protection Service	**Dir 92/43 EEC** Habitat
DL 96/1993 Management of water works **DLgs 275/1993** Modifications of water license rules			
L 36/1994 (Galli Act) Reform of municipal water supply (Regulations for tariff in DM 1/8/1996, other regulations in DPR238/1999	**L 61/1994** Establishment of Environmental Protection Agency		
	L 172/1995 Revision of L 319/1976 procedures	**DPR 18/7/1995** Approval of address act for the definition of the basin plan	
DPCM 4/3/1996 Directives on water resources	**DPR 12/4/1996** EIA procedures enlargement		**Dir 96/61 EEC** Modifications of Dir 91//271 EEC
D LLPP 99/1997 Criteria for losses evaluation in aqueducts and sewers		**D LLPP 14/2/1997** Directives to identify areas under hydrogeological risk	

(continued)

Table 3.2 (continued)

Italian acts on water resources and soil conservation issues			European Directives (Italian acknowledged acts)
Water resources uses	Water pollution control	Soil conservation and flood risk mitigation	
DLgs 112/1998 Transfer of functions to the regions and local authorities		**L 267/1998** (Sarno Act) Hydrogeological Asset Plan (HAP) (Basin Authority) and urgent plan for emergency (Civil Protection Service)	**Dir 98/15 EEC** Measures for pollution prevention and reduction **Dir 98/83 EEC** Quality of water for human consumption (D.Lgs 31/2001)
DRP 238/1999 Declaration of all water as public **DPM 29/4/1999** Chart for water supply service	**DLgs152/1999** Directives on pollution control and regional plans for water protection (Integrated by **DL 258/2000**)	**L 226/1999** Extraordinary plan for areas under high risk (Basin Authority) **DelCIPE 229/1999** National program to fight drought and desertification	
		L 365/2000 (Soverato Act) Procedures for HAP	Water Framework Directive **2000/60/EC**
		L 401/2001 Coordination of civil protection structures	
DM 185/2003 Regulations of reuse of wastewater	**Decree 19/8/2003** of Ministry for Environment on the quality of water bodies		**Dir 2003/40/EC** Natural mineral and spring waters (D.M.29/12/2003)
DM 30/6/2004 Management of reservoir storage **DM 28/7/2004** Guidelines for water balance of river basins		**DPCM 27.2.2004** Warning system for hydrogeological and hydraulic risk and emergency plan (Civil Protection Service responsibility)	
DLgs 152/2006 (Environmental Code) On the basis of Law 308/2008 which entrusted the government to reorganize legislation in environmental field, acknowledges Dir 2000/60/ EC, abrogates L183/1989 and establishes authorities of districts (Regulations in DM 131/2008)			**Dir 2006/7/EC** New rules on water bathing (D.Lgs 116/2008) **Dir 2006/118/EC** Pollution protection of groundwater (D.Lgs 30/2009)
			Floods Directive **2007/60/EC**

(continued)

Table 3.2 (continued)

Italian acts on water resources and soil conservation issues			European Directives (Italian acknowledged acts)
Water resources uses	Water pollution control	Soil conservation and flood risk mitigation	
DLgs 4/2008 Modifications to Environmental Code			
L 133/2008 (Ronchi Act) Preference to private companies for Integrated Water Service			**Technical Report 2008-023** Water Scarcity and Drought Expert Network Report
L 13/2009 Responsibility to national basin authorities of drafting district management plans			**Dir 2009/54/EC** Exploitation and marketing of natural mineral water
L 166/2009 Change in management of Integrated Water Service			
L 42/2010 Elimination of OTAs authorities and entrusting to the regions the regulation responsibilities **DLgs 85/2010** Transfer of the state's properties (including water and coasts) to local organizations		**DLgs 49/2010** Acknowledges the European Dir. 2007/60 about the Flooding Risk Management Plan (FRMP) and confirms the HAP and plan for emergency **DLgs 219/2010** Shares responsibilities among District Authority and region for FRMP	
DL 214/2011 (Art.21) Transfer of responsibility on regulating water municipal services to Authority for Electric Energy, Gas and Water System (AEEGSI)			
		DL 59/2012 Change of Civil Protection Service tasks	

(continued)

Table 3.2 (continued)

Italian acts on water resources and soil conservation issues			European Directives (Italian acknowledged acts)
Water resources uses	Water pollution control	Soil conservation and flood risk mitigation	
L 164/2014 Change of the rules for the management of Integrated Water Service in order to foster the implementation of the regional rules on governing body and company responsibilities		**L 116/2014 (Art.10) and L 164/2014 (Art.7)** Acceleration of the use of financial resources for flood risk mitigation **DPCM 27/5/2014** Establishment of a Mission Structure against hydrogeological risk disasters **L 164/2014** Financial resources for hydrogeological risk mitigation require (i) an agreement between regions and Ministry of Environment, (ii) priority to the actions which reduce the risk and improve ecosystems and biodiversity and give substitute power for defaulting cases	
L 221/2015 Amends the Environmental Code for districts territories and responsibilities for urban water supply; establishes a fund for municipal water infrastructures and guarantees drinking water access to low-income users			
Deliberation 2015 AEEGSI updating of the 1999 chart for water supply service			
L 205/2017 Change of the AEEGSI into Authority for Regulating Energy Networks and Environment (ARERA), due to the enlargement of responsibilities also to solid wastes			

DL Decree Law, *DLgs* Legislative Decree, *DM* ministerial decree, *DPR* Decree of the President of Republic, *DPCM* Decree of President of the Council of Ministers, *DLLPP* Decree of Public Works Ministry, *Dir* European Directive, *L* law, *DelCIPE* Deliberation of the Inter-ministerial Committee for Economic Planning, *EC* European Communities, *EEC* European Economic Communities

• Better functioning is pursued by separating the duty of strategic direction and control of the service (performed by a public OTA's authority) and the duty of the management of Integrated Water Service, to be performed by a management company, either private, public or in public-private partnership.
• A full cost recovery pricing criterion was envisaged in order to ensure that the tariffs charged to users should cover all the costs of the services, including operation, replacement and investments costs.

The law established a committee supervising the use of water resources (Co. Vi.R.I.), based at the Ministry of Public Works with the function of overseeing the management of OTAs and tariff and of guaranteeing service quality. The committee was composed by seven members in charge for 5 years. The committee was entrusted of reporting every year to the parliament on the current state of the service with the assistance of an observatory of water services, located at Ministry of Public Works. The committee should guarantee water users on effectiveness, efficiency and cost saving of the service, in cooperation with regional bodies established for the same purpose. The law also identified the consortia for land reclamation as responsible for construction and management of water networks for irrigation, for reuse of treated wastewater and for rural aqueducts to provide drinking.

The rules of Law 36/1994, at the time of its abrogation, were maintained with minor changes in the Part III, Section III, Title II (Integrated Water Service) of DLgs 152/2006 and modified with further minor changes by DLgs 4/2008.

The implementation of the Galli Act presented many differences among the various regions, both in the definition of the Optimal Territory Area (often coinciding with the territory of the provinces or of the region) and in the assignment of the Integrated Water Service to public companies, mixed companies with majority of public capital and the participation of private capital (allowed by Law 142/1990) or also with joint stock companies with public capital less than 20% (according to Law 498/1992). The implementation has been very slow, due to technical reasons such as inadequate knowledge of existing water infrastructures and also low quality of a few OTA plans, as well as to juridical reasons such as legal disputes concerning the assignment of the service.

Also political reasons had a key role. In particular, the opposition to Law 133/2008 (Ronchi Decree) and Law 166/2009, which had favoured the private companies, led to a referendum in June 2011. The referendum resulted in the elimination of the preference to private companies and of the obstacles to the direct assignment to public companies (*in house* operation) and in the elimination of the part of the tariff allocated to cover the capital costs. Also the principle of full cost recovery, introduced by the Galli Law, found significant implementation difficulties, especially in the South regions where a set of new plants for completing the needed infrastructures for water supply, sewage and wastewater treatment were required, but the new investments did not access to financial resources from the national fiscal system or from European funds.

Later, Law 42/2010 cancelled the OTA authorities and entrusted the regions of the regulation of all duties about the municipal services, namely, planning, tariff and surveillance on management companies. This act produced non-homogeneous rules in different regions. An opposite change occurred with DL 214/2011 (Art. 21), which transferred the tasks performed by the National Commission supervising the use of water resources (Co.N.Vi.R.I) (formerly by the committee supervising for the use of water resources, Co.Vi.R.I.) to the Authority for Electricity Energy and Gas, whose name was so completed with reference to water systems ("Sistema Idrico" in Italian), thus changing the acronym in AEEGSI.

Significant changes to management of municipal water services were introduced by Law 164/2014, which established (i) the mandatory participation of the municipalities to the governing body of the territorial unit; (ii) the preference to an unitary company for the whole territory to be selected with European rules referring to local services; (iii) the direct assignment to an in-house company only if it is formed by local bodies of the territorial unit; (iv) the responsibility of AEEGSI for drawing up standard contracts for service management; and (v) a substitutive power given to the region's president in case of default of the governing body of the territorial unit and to the President of the Council of Ministers in case of default of region.

Later, the rules for municipal water services were changed by Law 221/2015, which provided a fund for improvement of water infrastructures (aqueducts, sewers and treatment plants) and entrusted AEEGSI for the task of adjusting the standard scheme of the tariff in order to allow water access to low-income users, according to a decree of the President of the Council of Ministers. Also the same authority was supposed to provide directives for reducing the delays in the payment of bills by the users, according to another decree of the President of the Council of Ministers. Recently, Law 2015/2017 changed the name of AEEGSI into Authority for Regulating Energy, Networks and Environment (ARERA), since its responsibility was extended to include also the regulation on solid waste management.

3.3.3 Productive Water Uses

The rules regarding the water withdrawals for agricultural use were established by RD 1775/1933, which distinguished between small and large withdrawals (identified by the threshold of 1000 l/s, with even more constraining exception when the irrigated surface is larger than 500 ha). The RD 215/1933 gave priority to the land reclamation consortia in the licensing system with reference to individual applications. This priority was limited by DLgs 275/1993, which introduced new criteria for granting new licenses, in order to account for the quantitative and qualitative characteristics of the water body and for the rational use of water. It established the conditions for the renewal of the irrigation licenses and established that new groundwater abstraction licenses should be given for uses other than drinking purposes only in case of lack of other sources of supply.

The same legislative decree introduced a minimum ecological flow to be guaranteed in the stream. While the operation and maintenance costs are normally paid by the irrigations users or by the land reclamation consortia (totally or partially), the investment costs for building infrastructure for irrigation water supply were covered in general by the state, e.g. through the Southern Italy Development Fund (Cassa per il Mezzogiorno) or the Green Plan (e.g. Law 27.10.1966 and following acts).

The rules to allow the reuse of treated wastewater for irrigation, and also for municipal uses such as washing of roads, supply of dual aqueducts networks or cooling systems or for industrial use (such as water anti-fire or for washing), have been established by DM 185/2003.

3.3.4   Water Pollution Control and Water Quality Protection

The Legislative Decree 152/1999 (modified by DLgs. 258/2000) rearranged the previous Italian legislative framework on pollution control and water quality improvement according to the European Directives 91/271 on urban wastewater treatment and 91/67 on protection of water from agricultural pollution and also according to the proposals of the new European Water Framework Directive, which was ongoing at that time.

The decree introduced the objectives of a minimum standard of water quality in the water bodies and a specific level connected to each particular use (production of drinking water, bathing, support to fish life, etc.). It modified the previous standards on wastewater effluents (irrespective of the specific water body characteristics). It distinguished the actions for protection and restoration on the basis of the exposure of the site to eutrophication and of its vulnerability to pollution by nitrates originating from agriculture.

It replaced also the regional plan for water restoration (Law 319/1976) with the plan for water protection, considered as a part of the river basin plan, established initially by Law 183/1989. It provided criteria and technical guidance to foster the application of new rules (e.g. on classification of water bodies, surveying river basin characteristics and human impacts, definition of sensible areas, etc.).

The above-mentioned rules on water quality, with some changes, are now included in the Part III, Section II of the Legislative Decree 152/2006 (*Environmental Code*), which has substituted and integrated many previous acts.

3.3.5   Soil Conservation and Flood Risk Mitigation

Although Law 183/1989 had been issued with the objective of a unitary approach to soil conservation (including flood defense), water quality protection and water supply for various uses, many of the following laws adopted a different orientation. In contrast to the results of the Parliament Commission (Veltri 1998) which had confirmed the validity of the coordination approach of Law 183/1989, the subsequent politics reversed to face the problems in separate ways for each sector, thus deeming to assure more effective results. In particular, this approach was adopted to overcome the delays in the preparation of the comprehensive plans and to timely solve the dramatic effects of frequent flooding and landslides events.

In fact, after the hydrogeological Sarno disaster (which resulted in 159 fatalities), the Decree 180/1998 (indicated as Sarno Act, approved by the parliament as Law 267/1998), introduced the *Hydrogeological Asset Plan*, aiming at identifying the areas with high risk and at defining the necessary mitigations measures and works (commitment of River Basin Authorities or regions) and the *urgent plans for emergency* (commitment of the Civil Protection Service). The latter institution had been already established by Law 225/1992 with a more general purpose of coping with all natural and man-made disasters. Law 226/1999 introduced the extraordinary plan for areas under high risk, while Law 365/2000, issued after another flooding disaster (Soverato, with 13 fatalities), defined the procedures for approval of Hydrogeological Asset Plans and for developing meteorological monitoring.

Despite the increase in planning tools, the implementation of the planned actions to fight flooding and landslides was very limited, due to the economic crisis and bureaucratic delays which affected the amount of investments and the timely construction of hydraulic works. In the same years, policy on flooding risk mitigation increased the role of the meteorological monitoring service and of the civil protection, also as a consequence of a conceptual shift from the structural measures for flood defense to the measures for reducing damages by means of early warning systems and improved actions of aid during extreme flooding events. In particular, the directive of the President of the Council of Ministers 27.2.2004 improved the organization of the multifunctional centres and established the warning system for hydrogeological and hydraulic risk.

The following step in the legislative process was the incorporation of previous tools into the Legislative Decree 152/2006 (*Environmental Code*), which covered all aspects of the soil defense and water resources use and protection (Maglia and Galotto 2009). However, this act missed the opportunity to simplify and improve the very complex system of planning tools and management responsibilities in the water field (Rusconi 2010; Rossi Paradiso 2013).

The publication of the European Directive 2007/60/EC on the assessment and management of flood risks increased the complexity of the planning tools, since it required drafting the new Flood Risk Management Plans. The 2007/60 proposal to evaluate mapping of flooding hazard and risk was actually very similar request to the tools envisaged already by the Italian legislation but limited to the flooding without considering landslides.

The DLgs 49/2010 acknowledging the European flood risk directive, gave responsibility to the District Authorities for the preparation of this plan and confirmed the duty of regions in providing the part of the plan for hydraulic risk mitigation system. This duty must be carried out in cooperation with the National Department of Civil Protection. The plan includes (i) a synthesis of the urgent plans for emergency, prepared on the basis of Art. 67 DLgs 152/2006; (ii) the monitoring, forecasting, surveillance and watching actions developed by the multifunctional centres; (iii) the defense actions entrusted to the hydraulic land assistance/units (*Presidi territoriali idraulici*); and (iv) the flood control guaranteed by the flood control plans. The same law confirmed the obligation of the territorial or urban plans to respect constraints given by the Flood Risk Management Plan; indeed this obligation had been prescribed already for Hydrogeological Asset Plan, but very often it has not been respected, thus contributing dramatically to an increase of the flooding risk and of the amount of damages in the areas of urban expansion.

Since the District Authorities, established by the DLgs 152/2006, had not started yet their activities, the DLgs 219/2010 entrusted the responsibility of drafting the plans required by European Directives to the National River Basin Authorities (besides to the regions responsible for districts located in main islands). These authorities completed the plans within the deadline (June 2015), in particular, providing the hazard maps and the risk maps. However, the planning process emphasized the difficulties of an effective coordination among the different bodies and among the different planning tools, thus confirming the necessity of a simplification of the too complex regulation.

The more recent efforts are oriented to foster the implementation of designed measures. The Law 164/2014 aimed at accelerating the use of financial resources for hydrogeological risk mitigation by agreements between regions and Ministry of Environment, by the revocation of unused funds and by the allocation of new financial resources to the regions for flooding mitigation in urban areas. The law also established priorities of actions with the joint purpose of reducing the risk and of improving ecosystems and biodiversity.

The DPCM 27/5/2014 established a mission structure against the hydrogeological risk. It was named "Italia Sicura" (Safe Italy) and was located at the Presidency of the Council of Ministries aiming at developing a set of initiatives to increase the quality of new projects (such as guidelines) and improve the coordination among the Ministry for Environment, regions and local authorities.

The DPCM 15/9/2015 has contributed to foster the actions for flood risk mitigation by means of a plan for urban areas with a large amount of population exposed to flood risk. The Law 221/2015, besides the innovations on municipal water service for drinking use, which were mentioned previously, amended the Environmental Code in district definition, structure of District Authorities and new rules for prevention of flooding risk. These rules include a program of sediment management in river basins and specific instructions to remove unauthorized buildings in high-risk areas as well as to reduce vulnerability of buildings to hydrogeological risk at municipal level.

3.3.6 Drought Management

The Italian legislation does not include a specific mandatory act for drought management. The DLgs 152/2006 (Art.65) makes mention of "the actions against the drought risk" only in a list of the contents of the district plan, which includes also the actions against the risk of flooding and landslide and the actions to pursue the economic and social objectives and the land protection; however, no indication is provided on the methodology to identify these actions. A few indications for defining objectives and contents of a drought management plan have been given in the following regulations documents (Rossi 2017):

- The DPM 4/3/1996, which contained directives and methods for the implementation of Law 36/1994, included directives and technical parameters to individuate the areas prone to water crises: these directives, although aiming at identifying which municipal water systems were vulnerable to drought, required to evaluate the water shortage in the uses of all sectors in order to define the short-time and long-term measures to avoid emergency conditions. Unfortunately, these directives have not been implemented in the plans for optimal territorial areas.
- The DLgs 152/1999 (Article 20) established that regions and River Basin Authorities must check for the presence of areas subject to drought, soil degradation and desertification and to provide specific measures to cope with these phenomena by means of the plans for water protection, according to the criteria of the national plan against drought and desertification (CIPE Deliberation 229/1998). Unfortunately, also in this case, only a few regions (such as Emilia-Romagna and Veneto) have identified these areas and provided measures to mitigate the risk.

Nowadays, the European Commission suggests the member states to adopt drought management plans, but this advice is not mandatory. In fact, the technical report of the Water Scarcity and Drought Expert Network (E.C 2007) has extended the objectives and criteria of the Water Framework Directive 2000/60 to fight the drought, recommending that the member states develop and implement measures aiming at preventing and alleviating drought and water scarcity, by adopting an approach of risk management instead of crisis management. The report envisages that the authority responsible for the district water planning provides a drought management plan, to be incorporated into the River Basin Management Plan as supplementary plan according to Article 13.5 of WFD. Its specific objectives are (i) to guarantee sufficient water availability to cover water human needs and to ensure the population's health and life, (ii) to avoid or minimize negative drought impacts on water bodies and (iii) to minimize negative effects on economic activities.

3.3.7 Rules on Water Works

Several laws have been issued in order to improve the technical criteria for specific water works and plants, which have important roles in water resources management and soil conservation.

Primary attention has been payed to the rules for design, building and operation of large dams fostered by a few dam accidents. The first national regulation (RD 2540/1925) was issued soon after the collapse of the Gleno Dam which implied 500 fatalities. Then, DPR 1363/1959 established a new detailed regulation for dams higher than 10 m and reservoir capacity greater than 1,000,000 m^3. The responsibility for small dams continued to be in charge of the Genio Civile offices. Afterwards, the regulation was updated by the decree of the Ministry of Public Works 24.3.1982 and recently by the decree of the Ministry of Infrastructures and Transports 26.6.2014. The DM 30/6/2004 established the criteria for the management of sediments in the reservoirs, as required by DLgs 152/1999 (Art.40).

In addition, the structure for surveillance changed, as the previous Service for dams, part of the Ministry of Public Works, was cancelled by DL 112/1998 Art.91, transferring the responsibility to the Italian Register for dams, which in turn was cancelled by Law 286/2006 which transferred the functions to the Ministry of Infrastructures and Transports. The organization on the territory has been modified by the DPR 211/2008, which has reorganized the structure of this ministry. Responsibility for small dams was transferred to regions due to the transfer of Genio Civile offices from state to regions in 1977. At a later time, the divide between state and regional responsibility was elevated by Law 584/1994 to dams higher than 15 m or with reservoir capacity larger than 1,000,000 m^3.

Other technical rules of interest regulate the criteria for loss evaluation in aqueducts and sewers (DLLPP 99/197), the chart for water supply service (DPM 29/4/1999) and the guideline for water balance of river basins (DM 28/2004).

3.4 Significant Innovations in the Development of Water Laws

3.4.1 Shift in Basic Principles

The analysis of the development of the Italian water legislation has shown that several factors can be considered as drivers for the innovations introduced by water acts during the last decades. The most important factors are the administrative structure of the Italian Republic (namely, the growing role of regional governments compared to state in many sectors of governance) and the debate on the role of private companies in the management of public services.

The innovation processes were fostered strongly by the European legislative requirements. Their role has been particularly important since the 1970s, due to the obligation of introducing the rules of the European Directives into Italian legislation within the assigned deadlines. However, in many cases, the provision of the Italian acts can be considered more advanced than the ones of the European Directives. For example, the Law on Soil Defense (183/1989) has introduced a comprehensive water resources and soil conservation planning, while the two issues, even at a later time, were addressed in separate ways by the Water Framework Directive 2000/60 and the Directive 2007/60. Indeed, this last directive was limited to flood risk, thus missing to consider landslide risk which had been taken into account already by the Italian Law 183/1989.

Most of the innovations in the objectives and procedures introduced by acts of the last decades derive from the shifts occurred in several basic principles. In particular, the main shifts can be considered as follows:

- The regulation role of the state in water resources utilization by means of the licensing system to individuals (or consortia) who enact the resource exploitation (according to RD 1775/1933) was modified by the transfer of the licensing role to regions and by a stronger role of the planning tools of central and regional governments (especially for drinking water supply).
- The fight against the pollutant discharges, based only on the activity of the penal jurisdiction, shifted towards specific rules on wastewater to reduce the threats of pollution of water bodies and to protect the ecosystems since the Merli Act of 1976.
- The goal of a comprehensive river basin planning, which includes water resources uses, pollution control and soil conservation (pursued by Law 183/1989), modified the previous approach of separate regulations for each sector. However, in the following decades, this comprehensive approach has been progressively abandoned due to the adoption of specific plans for each water problem, under the push of urgent interventions as a consequence of the orientation towards a better management of the urban water services and of the urgency of actions aiming at reducing the risk of severe flooding events.
- Moreover, the focus on structural measures (new hydraulic works for water supply and river regulation) shifted to nonstructural measures, such as water demand management and a better management of water services on the water resources side, and development of warning systems on the flood risk control side.
- In the municipal water management sector, the objective of achieving a more efficient and effective management required the unification of the services for water supply and collection and treatment of wastewaters and the separation of the operational commitment given to a company with industrial approach from the responsibility of the public control, as well as the adoption of the full cost pricing.
- In the sector of the flood risk mitigation, a new awareness rose on the necessity of improving the cooperation between national, regional and local bodies and reinforcing the links between the structural measures and the prevention actions carried out by civil protection.
- The enhanced role of the citizens in all stages of the decision process – required by the so-called civil society and by the European Commission (e.g. by the Aarhus Convention of 1998) – has pushed for a greater transparency of the activity of the public administration and a greater participation of stakeholders in the water management.

Two major trends can be identified in the law innovation process, namely, (i) the increasing awareness of the necessity for a sustainable water management and (ii) an effort to improve access to information and increased participation of citizens and stakeholders in water planning and operation. The following sections will give some details on these trends.

3.4.2   Towards a Sustainable Water Resources Management

The sustainable development principle, which has been progressively accepted since the Brundtland Report (1987) and the UN Conference on Environment and Development (Rio de Janeiro, 1992), is becoming the new paradigm for ecosystem protection and the consideration of the rights of future generations. It requires a more cautious assessment of the impacts of human actions on natural resources and an effort to achieve equity in territorial and social sharing of benefits in the use of these resources (E.C 2000). The principle of considering the needs of future generations modifies substantially the previous approach to the planning, design and management of water resources systems. The sustainable development in water fields implies, in particular, an enlargement of the objectives, of the long-term impacts to be considered and of procedural policies that would require institutional changes, improved information sharing and stakeholders' participation (Rossi and Ancarani 2002).

In order to understand the difficult process of the implementation of the sustainability principle in water legislation, it is necessary to point out that the RD 1775/1933, which for many decades had regulated the water abstractions from surface resources and groundwater, did not mention the ecological requirements. In addition, the report of the National Water Conference (1972), which expanded the list of traditional uses (namely, hydroelectric, domestic supply, irrigation and land reclamation) to industrial and navigation uses, did not include recreational and ecological needs as actual uses, although some environmental, landscape and recreational needs were recognized as constraints.

However, a few years later, the technical directives, following Law 319/1976 on pollution control, introduced recreational uses and maintenance of fish and wildlife into the list of water uses. The Law 183/1989 for the first time referred explicitly to a "minimum constant in-stream flow", as the necessary streamflow for aquatic life protection and water quality improvement. The Legislative Decree 275/1993 confirmed the criterion that a minimum in-stream flow must be guaranteed in order to ensure water quality and ecosystem equilibrium of watercourses. The Legislative Decree 130/1992, in acknowledgement of EC Directive 78/659, considered the needed flow to preserve fish life as an "in-stream use". This objective was confirmed also by Law 36/1994.

As it can be seen from the above review, the need of an in-stream use, in the sense of a preserving function for water ecosystem and life, in contrast with the off-stream uses (municipal, irrigation, industrial, hydropower), has been accepted progressively in Italian legislation. However, the implementation of the concept has been delayed, and it was someway variegated since several different technical regulations for estimating in-stream flow were issued by several organizations (provinces, River Basin Authorities, regions). Only in 2004 the Ministry of Environment and Protection of Territory and Sea issued guidelines to assess "minimum in-stream flow" MIF, within the Decree 28/7/2004dedicated to the criteria for the census of water uses and the assessment of the water balance at basin scale.

The decree established that the plan for water protection has to define the minimum in-stream flow (MIF) in each homogeneous reach of a watercourse with the objective of *maintaining the typical biocenosis of local natural conditions*, safeguarding the *physical characteristics of the water body* and the *chemical-physical characteristics of water*. Recently, this decree has been replaced by new guidelines, issued by the Ministry of Environment (DDG 13/2/2017 n.30/STA) which refers to the European guidelines (CIS Guidance Document 31/2015) and introduces the new concept of "ecological flow" as the hydrological regime to be assured, in a hydraulically homogeneous reach of a watercourse, in order to achieve the environmental objectives of protection, improvement and restoration of surface water bodies and to obtain a good state, as defined by the Art. 4 of the Water Framework Directive 2000/60.

These guidelines entrust the District Basin Authority with the commitment of adapting the methodological approaches for the determination of the MIF within their districts to the guidelines, while the operating rules are still entrusted to the regions. The decree also assigned the *National Catalogue of MIF Calculation Methods* to ISPRA and established a *National Technical Table (NTT)* for the verification of technical-scientific congruity of the proposed methods, the updating of the catalogue and the experimental implementation of the guidelines.

The regulation of the conflicts among in-stream and off-stream uses of water, according to EC Dir 85/337, is assigned to the Environmental Impact Assessment (EIA) procedure, as a tool aiming at reducing negative impacts on environment components. The EIA procedure has been established as mandatory for most relevant hydraulic works, while the decision whether to require the procedure for other works has been left to each state. In Italy, the directive has been acknowledged in two steps. In the first step, after the creation of the Ministry for Environment (Law 349/1986), the EIA was required for main works (DPCM 337/1988), regulated by the technical directives issued by DPCM 27/12/1988. In the second step (DRP 12/4/1996), the regions have been entrusted with the decision whether to require the EIA for minor works or not.

The present legal framework imposes a constraint to the abstraction uses in order to satisfy the in-stream flow for ecological uses. The EIA procedure is entrusted with the solution of the conflicts among the different uses.

3.4.3   Towards an Improved Information Access and Stakeholders' Participation

Another important issue is the increasing will of legislators to guarantee better access to information and to encourage the actual participation of citizens in the decision-making process on water systems planning and operation (Massarutto 2008).

Transparency and participation are the two main requirements for maintaining relationships between public administration and citizen. They differ for the direction of the information flow between organizations and relevant stakeholders.

Transparency, as highlighted by Law 142/1990 and by Law 241/1990, refers to the information distributed by a public organization in order to make clear how it operates in terms of objectives, strategies, technologies and procedures. Participation refers to the possibility of citizens and stakeholders to take part in the decision-making process, either by expressing expectations and proposals or by proposing adjustments of the decisions.

Transparency should guarantee the free access to administrative documents and to the information about the process carried out by the organization in charge of the planning, design and service tasks. A typical tool for transparency in public activities is the establishment of offices for relationships with customers. Public hearings and procedures for oppositions are instruments for participation. Today most of information travels, in both senses, through the Internet.

Law 183/1989 considered the need for sharing information on river basin planning among the population by posting the plan in regional and provincial offices. The opportunity to consult the plan during a 45-day period and then either make written suggestions to the region or write suggestions directly in a special register should ensure citizen's participation in the plan drafting.

Law 36/1994 prescribed publication of the demands for water licenses in the *Gazzetta Ufficiale*, in a national newspaper and in a local newspaper, while the documents referring to the projects of hydraulic works must be available for consultation in the offices of Ministry of Public Works, regions and autonomous provinces. The same law prescribed that the managers of water services provide information to customers, promote activities to share water culture, promote water-related education and guarantee the free access to information on technologies applied for service management, plants operation and quantity and quality of supplied water. The same Law 36/1994 aimed at ensuring the citizen's right in the water municipal services, by checking the compliance of the service to the agreement between the provider and local authorities (Art.11). Furthermore, regional bodies can be established with the purpose of assuring citizen's rights (Art. 21).

In particular, the scheme of the chart for municipal water service, as reported in DPM 29/4/1999, includes criteria for service provision aiming at guaranteeing equity and impartiality in dealing with users of services, efficiency and effectiveness in service provision, continuity of service, courtesy of front-end personnel and clearness and intelligibility of messages to customers. It is confirmed that customer satisfaction is the target in addressing the efforts of management organization, even in the public sector. The chart for service also provided a penalty to be refunded to users in cases of inadequacy of the service. The deliberation of the Authority for Energy, Gas and Water Systems (2015) updated the 1999 chart for water supply service by rising the expected standards.

The DRP 12/4/1996 on Environmental Impact Assessment envisaged a broader participation of citizens in the decision-making process, the publication of related information in regional and provincial newspapers and the possibility of consulting the documentation in the pertinent offices. Anyone who wishes to give information on possible impacts of a work under an evaluation process can present written comments to the offices that must consider them together with the comments made by the public administration in order to evaluate the environmental compatibility of the project. Also, a public audit for examining the environmental impact studies as well as comments and suggestions by public administration is envisaged. Similar rules for citizen participation have been introduced in the decision-making process with reference to other planning tools.

The above analysis demonstrates an increased attention to public participation in the formulation of acts issued in recent years, due to the belief that the role of the citizen is invaluable in a democratic politic system and can improve environmental protection and equity in sharing benefits generated from the realization of water infrastructures.

3.5 Lights and Shadows of Italian Water Legislation

Recent Italian legislation on water resources protection and use and on flood defense is quite advanced, as it has incorporated many general principles which today are required by the socio-economic changes in the developed countries. The main principles are as follows:

- A more effective control of water pollution to guarantee human health and environment protection
- A shift from the focus on structural measures, such as new hydraulic works for water supply and river regulation, to nonstructural measures, such as water demand management and improved governance of water services for water resources utilization and early warning system for flood risk reduction
- The effort to improve municipal water service through a more efficient management and to the attention given to citizen's rights in terms of transparency and participation

Generally, the Italian legislation has followed (with some delay) the European Directives, particularly on water quality protection and Environmental Impact Assessment. However, some Italian acts had a pioneering role with respect to the European Directives (e.g. for acts dealing with flood risk mitigation) or have adopted a more comprehensive strategy (e.g. by taking into account a joint water-land approach for river basin planning and by developing tools based on prevention besides emergency actions).

Unfortunately, the positive aspects of the Italian legislation are suffering from by a great risk of failure due to the low level of efficiency of the public administration, which, in turn, is negatively affected by the disputes or conflicts between the central government and regional governments after the transfer of several competences from state to regions.

In addition, two pillars of the European legislation, such as environmental protection and public participation in the decision-making process, had a very slow and contradictory development. In particular, many weaknesses affect the procedure of Environmental Impact Assessment (EIA) and the criteria for evaluating the ecological flow requirements for river ecosystems conservation. Furthermore, citizen participation in the water system planning and operation can be considered as implemented only in a formal sense, since transparency of information to public participation is not yet assured in all the stages of the process and only partially accomplished.

With reference to extreme hydrological events mitigation, the meagre amount of financial resources has not facilitated a constant flow of investment to ensure an effective flood defense by means of the timely implementation of the planned prevention measures. Until now, the requirements of the European report on drought management plan are not acknowledged, although several severe droughts have struck many Italian regions in the last decades. However, the civil protection has carried out a worthwhile action to fight in emergency both flood and drought risks, particularly by means of help and assistance to population, of the development of monitoring and early warning systems, of the mitigation of damages and of the reconstruction of destroyed goods by hydraulic and geomorphological disasters.

Definitions in the European Directives and DLgs 152/2006

Directive 2000/60/EC

"Surface water"= inland waters, except groundwater; transitional waters and coastal waters

"Groundwater"= all water which is below the surface of the ground in the saturation zone and in direct contact with the ground or subsoil

"Inland water"= all standing or flowing water on the surface of the land and all groundwater on the landward side of the baseline from which the breadth of territorial waters is measured

"River"= a body of inland water flowing for the most part on the surface of the land but which may flow underground for part of its course

"Lake"= a body of standing inland surface water

"Transitional waters"= bodies of surface water in the vicinity of river mouths which are partly saline in character as a result of their proximity to coastal waters but which are substantially influenced by freshwater flows

"Coastal water"= surface water on the landward side of a line, every point of which is at a distance of one nautical mile on the seaward side from the nearest point of the baseline from which the breadth of territorial waters is measured

"Artificial water body"= a body of surface water created by human activity

(continued)

"Body of surface water"= a discrete and significant element of surface water such as a lake, a reservoir, a stream, river or canal, part of a stream, river or canal, a transitional water or a stretch of coastal water

"Aquifer"= a subsurface layer or layers of rock or other geological strata of sufficient porosity and permeability to allow either a significant flow of groundwater or the abstraction of significant quantities of groundwater

"Body of groundwater"= a distinct volume of groundwater within an aquifer or aquifers

"River basin"= the area of land from which all surface run-off flows through a sequence of streams, rivers and lakes into the sea at a single river mouth, estuary or delta

"Subbasin"= the area of land from which all surface run-off flows through a series of streams, rivers and lakes to a particular point in a watercourse

"River basin district"= the area of land and sea, made up of one or more neighbouring river basins together with their associated groundwaters and coastal waters

"Good surface water status"= the status achieved by a surface water body when both its ecological status and its chemical status are at least good

"Good groundwater status"= the status achieved by a groundwater body when both its quantitative status and its chemical status are at least good

"Pollution"= the direct or indirect introduction, as a result of human activity, of substances or heat into the air, water or land which may be harmful to human health or the quality of aquatic ecosystems or terrestrial ecosystems directly depending on aquatic ecosystems

"Pollutant"= any substance liable to cause pollution

"Environmental quality standard"= the concentration of a particular pollutant or group of pollutants in water, sediment or biota which should not be exceeded in order to protect human health and the environment

"Available groundwater resource"=the long-term annual average rate of over-all recharge of the body of groundwater less the long-term annual rate of flow required to achieve the ecological quality objectives for associated surface waters to avoid any significant diminution in the ecological status of such waters and to avoid any significant damage to associated terrestrial ecosystems

"Water services"= all services which provide for households, public institutions or any economic activity: (a) abstraction, impoundment, storage, treatment and distribution of surface water or groundwater, (b) wastewater collection and treatment facilities which subsequently discharge into surface water

"Water use"= water services together with any other activity having a significant impact on the status of water

(continued)

"Emission limit values"= the mass, expressed in terms of certain specific parameters, concentration and/or level of an emission, which may not be exceeded during any one or more periods of time

Directive 2007/60 EC
- "Flood"= the temporary covering by water of land not normally covered by water. This shall include floods from rivers, mountain torrents, Mediterranean ephemeral watercourses and floods from the sea in coastal areas and may exclude floods from sewerage systems.

"Flood risk"= the combination of the probability of a flood event and of the potential adverse consequences for human health, the environment, cultural heritage and economic activity associated with a flood event.

DLgs 152/2006 (Beside the Definitions of the 2000/60/EC Directive)
- "Soil" = the land, soil, subsoil, built-up areas and infrastructures

"Hydrographic network" = the system of drainage in the basin

"Hydrogeological calamity" = a condition of risk over the territory due to natural or anthropogenic processes referring to water bodies, soil or side.

"Soil defense" = the set of actions of protection of the land, rivers, canals, lakes, lagoons, coastline, groundwaters and connected territory with the aim to mitigate the hydraulic and hydrogeological risks, to optimize the management of the water resources and to enhance the environment and landscape features

References

Brundtland GH (1987) Our common future. Oxford University Press, Oxford

Canelas de Castro P (2008) European community water policy. Chapter 14. In: Dellapenna JW and Gupta J (eds) The evolution of the law and politics of water. Springer, pp 227–244

CNA (1972) Conferenza Nazionale delle Acque, I problemi delle acque in Italia. Relazioni e documenti, [The Water Problems in Italy: Reports and Documents] (in Italian). Tipografia del Senato, Roma, p 815

Commissione Interministeriale per lo studio della sistemazione idraulica e della difesa del suolo (1970) Relazione conclusiva [Inter-ministerial Commission for the study of hydraulic regulation and soil defense] (in Italian). Poligrafico dello Stato, Roma, p 900

E C (2000) Towards a sustainable and strategic management of water resources. Italy. Official Publications of the European Communites, Luxembourg, pp 203–233

E C (2007) Drought management plan report including agricultural, drought indicators and climate change aspects. Water Scarcity and Drought Expert Network Technical Report, 2008-023. Luxembourg, p 108

Greco N (1983) Le acque [Waters] (in Italian). Il Mulino, Bologna

Maglia S, Galotto G (2009) Il codice delle acque [Code of waters] (in Italian). IPSOA, Milano, p 1905

Massarutto A (2008) L'acqua [Water] (in Italian). Il Mulino, Bologna, p 140

Rossi G (2017) Policy framework of drought risk mitigation. Ch.28. In: Eslamian S, Eslamain F (eds) Handbook of drought and water scarcity, vol 2. CRC Press, New York, pp 569–588

Rossi G, Ancarani A (2002) Innovation in water legislation in Italy: ecosystem protection, and stakeholders participation. Int J Water 2(1):17–34

Rossi Paradiso G (2013) Gli strumenti di pianificazione delle acque e della difesa del suolo in Italia. Quali priorità per un approccio integrato? [The water and soil planning tools in Italy: what priorities for an integrated approach?] (in Italian). L'Acqua 5–6:53–68

Rusconi A (2010) Acque e assetto idrogeologico. [Water and hydro-geological asset] (in italian) DEI, Roma

Veltri M (1998) Indagine conoscitiva sulla Difesa del suolo, Atti e Documenti [Survey on Soil Defense. Proceedings and Documents] (in Italian) Senato della Repubblica, Roma

Chapter 4
Institutional Framework of Water Governance

Giuseppe Rossi

Abstract The chapter presents the current institutional framework concerning water resources and soil defense in Italy. The roles and functions of the organizations at central government level and river basin districts are described considering responsibilities in policy making, coordination, advisory, implementation of policy and surveillance over management bodies. Similar analysis is developed at regional and local levels. With the aid of the list of significant gaps of water governance in Italy, identified by the OECD (Organization for Economic Co-operation and Development), an evaluation of weaknesses and strengths of the institutional framework is carried out. The fragmentation of responsibilities between state and regions, the difficulty of coordination among the several bodies at national and subnational levels, the inadequate basin planning in the river basin districts, the preference for emergency actions instead than prevention measures to face flood and drought risks and the slowdown of the municipal water services reform can be identified as the main weaknesses. Conversely, positive results include the skill of civil protection for reducing the damages of the water-related disasters, the role carried out by of the Institute for Environment Protection and Research (ISPRA) for homogenizing the working rules of subnational authorities and regional bodies, as well the improvement of the supervising of municipal water services due to the new guidelines established by the Authority for Regulation Energy Networks and Environment (ARERA), e.g. on tariff computation and service quality monitoring.

Acronyms

AATO	Autorità dell'Ambito Territoriale Ottimale/Optimal Territorial Area Authority
AEEGSI	Autorità per l'Energia Elettrica, Gas e Sistemi Idrici/Authority for Electric Energy Gas and Water Systems

G. Rossi (✉)
Department of Civil Engineering and Architecture, University of Catania, Catania, Italy
e-mail: grossi@dica.unict.it

© Springer Nature Switzerland AG 2020
G. Rossi, M. Benedini (eds.), *Water Resources of Italy*, World Water Resources 5,
https://doi.org/10.1007/978-3-030-36460-1_4

ANBI	Associazione Nazionale delle Bonifiche, delle Irrigazioni e dei Miglioramenti Fondiari/Association of Land Reclamation Consortia
ANCI	Associazione Nazionale Comuni Italiani/Association of Italian Municipalities
ANPA	Agenzia Nazionale per la Protezione dell'Ambiente/Agency for Environment Protection
APAT	Agenzia per la Protezione dell'Ambiente e per i Servizi tecnici/Agency for Environmental Protection and Technical Services
ARERA	Autorità per la Regolazione di Energia, Reti e Ambiente/Authority for Regulation Energy, Networks and Environment
ARPA	Agenzia Regionale Protezione Ambiente/Regional Agencies for Environment Protection
ATO	Ambito Territoriale Ottimale/Optimal Territorial Area
CF	Centri Funzionali di Protezione Civile/Functional Centers of Civil Protection
CNR	Consiglio Nazionale delle Ricerche/National Research Council
Co.N.Vi.R.I.	Commissione Nazionale di Vigilanza sull'Uso delle Risorse Idriche/National Commission Supervising the Use of Water Resources
Co.Vi.R.I	Comitato di Vigilanza Risorse idriche/Committee Supervising the Use of Water Resources
DSTN	Dipartimento dei Servizi Tecnici Nazionali/Department of National Technical Services
ENEA	Ente Nazionale per le nuove tecnologie, energia e sviluppo economico sostenibile/Technology Energy and Environment Agency
INGV	Istituto Nazionale di Geofisica e Vulcanologia/National Institute for Geophysics and Vulcanology
ICRAM	Istituto della Ricerca applicata al mare/Institute for the Research Applied to the Sea
INFS	Istituto Nazionale Fauna Selvatica/Institute for Wildlife
ISPRA	Istituto Superiore per la Protezione e Ricerca Ambientale/Institute for Environmental Protection and Research
ISS	Istituto Superiore di Sanità/National Institute of Health
OECD	Organization for Economic Cooperation and Development
SII	Servizio Idrico Integrato/Integrated Water Service

4.1 Introduction

"The current water "crisis" is not a crisis of scarcity but a crisis of mismanagement, with strong public governance features. Key obstacles to improve water management are institutional fragmentation and badly managed multi-level governance. [...] Water policy involves a range of public stakeholders across ministries, departments and public agencies, and between various levels of government. In addition to the policy makers, citizens, private actors, end users, investment banks, and infrastructure and service providers have a stake in the outcome" (OECD 2011).

This quotation from a study of the Organization for Economic Co-operation and Development (which does not refer specifically to the Italian case) highlights the importance of the institutional framework on the water management and, in particular, the key role of government at various levels, together with civil society and private sector, to guarantee effective water governance. Nowadays, the term *governance* is generally used beyond its meaning of government "to encompass all the mechanisms, processes, relationships and institutions citizens and groups use to articulate their interests and exercise their rights and obligations" (OECD 2011). The term should be distinguished from *water management*, as the latter refers directly to the operational activities for meeting specific targets in the water services, while *water governance* refers to the set of administrative systems and focuses on formal and informal institutions as well as on organizational structures and their active performance in terms of legitimacy to govern, including also transparency in the decision-making process, accountability of the responsible bodies and the inclusiveness of stakeholders.

The institutional setting of policy making, planning, design, construction, operation and control of systems for water resource management and soil conservation in Italy is very complex, due to a historic process which has developed specific bodies for regulating and managing each specific sector, often supported also by a scientific community and by a sectorial approach (e.g. taking into account only agricultural or hydraulic engineering views) unable to produce real coordination.

The Italian legislative framework (Greco 1983; E.C. 2000; Rossi and Ancarani 2002; Maglia and Galotto 2009) has contributed to this complexity. The transfer of numerous tasks and relevant responsibilities from central government to the regions has worsened the situation in many instances. Besides the difficulties of communication at national level among the ministries, new difficulties have arisen between national and subnational organizations, particularly between state and regions and at regional level between region and local bodies. The reform of the water legislation, driven partially by European Directives, did not improve the confusing situation concerning the roles and responsibilities created by the overlapping of sectorial acts. In fact, the Legislative Decree (DLgs) 152/2006, aiming at incorporating almost all of the previous acts in a comprehensive and concise Environmental Code, failed this ambitious goal as it did not accomplish a judicious revision of the general structure of the system of governance. Furthermore, the DLgs 49/2010, acknowledging the Directive 2007/60 EC, did not resolve the difficulties in flood risk mitigation, due to the inadequacy of the planning provisions and to the lack of an effective coordination among prevention and emergency measures.

The chapter presents the institutional framework of water resource use, water quality protection and soil defense, distinguishing the roles and responsibilities both at national (central government) (Sect. 4.2) and at subnational levels (Sect. 4.3). Several considerations on the governance weaknesses along with some recent positive innovations are the subject of concluding remarks (Sect. 4.4).

4.2 Roles and Responsibilities in Water Governance at National Level

A simplified draft of the bodies, which cover specific roles in water governance at national level, is shown in Fig. 4.1. The sketch includes the central government and the districts defined by the DLgs152/2006 (Environmental Code) according to the European Directive 2000/60 EC.

The functions considered in the institutional mapping are (i) policy making and planning, (ii) coordination role, (iii) implementation of policies and advisor role, and (iv) vigilance over management bodies and citizen's participation. The current roles and responsibilities of various organizations are identified on the basis of the rules issued by the Environmental Code, as reformed by the following acts, particularly by Law 221/2015.

The first function refers to the upper level of water policy, i.e. to strategic choices in legislation and planning and to the coordination of decisions at other levels. These tasks are carried out by the President of the Council of Ministers (Prime Minister) according to the proposals done by the Minister for Environment, Land and Sea Protection and by the Committee of Ministers for Soil Defense Interventions. The Committee, according to Art.57 of Environmental Code, besides the Minister of Environment, which has a key role as delegate of the President to preside the

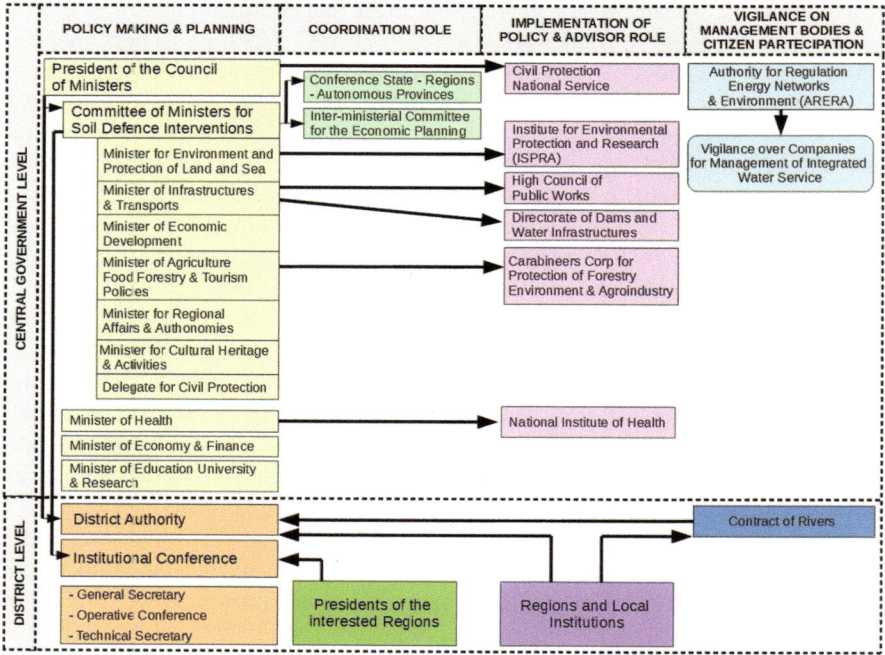

Fig. 4.1 Institutions responsible of water governance at national level

committee, includes several ministers, which perform different functions within water resources and soil defense fields: Minister of Infrastructure and Transport (formerly Minister of Public Works), Minister of Economic Development (formerly of Productive Activities), Minister of Agricultural Food and Forestry Policies and Tourism, Minister of Regional Affairs and Autonomies, and Minister of Cultural Heritage and Activities. Besides, the Committee includes the Delegate of the President of the Council of Ministers for civil protection.

According to the Environmental Code, **the President of the Council of Ministers**, who had a greater role under Law 183/1989, is responsible for the approval of the decrees concerning methods and criteria for basin planning and implementing water and soil management measures, for the approval of the river basin plans and of other acts concerning the direction and coordination of intervention at the lower level. However, the responsibility for several decrees, which had been entrusted to the Prime Minister, has been transferred to the Ministry of Environment by Law 221/2015. The same President of the Council, through the Department of Civil Protection, controls the Civil Protection Service which was reorganized by the Law 100/2012. In addition, for a short period, the new Mission Structure to cope with Hydrogeological Risk (*Safe Italy*) established in 2014 (Decree of President of Ministers Council 27/5/2014) was located at the Presidency of the Minister Council, until its cancellation in 2018.

The **Minister for Environment, Land and Sea Protection** is responsible for the enforcement of most of the rules on water resource planning, soil defense planning, water use (in particular domestic supply) and water quality standards. The Ministry is divided into four directorates (besides that of General Affairs and Personnel), devoted to Territory and Water Resources, Protection of Nature and Sea, Sustainable Development, Climate and Energy and Environmental Assessments. A very important responsibility is the duty to apply European Directives and approve the plans prepared by the District Authorities according to the Directives 2000/60/EC and 2007/60/EC. The Ministry is responsible for surveillance over the Institute for Environmental Protection and Research (ISPRA), which cooperates in preparing the biennial report on environment to be presented to Parliament, according to the Law 349/1986 which established the Ministry of Environment.

The **Minister of Infrastructures and Transport** continues to carry out part of the activities of the previous Minister of Public Works. It includes, at the national level, the directorate responsible for dams and water infrastructures.

The **Minister of Agricultural Food and Forestry Policies and Tourism** has responsibility for agricultural development and irrigation programs. It manages the forest services through the State Forestry Body that is dedicated to the protection of forests, environment and agriculture. Today such body is incorporated into the Carabineer Corp.

The **Minister of Health** has responsibility for the quality of drinking water and, in cooperation with the Ministry of Environment, establishes rules for the control of water quality through all stages of the process from withdrawal to delivery to the consumer. In the year 2017, it has prepared the decree establishing the programs of water quality control (including chemical and microbiological parameters and sampling frequency). This decree modified the annexes to the Directive 98/83 EC acknowledged through the DLgs 31/2001. Besides, together with the Minister of Agricultural Food and Forestry Policies, the Minister has responsibility for the quality of water used for irrigation.

The other Ministers, listed in Fig. 4.1 (particularly the Minister of Economy and Finance and the Minister of Education, University and Research), even though without specific functions in water-related decisions, play a role in the general process of planning, funding, implementation and control of investments for public works and research activities in water resources and soil conservation fields.

The main coordination role at national level is attributed to the **Conference State-Regions and Autonomous Provinces** (Trento and Bolzano). It has a key role in resolving frequent conflicts among central government and one or more regional governments also in the matter of water and soil management measures. The Conference can develop proposals on criteria and methods for district planning, and it can also express remarks on the river basin plans and on the sharing of funds for the 3-year programs of public works among the organizations responsible for the implementation.

Another body which is very active, especially with concerning shared financial resources provided by different investment sources, is the **Inter-ministerial Committee for Economic Planning** (CIPE), now Inter-ministerial Committee for Economic Planning and Sustainable Development. The Committee was established in 1967 with the participation of the main Ministries and is chaired by the Prime Minister. The Committee provides directives on national economic policy, State's balance and the relationship of the State with European economic policy. In the last few years, it has funded hydraulic works for water supply, sewage and wastewater treatment, with the aim of facing the infringement procedures regarding water pollution that the European Union established against Italy.

Among the bodies responsible for implementing national policy with regard to water-related disasters, an important role is assigned to the **Civil Protection Service,** established by Law 225/1992 and reorganized by the Law 100/2012 under the control of the Department of Civil Protection (Presidency of the Council of Ministers). The "civil protection" includes all the activities aimed at protecting life, property and settlements from risk of damage arising from natural and anthropogenic disasters (including flooding, landslides and drought). The focus of the Service is on prevision and prevention, the rescue of affected communities, the tackling and resolution of emergencies, together with the mitigation of risks. The civil protection duty is shared among national service, regions, provinces and municipalities. This requires the collaboration of National Fire Department, Armed Forces, Forestry Corp, Italian Red Cross, Mountain and Alpine Rescue Corps together with voluntary organizations. The Service operates at central, regional and local level under the principle of subsidiarity. The regions provide regulation of the regional organizations.

Institute for Environmental Protection and Research (ISPRA), under the authority of the Ministry for Environment Land and Sea, has the role of helping the public administration in many commitments (e.g. development of environmental data bases, synthesis of the planning provisions provided by the river basin authorities, guidelines for meteorological and hydrographic monitoring). ISPRA is a research organization established in 2008 through the unification of the Agency for Environmental Protection and Technical services (APAT), the Institute for the Research Applied to the Sea (ICRAM) and the Institute for Wildlife Studies (INFS). The APAT, established in 1999, had continued the activities developed by both the ANPA (Agency for Environmental Protection) and the DSTN (Department of National Technical Services) which, in turn, assumed the duties of the Hydrographic Service (in operation since 1917), and the other historic services such as Seismic and Geological Services. It seems the case to underline that the transfer of the hydrometeorological gauge networks from the Hydrographic Service to the regions and provinces (in most cases within the functional centres of the civil protection) since 2002 had negative consequences due to fragmentation in the different parts of the country of the previous unitary criteria adopted by the Hydrographic Service. The primary duty of the civil protection in mitigating water related disasters led to privilege meteorological forecasting and early warning systems, with negative impact on the role of hydrometeorological monitoring oriented to supply databases of long-term series. Also the hydrographic measures were strongly reduced in some regions, cutting off the very important heritage of long stream-flow series in the main Italian rivers. ISPRA developed in 2013 an initiative to homogenize the activities of the regional hydrological services and established a national table for the operative hydrology services which also includes Air Force and Civil Protection Department. Lately (2016), ISPRA has promoted the observatories for the use of water resources and voluntary structures to support the seven river basin districts, aiming to provide basic information for regulating withdrawals and uses, particularly during drought events in accordance with the objectives of the Management Plans of the Districts.

High Council of Public Works (HCPW), established by Law 1460/1942 within the Ministry of Public Works, (today Ministry of Infrastructures and Transport) with regulations updated by the DPR 204/2006, is a self-governing body with the following main functions: (i) to provide and to update technical rules and guidelines regulating design, building and maintenance of public works; (ii) to assess projects of relevant public works (amount greater than 25 million of euros), through the General Assembly and Sections; (iii) to provide official certification to building materials through the Central Technical Service of Ministry, chaired by the President of the HCPW; and (iv) to contribute to the activities of national and international organizations responsible of safety of building materials. It includes officials from national ministries and expert in civil engineering. The sections of HCPW responsible in water field include Section 4 (river basin plan, dams, soil conservation works, wastewater treatment plants, etc.) and Section 3 (harbours, coastal defense works, etc.).

A key role in the security of dams and the correct operation of reservoirs is played by the **Directorate for dams and water infrastructures**, part of the Ministry of Infrastructures and Transport, established by the DPR 254/2007. The Directorate, which replaced the Register of Large Dams, approves the projects of dams higher than 15 m and capacity greater than 1 million m^3. The Directorate has also the role of surveillance over building and operation of high dams. Its head office is in Rome, and nine technical offices are located in various cities (Torino, Milano, Venice, Florence, Perugia, Napoli, Cosenza, Palermo and Cagliari).

Since the unification of Italy in 1861, the State Forestry Body manages forest affairs, in particular in the State's propriety areas. In the year 2017, it became part of the **Carabineer Corp**, with the objective of guaranteeing the protection of forest, environment and agro-industry.

In addition, the **Association of Italian Municipalities (ANCI)** cooperates to monitor and elaborate data on water pollution, municipal water services and environmental costs and investments, as well as disseminating environmental information to the public in accordance with agreements drawn up with the Ministry for Environment, Land and Sea Protection.

The **Mission Structure against Hydrogeological Disasters** "Safe Italy", established by the DPM 116/2014 and located at the Presidency of the Ministers Council, had the function of accelerating the implementation of hydraulic works for mitigating flooding risk and landslide risk. It was cancelled in 2018 by the new government, and its functions have been transferred to the Ministry for Environment, Land and Sea Protection.

Under the surveillance of Ministry of Health, the **National Institute of Health** (established in 1934, as Institute of Public Health) is a technical-scientific body responsible of research, advice, control and training on public health, including water quality issues.

An important function of surveillance on the companies for management of municipal water services is carried out by the **Authority for Regulation Energy Networks and Environment (ARERA).** The Authority is independent from the government and has five members elected by the Parliament for 7 years. It was established in 1995 (Law 481/1995) under the name of Authority for Electric Energy Gas (AEEG), with the purpose of regulating and controlling the service of delivery of electric energy and gas. Since the year 2011 (by means the Law 214/2011), the task has been extended to the regulation of water services (AEEGSI). Since January 2018 (Law 205/2017), the Authority has the new name ARERA and is committed also in the regulation of the solid wastes management. Today, ARERA provides guidelines and control of tariffs for the municipal water supply, which are determined by the companies managing the Integrated Water Service under the vigilance of the Territorial Area Governments. In previous years, the task of regulating the municipal water services was carried out by the National Commission supervising the use of water resources (Co.N.Vi.R.I.) from 2009 to 2011, earlier Committee supervising the use of water resources (Co.Vi.R.I.) from 1994 to 2009, both located at the Ministry for Environment and Protection of Land and Sea.

The list of the institutions which have significant impact on water and soil issues at national level should include the bodies that are in charge of planning, funding and managing research and development programs (IPTS 1997). The main institutions for research in Italy are the universities, the National Research Council (CNR), as well as research institutes funded by other ministries. In particular, the CNR, founded in 1923, since 1989, is under the Ministry of Education University and

Research. The CNR's mission is to carry out research in its institutes, to promote the internationalization of the national research system, to provide technologies and solutions to emerging needs, to advice government and other bodies and to promote innovation of the national industrial system. Among the seven departments of the CNR, established by the reform of 1999, the Department of Earth System Science and Environmental Technologies contributes mainly to research in water resources and soil defense by means of several institutes. Other institutes carry out researches linked to water issues. The main institutes of CNR, operating in water field, are listed in the first part of Table 4.1, which also lists other institutes not connected with CNR, but funded by other ministries.

In particular, the researches of the **IRSA** (established in 1968) refer mainly to (i) management and protection of water resources and (ii) development of methodologies and technologies for water and wastewater treatment. The researches of the **IRPI** regard the fields of natural hazards with emphasis on geo-hydrological hazard, environmental protection and sustainable use of geo-resources.

Table 4.1 National research institutions that carry out activities on water-related issues

Reference body	Institution name	Location of sections
CNR Department of Earth System Science and Environmental Technologies	Water Research Institute (IRSA)	Rome, Bari, Brugherio, Taranto, Verbania
	Research Institute for Geo-Hydrological Protection (IRPI)	Perugia, Bari, Cosenza, Padua, Turin
	Institute of Atmospheric Sciences and Climate (ISAC)	Bologna, Lamezia Terme, Lecce, Rome, Turin
	Institute of Environmental Geology and Geo-Engineering (IGAG)	Roma, Cagliari
	Institute of geosciences and earth resources (IGG)	Pisa, Florence, Padua, Pavia, Turin
CNR Department of Biology, Agriculture and Food Science	Institute for Agricultural and Forest Systems in the Mediterranean (ISAFoM)	Ercolano Catania, Cosenza, Perugia
	Institute for Biometeorology (IBIMET)	Florence, Bologna, Sassari, Rome
CNR Department of Engineering, ICT and Technologies for Energy and Transportation	Institute for System Analysis and Computer Science (IASI)	Rome
	Institute for Applied Mathematics and Information Technologies (IMATI)	Pavia, Genoa, Milan
Ministry for Economic Development	Technology Energy and Environment Agency (ENEA)	Rome
Ministry of Education, University and Research	National Institute for Geophysics and Volcanology (INGV)	Rome, Milan, Bologna, Pisa, Naples, Catania, Palermo
Ministry of Agricultural, Food, Forestry and Tourism Policies	Council for Agricultural Research and Economics (CREA)	Rome

Italy hosts some international institutions, the most important of which is the Joint Research Centre of the European Union, located at Ispra, Varese. In the water problems, it has many contacts with the national universities and research institutions.

Research in specific sectors of water resources is also carried out by private companies, especially those dealing with chemical products.

In the field of hydroelectric power, the Italian National Electricity Board (ENEL) has proper laboratories and devoted structures, in close contact with the scientific community.

At the district level, the **District Authority** has the following main responsibilities: (i) drawing the district plan and the plans required by the European Directives and the programs of actions, (ii) checking the coherence between the objectives of the district plan and the measures of planning and programming at European, national, regional and local levels on soil defence, fight to desertification, water resources protection and management and (iii) analysing the impacts of human activities on surface and groundwater resources as well as an economic analysis of water uses. According to the modifications of Law 221/2015 to the Environmental Code, the District Authority is formed by the following bodies:

- The **Institutional Conference**, including the presidents of the regions which belong to the district, the Minister for Environment Land and Sea, the Minister of Infrastructures and Transport, the Chief of the Department of Civil Protection and also the Minister for Agriculture, Food and Forest Policies and Tourism and the Minister for Goods and Cultural Activities, if the topics to be discussed require their intervention. The main undertaking of the Conference is to deliberate the "Statute", to draw up and to adopt the planning tools (including the District Water Management Plan and the Flood Risk Management Plan (required by the European Directives). Also, the Conference has responsibility of drawing up the Hydrogeological Asset Plans and the Extraordinary Plans for areas under high risk (except for Autonomous Regions). It is entrusted with the protocols of agreements for water inter-basin transfers.
- The **General Secretary**, who provides the necessary actions for the functioning of the Authority, for implementing the directives of the Operative Conference and the data collection.
- The **Operative Conference**, in addition to the representatives of the public administrations which are already members of the Institutional Conference, including also the representatives from agricultural organizations and from the Association of Land Reclamation Consortia (ANBI). Experts (without voting rights) may be included to give advice on the plans and on the programs to be carried out.
- The **Technical Operative Secretary**.

According to the changes introduced by the Law 221/2105 to the DLgs 152/2006, the following districts have been defined: (1) Eastern Alps, (2) Po valley, (3) Northern Apennines, (4) Central Apennines, (5) Southern Apennines, (6) Sardinia and (7) Sicily. The boundaries of the districts are indicated in Fig. 4.2.

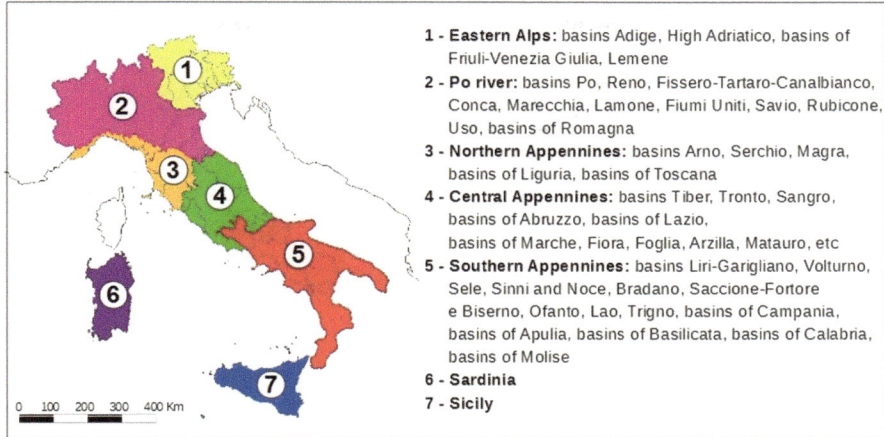

1 - **Eastern Alps:** basins Adige, High Adriatico, basins of
    Friuli-Venezia Giulia, Lemene
2 - **Po river:** basins Po, Reno, Fissero-Tartaro-Canalbianco,
    Conca, Marecchia, Lamone, Fiumi Uniti, Savio, Rubicone,
    Uso, basins of Romagna
3 - **Northern Appennines:** basins Arno, Serchio, Magra,
    basins of Liguria, basins of Toscana
4 - **Central Appennines:** basins Tiber, Tronto, Sangro,
    basins of Abruzzo, basins of Lazio,
    basins of Marche, Fiora, Foglia, Arzilla, Matauro, etc
5 - **Southern Appennines:** basins Liri-Garigliano, Volturno,
    Sele, Sinni and Noce, Bradano, Saccione-Fortore
    e Biserno, Ofanto, Lao, Trigno, basins of Campania,
    basins of Apulia, basins of Basilicata, basins of Calabria,
    basins of Molise
6 - **Sardinia**
7 - **Sicily**

Fig. 4.2 Current Italian Hydrographic Districts (as modified by Law 221/2015)

Recently, a body for citizen participation has been introduced through the
Contract of River, which can contribute to the definition and implementation of the
district plan in the fields of protection and management of water resources, development
of river basin and defense from hydraulic risk, as regulated by Article 59 of
Law 221/2015.

4.3   Roles and Responsibilities in Water Governance at Subnational Level

Figure 4.3 shows a simplified scheme of the institutions responsible for water governance
at subnational level, i.e. at regional, inter-municipal and local levels.

The functions here considered include (i) policy making and planning, (ii) executive
role, (iii) monitoring and advisor role and (iv) water resources management in
specific sectors, such as municipal, agricultural and industrial.

The *policy making and planning* role at regional level is carried out by the
regional government, including the president of the region and several regional
councillors (*Assessori*), which share the responsibility of regulating the sectors of
water resources and soil defense through many departments. The complexity of
coordination is similar to that of the central government level. Concerning water,
regional governments provide (i) the drawing up of the Regional Plan for Water
Protection, (ii) the definition of the Optimal Territorial Areas and the procedures for
the choice of the company managing the Integrated Water Service, (iii) the organization
of regional services, (iv) the creation of bodies to guarantee the quality of the
water service to the citizens of the region and (v) the regulation of the duties of the
bodies for water resources management in agriculture and industry sectors.

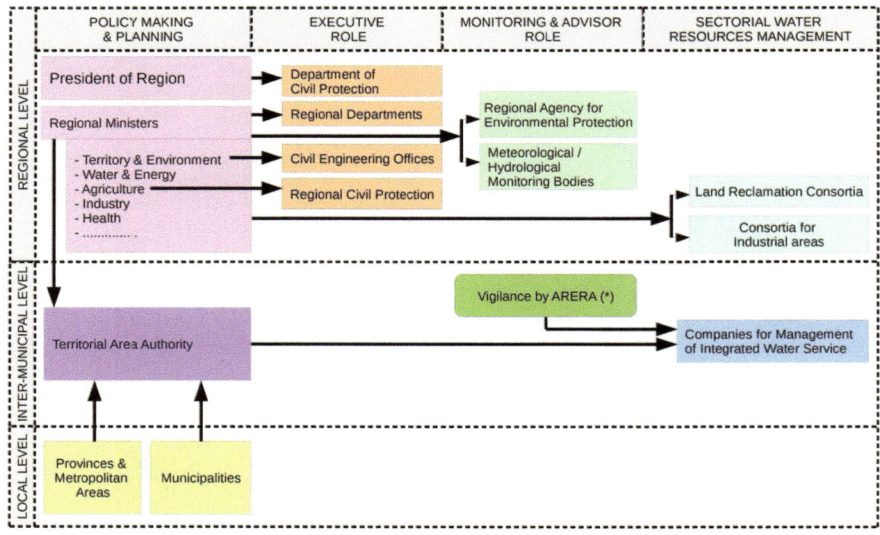

Fig. 4.3 Institutions responsible for water governance at subnational level

Most of these tasks are generally carried out by the **Department for Territory and Environment** (or other equivalent name). **Departments for Water and Energy, Agriculture, Health, Industry** exist almost in all the regions, although their names may slightly change.

Among the several offices, which have an *executive function* at the regional level, the Civil Engineering Offices (**Genio Civile**) have also a specific role for water resources and watercourse issues. From the establishment of the Italian Kingdom (1861), they carried out many responsibilities as organs of the Ministry of Public Works. Since the establishment of the ordinary regions, the civil engineering offices have been transferred to the regions and have dealt with all functions referring to public works at provincial level (Decrees of Republic President (DPR) 8/1972 and 616/1977). In particular, they are responsible or cooperate for granting abstraction licenses, for permits of occupation and crossing in watercourses, for maintenance of river public property and for hydraulic evaluation of the projects referring to the mitigation of flooding and landslide risks.

The regions are responsible for the functions and duties of civil protection for disasters of normal level, in particular for the flood events of normal level in the river basin of regional interest, while exceptional events which require extraordinary measures are under the responsibility of the National Department of Civil Protection. The first response to an emergency, whatever the nature and extent of the event, is guaranteed at the local level from the municipal structure, which is the closest institution to the citizen. The first head of civil protection in each municipality is, therefore, the mayor. However, when the event cannot be met by the means available to the municipality, the higher levels are activated through an integrated and coordinated action: from the province, the prefecture, the region, up to the involvement of the state in the event of a national emergency.

The network of regional functional centres for civil protection manages the system of alert with regard to hydrogeological and hydraulic risks that is made of two stages: first, the forecasting of severe meteorological events and, secondly, the stage of monitoring and watching. The regional structure utilizes the cooperation of voluntary organizations and organizes services of territorial hydrogeologic assistance.

The functions of *monitoring and advisor roles* on different aspects of water resources and hydrogeological risk are in charge of the Regional Agency for Environment Protection (ARPA) together with other bodies responsible for meteorological and hydrologic monitoring, established in various regions also with different names. The fragmentation of the hydrometeorological monitoring is evident from the list of the regional or provincial bodies, which have this duty, as reported in Table 4.2.

Table 4.2 Regional or provincial bodies responsible of the hydrometeorological monitoring

Region/province	Institution	Specific body
Abruzzo	Abruzzo Region	Hydrographic Office
Aosta Valley	Aosta Valley Region	Functional Centre
Apulia	Apulia Region	Functional Centre/Civil Protection
Basilicata	Basilicata Region	Department Infrastructures and Public Works/ Civil Protection
Bolzano	Bolzano Province	Hydrographic Office
Calabria	ARPA∗ Calabria	Functional Multi-risks Centre
Campania	Campania Region	Directorate Public Works Functional Centre Civil protection
Emilia-Romagna	ARPA ∗ Emilia Romagna	Hydro-Meteo-Climate Service
Friuli Venezia Giulia	Civil Protection	Functional Centre
Lazio	Lazio Region	Functional Centre
Liguria	ARPA∗ Liguria	Functional Meteo-Hydrologic Centre of Civil Protection
Lombardy	ARPA∗ Lombardy	Regional Meteorologic Service and Snow-Metorologic Centre
Marche	Marche Region	Functional Centre
Molise	Regional Agency of Civil Protection	Functional Centre
Piedmont	ARPA∗ Piedmont	Department of Forecasting Systems
Sardinia	Sardinia Region	Agency Hydrographic District Sardinia
	ARPA ∗Sardinia	Department Hydro-Meteo-Climate
Sicily	Sicily Region	Observatory of Water
Tuscany	Tuscany Region	Functional Centre/Regional Hydrographic Service
Trento	Trento Province	Dam Office
Umbria	Umbria Region	Hydrographic Office/Functional Centre
Veneto	ARPA∗ Veneto	Hydrologic Service/Functional Centre

Source: ISPRA (2018)
∗ARPA = Regional Agency for Environment Protection

The **Regional Agencies for Environment Protection (ARPA)** have been established by the Law 61/1994, together with the National Agency for Environment Protection after the reform of the Department of the Health Prevention, which has limited its role to specifically sanitary matters. Today the ARPAs operate in all 19 Italian regions and in the autonomous provinces (Trento and Bolzano). They monitor aspects of the environment (air, water, soil) to control pollution, and they both support the local organizations (municipalities, provinces) and maintain an information system on the environment. In particular, ARPA monitors climate, quality of drinking, bathing and coastal water, polluted discharge, treatment plants, etc. and develops campaigns for environmental education.

Based on the functions carried out by the ARPA, some regions have other **bodies for meteorological and hydrologic monitoring**, which replace the local offices of the old National Hydrographic Service. The latter were established on the Italian territory in 1917, based on the experience of the Office of the *Magistrato delle Acque* of Venice (1907) and Hydrographic Office of the Po river (1912) and operated until 1989 under the authority of the Ministry of Public Works. Afterwards, they were part of the National Technical Services, under the Presidency of Ministers Council until 1998, when the offices were transferred to the regions by the DLgs112/1998 (Bassanini Act).

The bodies for sectorial water resources management under the direct control of the regions include the Land Reclamation Consortia and Consortia for Industrial Development Area. The companies for management of municipal water services are under the direct control of the government bodies of the Optimal Territorial Areas and the vigilance of a regional department.

The **Land Reclamation Consortia** have been established by Royal Decree 215/1933 and are bodies formed by the private owners of the land with the function of undertaking public works in the sphere of land reclamation, irrigation and soil conservation. They are under the jurisdiction of the Ministry of Agriculture Food and Forestry Policies, which guarantees their special role in the implementation of the national plans for irrigation, periodically issued within the programs for agricultural development of the nation.

The **Consortia for Industrial Development Area** are public economic bodies, established by the Law 634/1957, which extended the mission of the Southern Italy Development Fund in order to promote the industrial development of the regions of the Southern Italy. They have been established by municipalities, provinces and chambers of commerce and were instrumental in building various infrastructures for industrial areas and managing related services (including water supply in some cases). Today some regions (e.g. Sicily) have cancelled the Consortia, with the establishment of Regional Agencies for the Development of Production Activities.

At inter-municipal level, the **Optimal Territorial Area Authority (AATO)**, as established by the Law 36/1994 (Law Galli), issues directives and controls the management of the municipal services (aqueduct, sewage, treatment plants) unified within the Integrated Water Service. The Law 42/2010 cancelled the authorities, transferring to the regions the responsibility of defining the boundaries of the ATOs and the rules to entrust the Integrated Water Service to management companies. A transition condition currently concerns the organization of the service, in particular, concerning the name and function of the government body of the territorial area, which exhibits non-homogeneous features in different regions.

At the same inter-municipal level, there is the body responsible for **managing the Integrated Water Service.** The rules employed for entrusting the task to a company (public, private or public-private) are defined by each region. The management of the service is regulated by the government body of the Optimal Territorial Area and is under the regulation of the Authority for Regulation Energy, Networks and Environment ARERA (particularly for tariff and service performance systems).

Finally, at local level the provinces and municipalities also have responsibilities in water matters. **Provinces** are responsible for controlling discharges into surface water bodies. However, the responsibilities of the provinces today are not clear. In fact, Law 56/2014 (Delrio Act) reduced the competences and eliminated the direct election (since they became bodies of secondary level), waiting for the approval of the constitutional referendum, which proposed the elimination of the province from the list of the bodies of Art. 114 of the Constitution. Since the referendum has not been approved, the provinces continue partially to operate, and the recent law for balance (2017) has given funds to make urgent works and to take on new personnel for covering retired people.

Municipalities, which carried out the services of drinking water supply and municipal uses, of sewage and of wastewater treatment until the Law 36/1994, have now the role of establishing the government body of the Optimal Territorial Area, according to the reform of municipal services. In several cases, municipalities continue to provide the water supply, sewage and waste-water treatment services, since the plants were not transferred to the management body of the Optimal Territorial Area.

4.4 Concluding Remarks

The survey shows the extraordinary complexity of the Italian institutional framework of water governance both at national and subnational levels. In order to examine the weaknesses of the system, it seems appropriate to review the analysis carried out in an OECD study (OECD 2011), made to evaluate the governance challenges of water policy in Italy. The evaluation was based on a list of seven proxy indicators chosen for the analysis of governance challenges in OECD countries (Table 4.3).

According to the responses of Italy to the 2010 OECD Survey on water governance, 5 out of 7 gaps have been identified as important or very important for Italy, namely, policy, administrative, information, capacity and accountability. Among the causes of these gaps, several obstacles have been cited. The main obstacles to horizontal coordination in water policy making at central level which can be considered very important are the following: (i) interference of lobbies, (ii) difficult implementation of central decisions at local level, (iii) difficulties related to implementation and (iv) lack of citizen concern with regard to water policy. Other obsta-

Table 4.3 Multi-level governance gaps in water policy

Multi-level governance gaps	Proxy indicator
Policy	Overlapping, unclear allocation of roles and responsibilities
Administrative	Mismatch between hydrological and administrative boundaries
Information	Asymmetries of information between central and subnational governments
Capacity	Lack of technical capacity, staff, time, knowledge and infrastructure
Funding	Unstable or insufficient revenues of subnational government to effectively implement water policies
Objectives	Competition between different organizations
Accountability	Lack of citizen concern about water policy and low involvement of water user's associations

Source: OECD (2011)

cles have been considered to be somewhat important: (v) overlapping, unclear, non-existing allocation of roles, (vi) absence of common information frame of reference, (vii) lack of high political commitment and leadership, (viii) lack of institutional incentives for cooperation, (ix) mismatch between ministerial funding and administrative responsibilities, (x) absence of strategic planning and sequencing and (xi) absence of monitoring and evaluation of outcomes. Many obstacles have been identified also in vertical coordination in water policy making and in coordination and capacity challenges. The most important obstacles include the asymmetries of information between urban areas and rural areas, the mismatch between hydro and administrative boundaries, the insufficient financial resources, the over-fragmentation of subnational responsibilities and a lack of synergies between policy fields at local level.

Although the evaluation seems too severe, as it does not consider the historic reasons which can partially explain the overlapping of some roles, most of the reported gaps in 2011 unfortunately persist today, in spite of the legislative reforms of the last years. Most of the difficulties continue to hinder the achievement of the objectives of a better water resources management and an effective soil defense. The fragmentation of responsibility at the national level and the difficulty of coordination between central and regional governments have not been overcome, yet. The appeal to the Constitutional Court is very frequent in resolving the conflicts among regions and state being reflected into delays in the implementation of laws and measures. In addition, the over-fragmentation of responsibilities at local level has negative impacts on the performance of services as well as on a timely implementation of water works.

The most severe deficiency is perhaps the delay in the functioning of the District Authorities, established in 2006 and which was supposed to operate according to Law 221/2015. These delays have likely contributed to reduce the quality of the planning provisions required by the European Directives, which have been prepared by the old National River Basin Authorities.

The modification to the Optimal Territorial Areas, introduced by the Law 42/2010, which has transferred to the regions the responsibility to define the boundaries of the OTAs and the rules to entrust the Integrated Water Service to management companies, contributed to add new difficulties in implementing the reform of the municipal water services (Law 36/1994) and increased the fragmentation of the regulation rules among the different regions.

Furthermore, considerable difficulties derive from the uneasy coordination between urban water use, agriculture, land use and energy policies. At central government level, many controversies rise between the Ministry for Environment Land and Sea and the Ministry of Agricultural Food and Forestry Policies and Tourism, which supports the farmers, which remain the largest water consumers. Conflicts of interest have also arisen at local level between energy and water supply entities in the use of stored volumes in the reservoirs. In addition, the constraints on the land use due to a high flooding risk are hindered by municipalities which are interested in enlarging urban areas and are averse to demolish buildings located in areas with a high risk of flooding.

On the other side of the evaluation of the institutional framework of water issues, one of the more appreciated positive experience can be considered the organization of the Civil Protection, which, since its establishment in 1992 and after its structural reform in 2012 at national, regional and municipal levels, has developed effective actions in mitigating the impacts of natural and anthropogenic disasters. The Civil Protection has achieved good results during the emergencies due to the water-related disasters (floods, landslides and droughts), in order to rescue the affected communities and reduce damages to property and settlement.

Other positive results have been obtained by the Institute for Environment Protection and Research (ISPRA) in its advisor role for Ministry for Environment Land and Sea, particularly to homogenize the working rules of subnational authorities and regional bodies responsible of hydrometeorological monitoring and of sharing available water resources during droughts events. The action of the Authority for Electric Energy Gas and Water Systems (AEEGSI), today Authority for Regulation Energy Networks and Environment (ARERA), has significantly improved the surveillance on the municipal water services through new criteria and methods established for management of the supply, sewage and wastewater treatment in particular for the tariff computation and quality service performance.

The Mission Structure against Hydrogeological Disasters "Safe Italy", established at the Presidency of Ministers Council in 2014 with the aim to accelerate the implementation of structural measures and improve the mitigation of flood risk, has shown a few positive results, in spite of the difficulties of the relationship with the Ministry of Environment. The new Italian government (on May 2018) cancelled the "Safe Italy" whose responsibilities have been now transferred to the Ministry for Environment.

Even though rather unpopular, the reordering of the Italian water governance appears an impellent challenging enterprise.

References

E C (2000) European Commission. Towards a sustainable and strategic management of water resources. Italy. Official Publications of the European Communities, Luxembourg, p 178

Greco N (1983) Le acque (Waters) (in Italian). Il Mulino, Bologna, p 481

Maglia S, Galotto G (2009) Il codice delle acque (Code of waters) (in Italian). IPSOA, Milano, p 1905

IPTS (1997) Institute for prospective technological studies, Seville, Research and development policies: Case study: Water. EC Joint Research Centre, p 198

ISPRA (2018) Linee guida sugli indicatori di siccità e scarsità idrica da utilizzare nelle attività degli osservatori distrettuali per l'uso della risorsa idrica [Guidelines on drought and water scarcity indicators to be used by the District's Observatories for water resources use]. (in Italian) ISPRA Report p 55

OECD (2011) Water governance in OECD countries: a multi-level approach. OECD Publishing. https://doi.org/10.1787/9789264119284-en

Rossi G, Ancarani A (2002) Innovation in water legislation in Italy: ecosystem protection, and stakeholders participation. Int J Water 2(1):17–34

Part II
Water Resources and Water Demands

Chapter 5
Conventional Water Resources

Giuseppe Rossi and Marcello Benedini

Abstract The assessment of surface water resources carried out in 1972 and updated in 1990 is compared with the results of an estimate relevant to the 1996–2015 period. The assessment of groundwater resources, also estimated in 1972, is compared with the results of hydrogeological studies developed in 1979 and with a more recent estimate (2017), based on hydrological balances in selected zones of the country. The analysis of all the available data stresses some discrepancy in the evaluation of the national water resources, due, in particular, to large uncertainties on river runoff and aquifer infiltration. The total potential freshwater availability of Italy can be per capita almost 2000 m^3/year, with reference to the population of 2015. The impacts of climate change on the water resources in the various regions are analysed according to the Plan for Adaptation to Climate Changes (2017). A decrease of the amount of usable water, in particular due to the emission of greenhouse gas, is expected in the next decades.

5.1 Introduction

The consistency and the problems of water in Italy have been analysed following the approach and the procedures currently applied for the goods necessary for human life, which lead to the concept of "resource". Consequently, the multidisciplinary activities for definition, exploitation, control and protection of water resources belong to the more general concern known as "water resources management", which has been considered the most appropriate way to approach and solve the complexity of Italian water problems.

G. Rossi (✉)
Department of Civil Engineering and Architecture, University of Catania, Catania, Italy
e-mail: grossi@dica.unict.it

M. Benedini
Italian Water Research Institute (Retired), Rome, Italy
e-mail: benedini.m@iol.it

© Springer Nature Switzerland AG 2020
G. Rossi, M. Benedini (eds.), *Water Resources of Italy*, World Water Resources 5,
https://doi.org/10.1007/978-3-030-36460-1_5

The aim of this chapter is to assess the available resources in the Italian territory, with particular attention to the surface and underground water naturally available, namely to the "conventional resources". The "unconventional resources", supplementary quantities of water that man can artificially provide to fulfil essential uses, will be dealt with in Chap. 6.

In line with the scientific and technical approach for the water resources management, the following fundamental definitions have been adopted:

- *Natural water resource:* the volume of water that flows through a river cross section or through an identified aquifer during a defined time interval.
- *Potential water resource:* the maximum volumetric portion of an identified resource that is available for different uses by means of artificial tools, like storage reservoirs.
- *Usable water resource:* the water volume really utilized for different purposes.

The potential resource takes into account the geographic, hydrographic, geological and technologic constraints, which limit the possibility of using the water of a surface or underground body. The usable resource has to face the economic and social constraints, which affect the amount of potential resources. An economic constraint can be a limited financial availability; a social constraint is the opposition of the interested populations towards the construction of a structure able to alter their usual living conditions.

The evaluation of the resources has requested long series of data. The starting point of an effective and sustainable water resources management has been the balance between the usable water and the requirements for human uses and environmental protection (AMBIENTEITALIA 1997).

5.2   Basic Data for a Water Resources Assessment

As already anticipated in Chap. 1, the assessment refers only to the renewable resources due to the precipitation on the Italian territory, since the inflow from nearby countries is negligible with the exception of that from Switzerland, estimated 3.10 km³/year, and that from Slovenia, estimated 5.00 km³/year. This contribution is measured at gauging stations located downstream the country's border and is therefore included in the amount of water belonging to the national territory.

Concerning the groundwater, the fossil amount is neglected, not only because of the shortage of reliable data, but also because until now it has no practical utilization. The attention is on the aquifers described in Chap. 1, for which sufficient information is available.

A first detailed evaluation of water resources in Italy was done by the National Water Conference, a countrywide initiative promoted by the Senate in 1967, with the participation of the scientific community and the institutions responsible to use, protect and control the water resources (CNA 1972). Scope of the initiative was to identify the main problems in the framework of the Italian life, considering also some feasible future outcomes. An action appreciated with favour even by foreign countries, the Conference was not only the opportunity to underline some positive aspects of the Italian water expertise, heir of long traditions, but also to focus some negative aspect, due, in particular, to the lack of overall coordinating structures. The Conference fostered fruitful activities, particularly among the scientific community, and even at the present time, it can be considered a reference point for the water resources management in Italy.

Several aspects of the main problems have been further examined 10 years after and are the subject of a document in Italian, issued by the Ministry of Agriculture and Forestry (MAF 1990). Other more up-to-date documents are also in the Italian specialized literature, among which very important is the review carried out by the Water Research Institute (IRSA 1999), mostly concentrated on the economic aspects.

The first assessment of the National Water Conference and those developed in subsequent years have the advantage to give estimates of natural, potential and usable resources, taking into particular account the role of the reservoir for increasing the amount of water available when the natural streamflow is lower than the requirement.

The methods adopted by the National Water Conference to assess the natural water resources, as above defined, used the data collected in several stations of the "Italian Hydrographic Service".

As anticipated in Chap. 3, the Service was established in 1917, extending to the whole national territory the activities already carried out by the "Magistrato alle Acque" in Venice since 1907 and by the "Ufficio Idrografico" of River Po in Parma since 1912, in their relevant area of jurisdiction. Twelve "Hydrographic Compartments" were then established and identified by the catchment area of the most important rivers or taking into consideration the extension of the regional administrations. Figure 5.1, deduced from historical documents of the national government, shows the boundary of the compartments, which, according to the geographic description of Chap. 1, were structured as summarized in Table 5.1. At that time, the Hydrographic Service was a specialized branch of the Ministry of Public Works with its main offices located in Rome. In order to activate its mandatory role, it was structured with detached offices and sections hosted in the various compartments.

The equipment of the Service was more than 3500 rain gauges, 1100 temperature gauges and 500 stream gauges, disseminated in the various compartments with the facilities for analysing and processing the collected data. Every year, published "Annals" made up a precious set of information, very useful to compile a detailed history of the Italian water resources. After many years of distinguished activity under the direction of the Ministry of Public Works, the Service was upgraded directly to the office of Prime Minister, in a way that confirmed the importance of the water problems in the development of the national policy.

At the end of the twentieth century, in the progressive transformation of the administrative structure, which now enforces the importance of the local government, several compartments were transferred to the regional administrations, which have now the responsibility to control their relevant water resources.

Fig. 5.1 The Hydrographic Compartments in 1932, according to official documents of the national government

5.3 Former Estimates of Surface Water Resources

With all the available data, the National Water Conference estimated annual average precipitation and total runoff values for all the Hydrographic Compartments. Since some compartment hosted also secondary streams without gauging equipment, their hydrologic terms were estimated applying simple similarity procedures with nearby harnessed rivers. These values helped to estimate the average values for the 1921–1960 period, for all the national territory.

The numerical results obtained for that period are not fully adequate to describe the actual situation. After that period, non-negligible changes in the hydrologic regimes occurred in Italy, affecting the climatic pattern and the soil utilization. A

Table 5.1   Main characteristics of the Hydrographic Compartments

Hydrographic compartment	Area (km²)	Location	Headquarter
1. Veneto	37,000	Part of Veneto, Trento and Bolzano Provinces, and the entire territory of Friuli-Venezia Giulia	"Magistrato alle Acque", Venice
2. River Po	67,100	Entire regional territory of Aosta Valley, Piedmont and Lombardy, and the portion of regional territory of Emilia, Veneto, Trentino and Liguria belonging to the tributaries of the River Po	Hydrographic Office of River Po, Parma
3. Liguria	4800	Remaining area of the regional territory	Hydrographic Section of Genoa
4. Romagna and Marche	23,000	Regional territories	Hydrographic Section of Bologna
5. Tuscany	20,700	Regional territory	Hydrographic Office of River Arno, Pisa
6. Umbria and Lazio	23,600	River Tiber catchment and the remaining part of the two regional territories.	Hydrographic Section of Rome
7. Abruzzi and Molise	13,200	Regional territories	Hydrographic Section of Pescara
8. Campania	19,300	Regional territory and part of nearby regions	Hydrographic Section of Naples
9. Apulia	20,000	Part of regional territory and part of nearby regions	Hydrographic Section of Bari
10. Basilicata and Calabria	23,700	Regional territory and part of nearby regions	Hydrographic Section of Catanzaro
11. Sicily	25,700	Regional territory	Hydrographic Section of Palermo
12. Sardinia	24,100	Regional territory	Hydrographic Section of Cagliari

review of the estimated values should be necessary for adapting them to the current conditions. Moreover, the methods adopted for evaluating the surface water resources, although scientifically valid, considered only the man's requirements in a time of strong economic development, without the necessary attention to the ecosystem protection. At that time, the environmental awareness requested a very poor attention.

Likewise, the updating of the National Water Conference values, carried out in 1990 by the Ministry of Agriculture and Forestry, as well as the revision made 10 years after by the Water Research Institute, cannot be fully acceptable for an actual estimate, because they do not consider the most recent meteorological and hydrologic records due to the climate change. An accurate analysis of the available data is therefore necessary.

From Table 5.2 it is possible to appreciate the high space variability of both precipitation and runoff. The average annual precipitation, which was estimated 980 mm overall the Italian territory, varies between the minimum of Apulia (600 mm) and the maximum of Liguria (1340 mm). Similarly, the average runoff, estimated

Table 5.2 Natural water according to the National Water Conference (CNA 1972)

Hydrographic compartments	Average annual precipitation		Average annual runoff		Runoff coefficient
	(mm)	(km³)	(mm)	(km³)	
Veneto	1160	42.80	810	30.00	0.70
River Po	1070	71.80	670	47.00	0.62
Liguria	1340	6.40	990	4.80	0.74
Romagna and Marche	940	20.60	460	10.10	0.49
Tuscany	1010	20.90	470	9.70	0.47
Lazio and Umbria	1020	24.10	440	10.30	0.43
Abruzzi and Molise	900	11.90	490	6.50	0.54
Campania	1200	23.20	670	12.90	0.56
Apulia	660	13.20	150	2.90	0.23
Basilicata and Calabria	1012	24.00	413	9.80	0.4
Sicily	730	18.80	190	4.90	0.26
Sardinia	780	18.30	250	6.10	0.33
Italy	**990**	**296.00**	**510**	**155.00**	

500 mm on the whole country, shows very low values in Apulia and Sicily (150 and 190 mm, respectively) and high values in Liguria and Veneto (990 and 810 mm).

The time variability that characterizes the rainfall distribution, particularly in the southern regions, gives rise to great variability of the runoff. This is particularly noticeable in some zones where the summer is often completely dry with up to 150 consecutive days with no rainfall.

Table 5.3, also deduced from the results of the National Water Conference of 1972, underlines the different runoffs in significant territorial aggregations of the country. The weather conditions are affected by the geological nature and by the geographical latitude of the catchment. Normally, the permeable catchments have more uniform runoff along the various seasons.

The winter runoff predominates in central and southern parts of the country and in the islands, while in the northern Alpine zones the highest values are in summer and autumn. In the catchment of River Po the runoff is almost constant all over the year.

5.3.1   *Natural Surface Resources*

An assessment of natural surface resources can start from the evaluations of National Water Conference shown in Table 5.2. It is acceptable to assume the surface runoff as the natural resource, with its partition among the hydrographic compartments shown in the preceding paragraphs. For the national territory, the estimated surface natural resource is assumed to be 155.00 km³/year. Improvements and adjournments can be done with the results of other investigations carried out in subsequent times.

Recent estimates of water resources have been promoted by the Central Institute of Statistics (ISTAT 2018), relevant to the 2001–2010 decade, but they do not take into account the usable water for in-stream and off-stream uses. For all the national terri-

Table 5.3 Distribution of runoff during the seasons in different parts of the country (CNA 1972)

Basins	Seasonal runoff (% of annual amount)			
	Winter	Spring	Summer	Autumn
Alps	12	21	42	25
River Po	21	28	24	27
Apennines (impervious catchments)	46	33	5	16
Apennines (permeable catchments)	33	33	14	20
Islands	57	29	3	11

tory, following such estimates, the average annual precipitation during the 1971–2000 period is 708 mm (241 km^3); the average annual runoff is 384 mm (116 km^3).

The estimate for the 2001–2010 decade gives a further increase of the annual average precipitation (245 km^3) and of the natural resources (123 km^3), due to a decrease of the estimated evapotranspiration. These figures are close to those proposed by the National Water Conference.

There is also an estimate carried out by National Institute for Environmental Protection and Research (ISPRA 2017), after developing a GIS-based hydrologic balance, but without any mention to the usable amount of water. Moreover, the Plan of Adaptation to Climate Changes (CMCC 2017) has conducted an estimate of the climate change that affects the Hydrographic Districts, the area of which is different from that of the Hydrographic Compartments.

Therefore, due to the uncertainties of the investigations carried out using different methodologies and different basic data, the values of Table 5.2 still provide an acceptable picture of the mean annual natural surface resource in the different parts of Italy, also if a precaution approach requires to consider a reduction of about 20% in the total amount.

5.3.2 Potential Surface Resources

For each Hydrographic Compartment, the National Water Conference proposed also an assessment of the potential water resources (CNA 1972). The estimated values were based on the assumption that the amount of natural water potentially utilizable corresponds to that suitable to be stored in a real or foreseeable reservoir.

For some significant hydrographic basins, the runoff in the area at higher elevation, having topographic and geological features able to host a reservoir, was considered. Such assumptions allowed obtaining some estimates of the potential amount according to foreseeable utilization criteria of the water stored in the reservoir. The series of monthly streamflow for the 1929–1963 period, measured in 23 hydrographic basins, have been representative of the hydrologic characteristics of the various compartments. An extension to the Hydrographic Compartments has given the values summarized in Table 5.4.

The procedure was adopted also in the investigation carried out in 1990 by the Ministry of Agriculture and Forestry (MAF 1990), using more reliable data con-

Table 5.4 Potential surface
water resources (CNA 1972)

Hydrographic compartment	(km³/y)
Veneto	16.33
River Po	46.05
Liguria	2.86
Romagna and Marche	5.53
Tuscany	6.03
Lazio and Umbria	8.23
Abruzzo and Molise	4.63
Campania	9.51
Apulia	0.75
Basilicata and Calabria	5.46
Sicily	1.80
Sardinia	3.23
Italy	**110.41**

cerning the existing reservoirs. The results were very close to those relevant to the previous investigations.

The available procedure was the most reliable at that time, in a complex situation characterized by lack of consistent data. Anyhow, the results have been recognized and accepted in successive investigations (Benedini 1996). The values of Table 5.4 can give, at least, an order of magnitude also for the actual amount of the potential surface resources.

5.3.3 Usable Surface Resources

The procedure proposed by the National Water Conference allowed also an estimate of the usable surface resources. After analysing the series of monthly streamflow of the 1929–1963 period, measured in the 23 hydrographic basins considered representative of the hydrologic characteristics of the various compartments, a relationship between the reservoirs capacity and the annual release has been developed. This method was initially proposed during the first attempt of the National Water Conference with the data available at that time, which underlined the important role of the reservoirs. Ten years later the Ministry of Agriculture and Forestry (MAF 1990) adopted the method using more complete data.

From the capacity-yield relationship of multi-year regulation, an estimate of the usable resources has been proposed according to two alternative possibilities, namely:

- "Hypothesis **A**" with reservoirs for an industrial demand with constant distribution fully satisfied during the year

Table 5.5 An estimate of the usable surface resource (Source: MAF 1990)

Hydrographic compartment	Potential resource P (km³/y)	Usable Resource (Hypothesis A) (km³/y)	A/P (%)	Usable Resource (Hypothesis B) (km³/y)	B/P (%)
Veneto	16.33	7.67	47	10.94	67
River Po	46.05	24.87	54	16.12	35
Liguria	2.86	0.4	14	0.37	13
Romagna and Marche	5.53	1.38	25	1	18
Tuscany	6.03	0.54	9	0.54	9
Lazio and Umbria	8.23	1.07	13	1.4	17
Abruzzo and Molise	4.63	2.45	53	2.45	53
Campania	9.51	0.86	9	1.24	13
Apulia	0.75	0.46	61	0.52	69
Basilicata and Calabria	5.46	2.84	52	2.51	46
Sicily	1.8	0.67	37	0.74	41
Sardinia	3.23	1.62	50	1.84	57
Italy	110.41	44.83	40.6	39.67	(36)

- "Hypothesis **B**" with reservoir for an irrigation demand concentrated only in 6 months from April to September, following the irrigation practice of various regions, with a 20% failure.

The resulting values are in Table 5.5, compared with the potential resource described in the above paragraphs.

Both **A** and **B** hypotheses can be close to the real situation, taking into account the existing and designed reservoirs, and eventually they give results with the same order of magnitude. In both cases the amount of total usable water is less than the 40% of the potential resources, as shown by the average values of the ratio A/P in the table. Once more, the discrepancy among the various zones of the country is evident, although the determinant factor is not the geographical latitude, but several constraints relevant to local conditions of feasibility.

As an overall consideration, the amount of 40 km³/year is acceptable for estimating the total usable water in the country.

5.4   Groundwater Estimates

The importance of groundwater arises from the fact that, at least up to the end of Second World War, most drinking water was that of wells and springs. Also in the subsequent years, when the high increase of urban demand used the surface water stored in reservoir originally designed for hydroelectric or irrigation purposes, the groundwater was preferable since it required less treatment for drinkability. This is now questionable because of the increasing pollution of many wells as an effect of uncontrolled seepage of wastewater strewn on the ground.

In the future, the aquifers will have a new role for artificial recharge, storing the surface water during the rain-excess periods and the treated wastewater. It is already a common practice in arid or semiarid countries, like in many Mediterranean countries such as Israel and Jordan.

As anticipated in Chap. 1, the groundwater resources are naturally contained in aquifers located in all the national territory. The characteristics of the aquifers are variable according to the hydrogeological and climatic regimes of the regions.

The assessment of groundwater resources has been more difficult than that of surface water, for the uncertain identification of the aquifers and the zones of seepage, as well as for the difficulty of measuring the amount exchanged between watercourses and subsoil. The coastal aquifers have losses into the sea. In many cases also the indirect estimation obtained through a water balance, subtracting from the rainfall both evapotranspiration and surface runoff, was hindered by the uncertain evaluation of the last two terms of the balance. Figure 5.2 shows the main hydrogeological patterns.

Following these considerations, in order to estimate the groundwater potentially usable, the National Water Conference promoted a first evaluation (CNA 1972) on the preliminary assumption that the amount of water withdrawn through wells matches the existing resource. This has been done updating a survey carried out by the National Hydrographic Service in 1966. The survey took into consideration 124,190 wells assumed to exist in the Hydrographic Compartments. The total annual volume in Italian territory was estimated in 9.45 km^3/year. This value was not in line with other investigations carried out during the same period, which considered unreliable the basic assumptions, not only because the available information was too scarce, but also because the water abstracted from the well can be lesser than the amount naturally stored in the subsoil. Accordingly, the National Water Conference, with the support of the Hydrographic Service, carried out another investigation, based on more reliable information and, where possible, on direct measure in the field. The results are in Table 5.6, and the total amount of estimated groundwater is of the order of 13 km^3/year.

A few years after, the Ministry of Agriculture and Forestry promoted another evaluation (MAF 1990), referred to the regional territories. The evaluation considered more reliable data inferred from an investigation promoted by the European Commission (Mouton 1980), based on the amount of groundwater withdrawn from well and natural springs. The results are in Table 5.7, with the amount of 12.15 km^3/year for all the country.

Mention must be done of an estimate of the freshwater resources carried out by the Euro-Mediterranean Centre on Climate Change using the information collected from some regional administrations (CMCC 2017), for which the total usable groundwater amounts could be 16.24 km^3/year.

On the other side, mention must be done also on an attempt carried out by the National Institute for Environmental Protection and Research by means of a GIS-based procedure (Braca and Ducci 2017), considering the aquifer recharge for the 1996–2015 period.

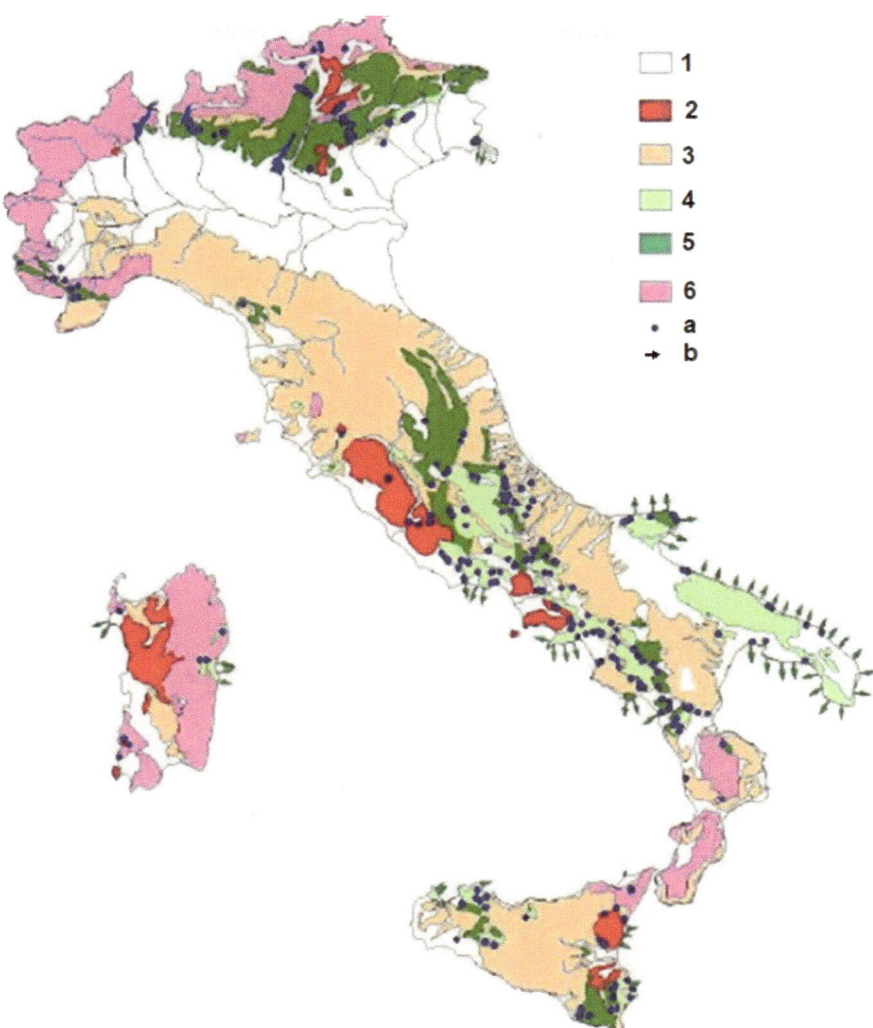

Fig. 5.2 Schematic hydrogeological chart of Italy: 1 = recent hydrogeological sequences of sedimentary nature; 2 = vulcanite and pyroclastic hydrogeological sequences; 4 = mostly terrigen Cenozoic hydrogeological sequences; 5 = continued carbonate hydrogeological sequences; 6 = vertically segmented hydrogeological sequences; 7 = metamorphic and magmatic hydrogeological sequences; a = main springs; b = main outlets and/or aquifer losses to the sea. (Source: Civita et al. 2010)

Table 5.6 Estimated groundwater resources by the National Water Conference (CNA 1972)

Hydrographic compartment	(km³/y)
Veneto	1.76
River Po	6.66
Liguria	0.20
Romagna and Marche	0.68
Tuscany	0.49
Lazio and Umbria	0.28
Abruzzo and Molise	0.18
Campania	0.55
Apulia	0.76
Basilicata and Calabria	0.35
Sicily	0.86
Sardinia	0.19
Italy	12.96

Table 5.7 Groundwater resources according to the Ministry of Agriculture and Forestry (1990)

Region	(km³/y)
Piedmont	1.49
Aosta Valley	0.05
Lombardy	2.59
Trentino-Alto Adige	0.20
Veneto	1.31
Friuli-Venezia Giulia	0.20
Liguria	0.31
Emilia-Romagna	0.66
Tuscany	0.44
Umbria	0.10
Marche	0.29
Lazio	1.03
Abruzzo	0.21
Molise	0.04
Campania	0.93
Apulia	0.33
Basilicata	0.18
Calabria	0.41
Sicily	1.15
Sardinia	0.22
Italy	**12.15**

The resulting average value for all the national territories should be of 6.7 km^3/year, much lesser than the previous estimates. Such a discrepancy can be justified also considering that only a portion of the total precipitation contributes to the aquifer recharge.

Likewise, an evaluation of the National Group of Defence from Hydrogeological Disasters on 1986 data (Ubertini 1989), oriented to identify the risk of aquifer pollution and situations of over-exploitation, has given useful information also on the estimates of usable volumes for specific aquifers located in different regions, but without any updated assessment for all the country.

The large discrepancy of the above results underlines the difficulty for a reliable evaluation of groundwater resources, in particular for a distinction between potential and usable water. Taking into account all the causes hindering the efforts for a reliable evaluation, it is worthwhile to call the attention to the estimates that agree on 12 km^3/year as usable resource, with the assumption that the actual exploitation of Italian groundwater is close to the upper limits imposed by environment preservation. The progressive deterioration of aquifers due to the discharge of polluted wastewater and the increasing costs of intervention can also reduce the practical feasibilities.

5.5 Updated Estimates of Italian Water Resources

Recent estimates of water resources have been included in the planning tools developed by the regions and the River Basin Authorities, almost oriented to water quality problems, while the subject is also in the Districts Management Plans recommended by the European Directives and relevant to the new structure of Italian water policy, as described in Chap. 1 (CEC 1991). Unfortunately, these tools adopt different methodologies, also due to the fragmentation of the national Hydrographic Service into regional offices. Therefore, it is not possible to compare the estimates of the Districts with the previous assessments relevant to the Hydrographic Compartments and revise the resource amounts for all the national territory.

The most recent assessment of natural water resources has been carried out by National Institute for Environmental Protection and Research (ISPRA 2017), with monthly values of precipitation and temperature over 1,0 km grid ("BIGBANG 1.0 model"), for estimating the evapotranspiration, the recharge of aquifers and the runoff. The assessment includes the average monthly values of the balance for the 1996–2015 period, as shown in Table 5.8.

However, this estimate is limited to the natural surface resources (runoff) and the recharge of aquifers (infiltration), without any mention to the amounts of potential and usable water resources, as the National Water Conference did.

Anyhow, a comparison between all these estimates, which differ for both basic data and methods of assessment, can infer some useful considerations. According to the National Water Conference of 1972 (CNA 1972), the total annual precipitation

Table 5.8 Average monthly values of the hydrological balance in the national territory (1996–2015) computed by ISPRA 2017

Month	Precipitation (mm)	Precipitation (km³)	Potential ETP (mm)	Potential ETP (km³)	Real ETP (mm)	Real ETP (km³)	Infiltration (mm)	Infiltration (km³)	Runoff (mm)	Runoff (km³)
January	81	24.40	10	3.10	10	2.90	33	10.00	28	8.50
February	69	20.90	11	3.40	11	3.20	29	8.70	25	7.40
March	76	22.90	26	7.90	25	7.40	30	9.20	26	7.80
April	82	24.80	46	13.90	43	13.00	25	7.50	23	7.00
May	74	22.40	84	25.40	78	23.40	13	3.90	17	5.00
June	60	17.90	119	35.90	87	26.20	4	1.10	7	2.10
July	47	14.20	140	42.30	61	18.50	2	0.60	4	1.10
August	55	16.50	131	39.50	51	15.40	2	0.70	4	1.20
September	85	25.50	85	25.60	60	18.10	4	1.30	7	2.10
October	103	31.00	54	16.40	46	13.70	13	4.00	14	4.20
November	133	40.00	26	7.80	24	7.30	37	11.00	32	9.50
December	97	29.10	12	3.60	11	3.40	35	10.70	30	9.10
YEAR	962	289.60	744	224.80	507	152.50	227	68.70	217	65.00

(990 mm, 296 km³) is shared between natural surface resources (510 mm, 155 km³) and groundwater (43 mm, 13 km³), with a total loss mostly due to evapotranspiration (437 mm, 128 km³), which is about the 44% of precipitation. The ISPRA results for the 1996–2015 period give a total annual precipitation very similar (962 mm, about the 3% smaller than the amount of CNA), while the real evapotranspiration is about the 53% of the precipitation, and the total surface and groundwater resource is 46% of the precipitation instead of the 56.4% estimated by the Conference. It is difficult to appreciate if these differences are due to the increase of temperature and different seasonal regimes or almost exclusively to the different methods adopted.

The various outcomes described above considered different geographic partitions, in a way that a comparison of the estimated values cannot be easily done. In fact, the National Water Conference referred to the Hydrographic Compartments, which have been the basic territorial entity also for other evaluations carried out during the first investigations, while the groundwater evaluation promoted by the Ministry of Agriculture and Forestry in 1989 considered the regional territory. Now, there is not a clear definition on the respective areas as concerns the hydrological aspects. Some approximations are therefore possible, as already done above for the natural surface resources, reliable only on the overall values relevant to all the national territory.

According to the above paragraphs, the total potential resource, resulting from about 110 km³/year of surface and 12 km³/year of underground, is of the order of 122 km³/year.

In line with the above considerations, the basic assumptions of the National Water Conference give an estimate of the total Italian usable resources of an order of 52 km³/year, of which 40 km³/year of surface water and 12 km³/year of groundwater. A recent estimate carried out by the Euro-Mediterranean Centre on Climate Change (CMCC 2017) gives an amount of 58 km^3/year, of which about 42 km^3/year from surface water and 16 km^3/year from groundwater. An estimate of the Ministry of Environment and Protection of Land and Sea (MATTM 2014) confirms the values proposed by the National Water Conference. The discrepancy of the various estimates can be due to the approximation of the basic data. Anyhow, a foreseeable increasing of the environmental constraints, which will certainly reduce the water withdrawal, enforces the above value of 52 km^3/year.

An interesting consideration can come recalling the values of the actual population, as described in Chap. 1, which at national level is more than 60 million. Consequently, the per capita amount of potential water resources is of the order of 2000 m^3/year, an average between the various parts of the country characterized by great difference of water resources and population density.

As already pointed out for the natural resources, some southern regions, which have scarce potential water but high population, can have a per capita less than 300 m^3/year, while some northwestern regions can rely on a per capita greater than 3500 m^3/year.

To complete these general considerations, the overall Italian situation, relevant to the potential resources, can be compared with that of some European countries tied geographically or politically to the Mediterranean basin, as shown in Fig. 5.3 (IRSA 1997; Suzenet 1997). Italy is among the countries in which the per capita availability is scarcer.

This explains why the water problems are so important, in a challenge that involves Italy's present and future life.

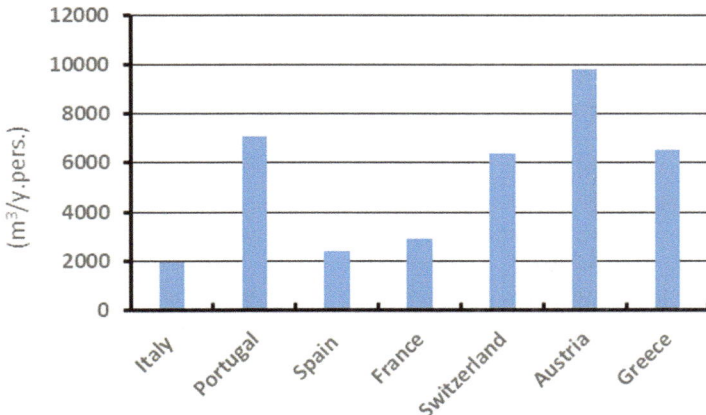

Fig. 5.3 The annual per capita potential water in some European and Mediterranean countries

5.6   Expected Modifications Due to Climate Change

The Southern European countries are among the most vulnerable areas affected by the climate change in the world. Increased temperature and frequency of meteorological and hydrological extremes (heatwave, drought and storm) and decreased annual precipitation will characterize the future (Portoghese and Vurro 2010). Following the Directives of the European Union (CEU 2000), Italy has developed a "National strategy for adaptation to climate change" (MATTM 2014) and a more detailed analysis in the "Adaptation Plan to Climate Changes". This Plan, supported by the Euro-Mediterranean Centre for Climate Change, (CMCC 2017), aims at identifying the impact on environment and society sectors of the future climatic conditions and defining feasible actions for adaptation. The set of analysed sectors includes an extensive list of natural systems and processes threatened by climatic change, such as flooding, landslide, drought and desertification, in the frame of the socio-economic sectors of agriculture, forestry, fishing, health, energy, tourism and transports.

The section of the plan concerning the water resources includes the analysis of the climatic conditions, existing now and expected in the future, for some climatic "macro-regions", identified in the Italian territory, by using the data of the 1981–2010 period (Cornes et al. 2018; Haylack et al. 2008). The six homogeneous "climatic macro-regions" are:

- Alps.
- Northern zones.
- Pre-Alps and Northern Apennines.
- Po plain, Adriatic and Tyrrhenian coast.
- Central and Southern Apennines.
- Islands and Southern areas.

For these macro-regions, the climate projection was obtained with the application of specific tools ("COSMO CLS Model" with 8 km resolution and "CMCC-CM Model" with 80 km resolution), for two scenarios of "Representative Concentration Pathways" (RCP), considering the emission of greenhouse gas. Two levels of "radiative forcing" (the difference between the insolation absorbed by the Earth and the energy radiated back to space, expressed in terms of Watt per square metre) were considered, identifying, respectively, a 4.5 W/m^2 medium-level scenario (RCP4.5) and a 8.5 W/m^2 high-level scenario (RCP8.5).

The indicator anomalies (differences between future and baseline values) for temperature and precipitation, obtained for the 2021–2050 period, have been grouped in homogeneous categories. The overlapping of climatic macro-regions with the anomaly clusters have allowed to define homogeneous areas with equal actual and future anomalies, in order to estimate the main expected climatic change and related impact on the selected sectors. The list of the indicators includes:

1. *Winter   precipitation,   WP*   (cumulated   precipitation   of December–January-February).

2. *Summer precipitation, SP* (cumulated precipitation of June–July-August).
3. *Mean annual temperature, T_{mean}.*
4. *Frost days, FD* (average number of days with temperature lower than 0 °C).
5. *Summer hot days, SU95p* (average number of days with temperature higher than 29.2 °C).
6. *Consecutive dry days, CDD* (average number of days with rainfall lower than 1 mm).

The values of the climatic indicator of the macro-regions and for the 1981–2010 period are in Table 5.9.

The anomalies identified comparing the 2021–2050 climatic projection with the 1981–2010 condition indicate a significant expected change in the climatic areas of the macro-regions.

In the Alps, the RCP4.5 scenario leads to a reduction of summer and winter precipitation, accompanied by a decrease of frost days and consecutive dry days. The RCP8.5 scenario confirms the reduction of summer precipitation and frost days, while an increase of winter precipitation is expected.

A significant reduction of summer precipitation is foreseen for northern zone in both scenarios, but different results are obtained for winter precipitation by RCP4.5 and by RCP8.5. High reduction of frost days is also expected.

In Pre-Alps and Northern Apennines, a considerable reduction of frost days and a medium increase of the summer hot days are expected for both scenarios. A prevalent decrease of summer precipitation and an increase of winter precipitation are

Table 5.9 Average values and standard deviations of climatic indicators in Italian macro-regions in the 1981–2010 period (Source: CMCC 2017)

	Winter precipitation	Summer precipitation	Mean temperature	Frost days	Summer hot days	Consecutive dry days
Macro-region	WP (mm)	SP (mm)	T_{mean} (°C)	FD (days/y)	SU95p (days/y)	CDD (days/y)
Alps	143(±47)	286(±56)	5.7(±0.6)	152(±12)	1(±1)	32(±8)
Northern zones	321(±89)	279(±56)	8.3(±0.6)	112(±12)	8(±5)	28(±5)
Pre-Alp and Northern Apennines	187((±61)	168(±47)	13.0(±0.6)	51(±13)	34(±12)	33(±6)
Po plain, Adriatic and Tyrrhenian coast	148((±55)	85(±30)	14.6(±0.7)	25(±9)	50(±13)	40(±8)
Central Southern Apennines	182((±55)	76(±28)	12.2(±0.5)	35(±12)	15(±8)	38(±9)
Islands and Southern areas	179((±61)	21(±13)	16.0(±0.6)	2(±2)	35(±11)	70(±16)

possible in all the regions with the exception of Tuscany, where an increase of seasonal precipitations is given by scenario RCP8.5 and a decrease of summer precipitation by the RCP4.5.

In the macro-region of Po and the geographically distant Adriatic and Tyrrhenian coasts, a significant decrease of summer rainfall and increase of winter rainfall are expected for the RCP4.5 scenario, while this feature is limited to the Po plain for RCP8.5. In Tyrrhenian coast a small increase will occur for seasonal precipitations for RCP 8.5, while non-homogeneous results will occur for RCP4.5. A general increase is possible for summer dry days and a negative trend for frost days.

In Southern Apennines a decrease of seasonal precipitation is indicated by RCP4.5, while non-homogeneous results are given by RCP8.5. Both scenarios show a high decrease of frost days and a limited increase of summer hot days.

An increase of summer hot days and a decrease of frost days will characterize Sardinia, Sicily and the extreme southern areas of Calabria and Apulia. Heterogeneous changes are relevant to winter and summer precipitation, since scenario RCP8.5 gives a small increase of summer precipitation, while the RCP4.5 gives a high decrease.

The indicator anomalies that characterize the temperature and precipitation have been analysed by means of clustering algorithms, which allowed to identify five areas (denominated from A to E), where the impact can be considered homogeneous. Figure 5.4 presents the results relevant to the RCP 8.5 scenario, with a quantification of the areal percentages for the Hydrographic Districts (Eastern Alps,

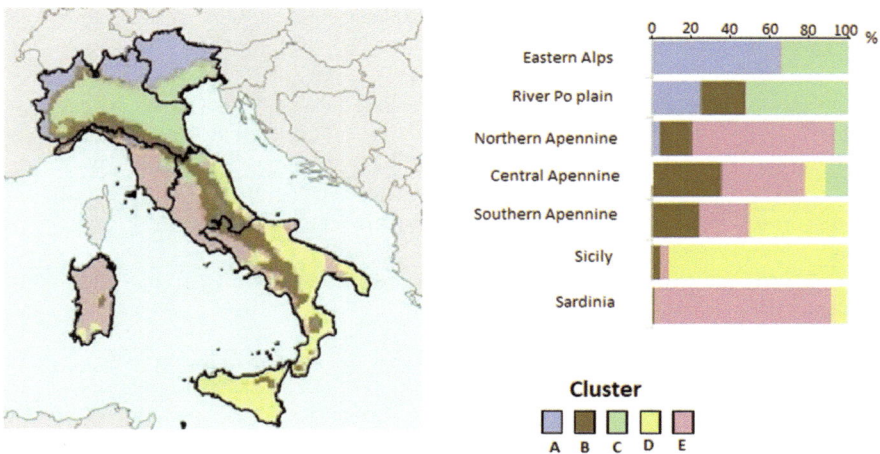

Fig. 5.4 Distribution of the cluster of climatic anomalies in the Hydrographic Districts according to the scenario RCP 8.5 (clusters A = wetter winter and dryer summer, B = hotter winter, C = wetter summer and winter, hotter summer, D = dryer winter and hotter summer, E = hotter-wetter winter and dryer summer) (Source: CMCC 2017)

River Po plain, Northern Apennine, Central Apennine, Southern Apennine, Sicily and Sardinia). For this RCP8.5 scenario, the clusters of the anomalies have the following meaning:

- *Cluster A* (wetter winter and dryer summer), affected by an increase of winter precipitation and by a significant reduction of summer precipitation, as well as a decrease of frost days (23 days/year) and snow cover (20 days/year).
- *Cluster B* (hotter winter), affected by a significant reduction of both frost days (28 days/year) and of the snow cover (18 days/year). In addition, there is a moderate reduction of summer rain.
- *Cluster C* (wetter summer and winter, hotter summer), affected by an increase of both winter and summer rainfall and by an increase of extreme precipitation, with a significant increase of summer days (12 days/year).
- *Cluster D* (dryer winter and hotter summer), with an overall reduction of winter rainfall and an increase of the summer rainfall accompanied by a noticeable increase of the summer days (14 days/year) and an overall reduction of evaporation.
- *Cluster E* (hotter-wetter winter and dryer summer), characterized by a significant increase of summer days (14 days/year) and of extreme rainfall events, with a reduction of summer rainfall and a significant increase of winter rainfall. The cluster presents also a noticeable reduction of the frost days (27 days/year).

The impact of climate change on the water resources will be significant all over the national territory, despite the great uncertainty that affects the assessment of climatic indicators over the different areas. As a general remark, the regions already denoting less availability of natural resources will have a further reduction of usable water. Consequently, the water resources problems will worsen in the near future, and the responsible people and institutions should be right now ready to look for suitable measures.

The main indications given by the Adaptation Plan to Climate Changes, relevant to the improvement of the territory and society resilience, are discussed in Chap. 14.

5.7 Conclusions

The above paragraphs have put into evidence the difficulty of identifying the actual situation of water resources in Italy. Several attempts have been done during the past decades, starting from different data and using different methods, but their results are strictly conditioned by the scarce information, which refers only to specific zones and limited observation periods, in which many changes occurred, affecting the basic criteria of a rational water management policy. Besides, the environment constraints, strictly tied to the water quality aspects, the increase of population, its migration from a region to the other and the improved living style had an impact on the water withdrawal and waste discharge. The progressing transformation of the national organizational structures about the meteo-hydrological data collection, which shifted from the Hydrographic Compartments to the Regions, and now to the Hydrographic Districts, greatly affected the possibility of achieving a fully comprehensive picture. The responsible institutions are now deeply committed to prove how much water is now available in the various parts of the country and what solution can be suitable for the actual and future problems.

References

AMBIENTEITALIA (1997) Collection of data concerning water resources use and management in Italy. Final Report, Contract n. 12515-F1PC I with the Joint Research Centre of the European Commission

Benedini M (1996) Some problems of the present Italian situation. In: Proceedings of the International Congress Metropolitan Areas and Rivers, Rome, Italy, 27–31 May. Quintily, Rome, vol 3, pp 5–18

Braca G, Ducci D (2017) Development of a GIS based procedure (BIGBANG 1.0) for evaluating groundwater balances at National scale and comparison with groundwater resources evaluation at local scale. In: Proceedings Congress on Groundwater and Global Change in the Western Mediterranean. Granada, Spain

CEC (1991) (Commission of the European Communities) Research and technological development for the supply and use of freshwater resources. Progress report and use, TECHWARE, Brussels, p 49

CEU (2000) (Commission of the European Union) Towards a sustainable strategic management of water resources: evaluation of present policies and orientation for the future. Study 31, ISBN 92, Luxembourg

Civita MV, Massarutto A, Seminara G (2010) Groundwater in Italy: a review. In: EASAC (European Academies Science Advisory Council) Groundwater in the Southern Member States of the European Union: an assessment of current knowledge and future prospects. Accademia dei Lincei, Rome

CMCC (2017) (Euro-Medit. Centre on Climatic Change) Piano nazionale di adattamento ai cambiamenti climatici [National Adaptation Plan to Climate Changes] (in Italian). http://www.climateadaptation.eu/italy/fresh-water-resources

CNA (1972) (National Water Conference) I problemi delle acque in Italia. Relazioni e documenti [The Water Problems in Italy: Reports and Documents] (in Italian). Tipografia del Senato, Rome. p 815

Cornes R, van der Schrier G, van den Besselaar EJM, Jones PD (2018) An ensemble version of the E-OBS temperature and precipitation datasets. J Geophys Res Atmos 123. https://doi.org/10.1029/2017JD028200

Haylack MR, Hofstra N, Tank AMGK, Klok EJ, Jones PD, New M (2008) A European daily high-resolution gridded data set of surface temperature and precipitation for 1950–2006. J Geophys Res D 113. https://doi.org/10.1029/2008jd010201

IRSA (1997) (Water Research Institute) Long range study on water supply and demand in Europe. Level A: studies at country level. Italy. ICWS, Final Report

IRSA (1999) Water Research Institute, Un futuro per l'acqua in Italia. [A Future for Water in Italy] (in Italian), Quaderni IRSA n.109, Roma, p 235

ISPRA (2017) (National Institute for Environmental Protection and Research) La procedura BIGBANG per il bilancio idrologico a scala nazionale [BIGBANG procedure for the national Italian water balance]. http://www.isprambiente.gov.it/pre_meteo/idro/BIGBANG_ISTAT (2018) Central Institute of Statistics. Ann report: The state of the Nation. https://www.istat.it/en/archivio/217955

MAF (1990) I problemi delle acque in Italia [Water Problems in Italy] (in Italian). Ministry of Agriculture and Forestry. Edizioni Agricole, Bologna, p 398

MATTM (2014) (Min. of Environ. and Protection of Land and Sea) Strategia nazionale di adattamento ai cambiamenti climatici. [National strategy of adaptation to climate changes] (in Italian). http://www.minambiente.it/notizie/strategia-nazionale-di-adattamento

Mouton J (1980) L'exploration des eaux souterraines en Italie [Exploration of groundwater in Italy]. Proc. of the 26th International Geological Congress, (in French), Paris, France

Portoghese I, Vurro M (2010) The challenge of water balances and run-off projections in the Mediterranean hydrology. Impact of climate change on water resources- 200 Years hydrology in Europe, Euraqua Symposium, Koblenz, Germany

Suzenet G (1997) R&D policies: Case Study: Water, Present Status in Greece, Italy, Portugal and Spain. Technical Report, EUR 17726EN

Ubertini L (1989) The activity of the Italian Research Group for Hydrogeological Hazard Mitigation" NSF-CNR, Proceedings selected papers from workshop on natural disasters in European Mediterranean Countries, Colombella, Perugia, Italy, pp 33–4

Chapter 6
Assessment of Non-conventional Water Resources

Giuseppe Rossi and Giuseppe Luigi Cirelli

Abstract In Italy, the non-conventional water resources represent a small percentage of the natural freshwater. On the basis of the limited available information, this chapter describes at first the development of the use of desalinated water. In a first stage, this has been achieved in industrial factories through two-purpose plants aimed at producing electric energy and desalinated water. Later, most of recent desalination plants have been built either to solve significant water shortages occurred during severe droughts in Southern Italian regions or to integrate or replace the water supply guaranteed by shipping service for small islands, affected by a very high increase of tourism.

The other significant non-conventional resource of interest, i.e. the reclaimed water, is still marginally used in spite of the advanced available research on their application and the project planning of several wastewater reuse systems in various Italian regions. Wastewater reuse has been proposed mainly to meet agricultural and landscape irrigation demands or industrial and urban demands. The implementation of the projects has been very limited, due either to the constraints of the legislation regulating the water reuse quality standards or to the high costs of additional treatment processes in comparison to the supply costs of other conventional resources. Finally, the chapter provides an estimate of the potential for wastewater reuse in Italy, based on a survey of urban wastewater treatment plants (WWTP) at regional level. The chapter also highlights potential benefits of reuse for irrigation purposes in a region (Sicily), through the analysis of the urban WWTPs in operation or under construction and the nearby areas of irrigation.

G. Rossi (✉)
Department of Civil Engineering and Architecture, University of Catania, Catania, Italy
e-mail: grossi@dica.unict.it

G. L. Cirelli
Department of Agriculture, Food and Environment (Di3A), University of Catania, Catania, Italy
e-mail: giuseppe.cirelli@unict.it

© Springer Nature Switzerland AG 2020
G. Rossi, M. Benedini (eds.), *Water Resources of Italy*, World Water Resources 5,
https://doi.org/10.1007/978-3-030-36460-1_6

6.1   Introduction

Non-conventional water resources have an important role to reduce the water gap and to promote sustainable water use in several countries affected by water stress, that is, when the demand for water exceeds the available natural resource or when poor water quality restricts its use (Tran et al. 2016; Ventura et al. 2019). Resources are currently limited to desalinated water and to municipal-treated wastewater, since the possibilities to boost water resources through the artificial rainfall by cloud seeding were not as effective as expected. Also the researches about the reduction of the evaporation from the surface of lakes, by using different substances (powders, emulsions aimed to form monolayers which avoid the evaporation), have had contradictory results in terms of costs to recharge the dispersed substances (due to the effects of the wind) as well as of negative biological impacts.

In Italy the non-conventional resources represent a small percentage of the total amount of useable natural resources, due to various reasons. First of all, a large part of the country has been able to meet water demands by using surface and groundwater and did not need to use desalinated water which had greater cost as compared to conventional sources. A limited number of desalination plants have been built in large petrochemical factories and/or in a small island where the touristic water demand required water shipment with high costs.

In spite of the advanced research activities and the project planning of several wastewater reuse systems in Italian semiarids regions, exploitation of reclaimed water has been mainly constrained by the concern about the potential public health and environmental impact risk; the complexity and restrictiveness of the required quality standards of the reclaimed water; and the economic and management constrains due to reclaimed water usually competing with undervalued and/or subsidized conventional water sources (Salgot et al. 2017).

In the next sections, the scarce available information on water desalination plants and on the experiences of wastewater reuse in Italy is presented (Sects. 6.2 and 6.3). The perspectives of wastewater reuse in Italy are discussed, by using the survey carried out by ISTAT (2015) on urban wastewater treatment plants (WWTPs) in Italian regions (Sect. 6.4). An example of the potential reuse for irrigation in a region is given through the synthesis of more detailed analysis of the WWTPs in operation or under construction and of the irrigated areas in Sicily (Sect. 6.5). Conclusive remarks (Sect. 6.6) close the chapter.

6.2   Desalinated Water

The survey carried out during the National Conference for Water (NCW) (CNA 1972) has identified five seawater desalination plants currently in operation in Italy, mostly located at industrial sites where it was possible to install two-purpose plants producing both freshwater and electric energy in a cost-effective way.

Table 6.1 Main desalination plants built until 1989

Plant location	Building year	Design productivity (m³/day)	Type of process
Gela	1964–1970	2 × 2150	MSF
Gela	1974	4 × 14,400	MSF
Taranto (Italsider)	1964–1970	2700 + 3000	MSF
Taranto (Shell)	1966	2250	MSF
Brindisi (Montedison)	1968–1974	2x4,400 + 8800	MSF
Porto Torres (Sir)	1972–1973	17,000 + 36,000	MSF
Linosa	1972	50	MVC
Lampedusa	1972	500	MVC
Capri	1975	2 × 2200	MED
Pantelleria	1975	500	EDR
Bari (IRSA, Enel)	1979	2000	MSF

Source: MAF (1990)
EDR electrodialysis, *MED* multi-effect distillation, *MSF* multi-stage flash evaporation, *MVC* distillation with thermocompression

On the basis of the trend analysis of the costs of desalination plants (particularly of the technology of the reverse osmosis), the NCW report estimated a strong reduction of the cost on a short-term period, with the optimistic hypothesis that already in the 1980s the cost of desalinated water would be lower than the cost of conventional water deriving from a complex supply infrastructure.

The survey carried out by Ministry of Agriculture and Forestry in the year 1989 (MAF 1990) updated the estimates of the NCW, reporting 12 desalination plants (Table 6.1).

As stated in the MAF report (1990), a few old plants had been dismantled, due to the out-of-date technology or the high costs. Further, the optimistic hypotheses about the progressive reduction of desalinated water costs were finally set aside due the growing cost of energy. The future of this non-conventional water resource was foreseen in the large industrial facilities and in the small islands with high touristic water demands.

This expectation has been confirmed particularly in Sicily, where several desalination plants are currently in operation within the petrochemical factories and in the small islands with very limited natural water resources and large touristic presence, particularly during the summer season. Other desalination plants were built for drinking supply in the western Sicily, which had experienced dramatic shortages during severe droughts of 1988–1991 period (Table 6.2).

In reality, though designed as emergency facilities to overcome drought, these plants have been used as primary sources even under normal climatic conditions in order to provide increased reliability than supply from spring or surface waters. However, plants resulted very expensive for the Sicily region, which supported the O&M costs.

The construction of small desalination plants for the municipal water supply of small islands of Sicily (Pantelleria, Lampedusa, Linosa and Lipari) and of Tuscany (Giglio and Capraia) can be considered as a positive experience. In these islands the

Table 6.2 Characteristics of desalinated plants in operation in Sicily

Plant – Process	Starting year	No. units	Daily productivity (m³/d)	Days in operation	Max annual productivity (m³/y)
Gela – MSF	1974	4	52,531	330	17,335,296
Pantelleria – EDR	1985	2	899	300	269,568
Pantelleria – RO	1985	1	199	300	59,616
Lampedusa – MVC	1990	2	899	300	269,568
Lampedusa – MVC	1990	1	50	300	15,034
Linosa – MVC	1990	2	501	310	155,347
Porto Empedocle – MVC	1992	3	4666	310	1,446,336
Gela – RO	1994	1	16,848	330	5,559,840
Trapani – MED	1995	4	34,560	330	11,404,800
Pantelleria – MVC		2	3214	310	996,365
Lipari – MVC		3	4795	310	1,486,512
Ustica – MVC		2	1002	310	310,694
Total			120,163		39,308,976

EDR electrodialysis, *MED* multi-effect distillation, *MSF* multi-stage flash evaporation, *MVC* distillation with thermocompression, *RO* reverse osmosis

Table 6.3 Desalination plants in Italian regions

Use	Region	Plants (no.)	Desalinated water (hm³/year)
Drinking supply	Sicily	5	3.400
"	Tuscany	3	0.384
"	Liguria	1	0.009
Petrochemical plants	Apulia	6	4.370
"	Sicily	1	2.751
"	Sardinia	1	1.775
Thermoelectric	Sardinia	4	0.546
"	Lazio	4	0.592
"	Sicily	3	1.852
"	Apulia	1	1.641
"	Liguria	1	0.428
"	Calabria	1	0.084

Source: ISTAT (2012)

alternative source for supply is the shipping service, which has higher costs and presents the risk not to guarantee safe supply due to sea conditions and, in some cases, the risk of not adequate hygienic conditions.

According to the ISTAT survey (2012) on the use of water resources on drinking use, 31 desalination plants were in operation in the year 2012, mostly located in the islands, in Apulia or in Tuscany (Table 6.3). The annual production of desalinated water was estimated as follows: 3.80 hm³ for drinking water supply, 8.89 hm³ for petrochemical factories and 5.14 hm³ for thermoelectric plants. The total amount of annual production for all the uses was estimated in about 17.83 hm³ and results smaller than the past estimates of annual productivity of the plants.

The volume of desalinated water for municipal use is very low (less of 0.1% of the total amount of the water volume supplied for this use).

The advances in desalination technology occurred in the last decades, and the expected cost reductions let foresee that also Italy could be interested by a significant growth of desalination plants in the next years as other Mediterranean countries.

6.3 Treated Wastewater

6.3.1 Strength and Weakness of Reuse

The use of reclaimed water (RW) constitutes an important solution for promoting and enhancing the sustainable use of the available water (Tran et al. 2016), especially in arid and semiarid regions, where the limited amount and the uneven distribution of water resources in time and space are pressing issues. In several countries of the South Mediterranean and Middle East (such as Spain, Greece, Tunisia, Israel, Jordan), RW are already used to meet, at least partially, water demand. In particular, RW are used for direct irrigation of crops or for aquifer recharge and subsequently for agriculture use after the withdrawal from wells.

However, many restrictions prevent full exploitation of the treated wastewater (TW). According to Salgot et al. (2017), the main restrictions for the various categories of water reuse are as follows:

(a) Agriculture and landscape irrigation: the health risk for workers, consumers and final users (public access); the increased treatment cost for removing or reducing chemical and toxic components of wastewater not adequate for soil and plants, such as sodium, or with too rich nutrient; the need of storage facilities to meet the seasonal demand; the cost of water delivery from the wastewater treatment plant to the final users
(b) Industrial for cooling or process water: health risk (i.e. aerosol dispersion; need for high water quality)
(c) Non-potable urban use for road washing or domestic use (i.e. toilet flushing): the health risk for the end users; the need for dual water distribution system
(d) Recreational/leisure, such as irrigation of golf courses: the health risk for end users; the social acceptance
(e) Recharge of aquifers: the need of hydrogeological studies; the need of advanced water treatment if groundwater is for withdrawal for potable purpose; site specific and not allowed by national legislation in several countries
(f) Replenishment of surface water bodies to maintain an ecological flow: the health risks, in particular if there is a possible indirect recharge of aquifer; the need for

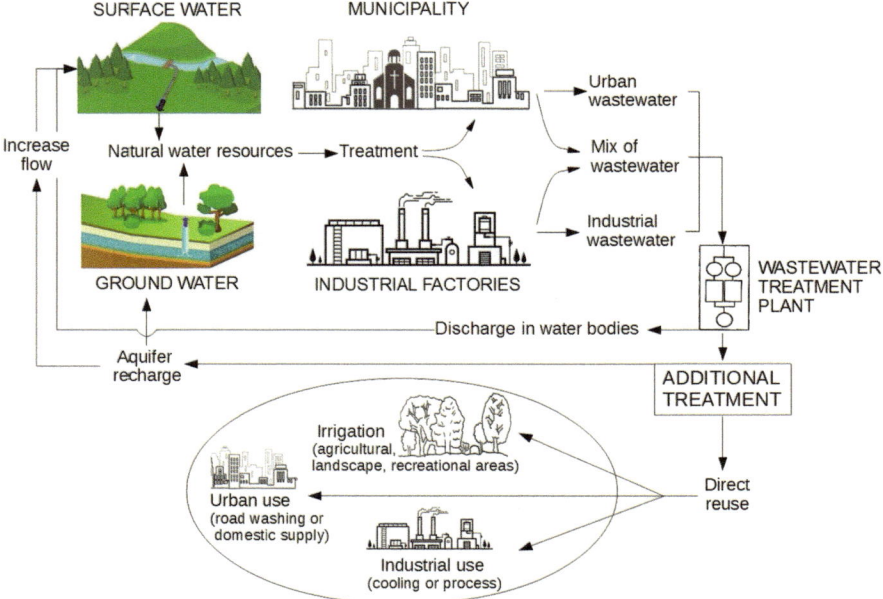

Fig. 6.1 Outline of the water cycle in the wastewater treatment and reuse system

    special permits (in many countries it is not permitted by the legislation on the use of RW); site specific and not allowed by national legislation in some countries

The water cycle for supplying municipalities and industrial factories and for wastewater treatment and reuse system is described in Fig. 6.1.

    In Italy, indirect reuse is very common and since long time practised both in the north and in the south. However, different approaches have been adopted among the different Italian regions. In the North and Central Italy, besides agriculture and landscape irrigation, several examples of industrial uses have been implemented. In the last years, the growing price of water supply and the increase in wastewater treatment costs have induced a strong effort for a more rational organization of water uses in the industry. Internal recycling has been implemented, and also some interest began to arise for wastewater reuse. In the Southern and Insular Italy, effluent reuse mainly regards irrigation of woody and horticultural crops in order to save freshwater for drinking purposes, thus reducing the competition among different users.

    In southern regions of Italy, the increase in water demand for drinking purpose and the deterioration of water quality in several coastal aquifers (due to the salinization connected to the groundwater overexploitation) have significantly reduced the available water resources for agricultural use. Also, the increased frequency of occurrence of severe droughts in the last decades (which have the greatest impacts on the agriculture) has played an important role to promote the use of treated wastewater.

Municipal wastewater can be considered as the most valuable non-conventional resource for a lot of reasons. The wastewater resource presents a high reliability, as it is slightly affected by drought, and it is available in WWTPs often located close to irrigation districts. Further a significant part of the reclaimed water cost is covered by the tariff paid by the municipal service users in order to meet the quality standards for effluent disposal required also in the case of no wastewater reuse. Furthermore, in many cases the additional treatment cost necessary for achieving the water quality for reuse could be less than the supply costs from other sources (e.g. deep groundwater).

Besides, in southern regions of Italy, many watercourses have long dry periods with very little natural flow; thus most of the water flow downstream of WWTP discharge point is made by wastewater (treated or partially treated). So, in these cases a planned and monitored wastewater reuse system that may assure the respect of the legislation standards can be a better and safety solution than a water withdrawal of "natural flows".

In all Italian regions, where the pollution of water bodies (both surface and groundwater) has determined very severe problems, wastewater reuse could play an important role for improving the water quality of water bodies, particularly in the Po river basin (Barbagallo et al. 2001).

Unfortunately, the reasons supporting a larger reuse of treated wastewater are not sufficient to overcome the severe constraints deriving from the health risk concern, which has characterized the Italian legislation.

Since 1977–2003, the use of RW for irrigation in Italy has been regulated in the frame of the 1976 Water Protection Act (Annex 5: CITAI 1977), which prescribed extensive treatment process of wastewater before its reuse. The approach was quite stringent, especially if we consider that surface water used for irrigation in Italy is normally characterized by a lower microbiological quality. No limits have been set for toxic or bioaccumulative substances, while a specific evaluation of the effluent which can be yearly applied was required, depending on soil and crop characteristics.

In 2003, within the frame of Legislative Decree (DLgs) 152/1999, which acknowledged the European Directives 91/271/EC (on urban wastewater treatment) and 91/676/EC (on water resources protection from nitrate pollution), new rules were issued for effluent reuse by the Ministry of Environment and Protection of Land and Sea (Ministerial Decree 185/2003). This Decree considers the same water quality for three main categories of water reuse:

(a) Irrigation of food and non-food crops, green areas, parks and sport fields.
(b) Urban non-potable uses such as street washing, heating or cooling systems, toilet flushing through dual networks, clearly separated from the drinking water.
(c) Industrial uses such as firefighting, processing, washing and thermal cycling of industrial processes; but it does not allow uses which may involve contact between the effluent and food or pharmaceutical products and cosmetics.

The Italian criteria follow a quite restrictive approach. In many cases, the quality limits are identical to those of the drinking water, or additional parameters are considered (Cirelli et al. 2008). In Table 6.4, the main parameters, usually detected in water monitoring, and the limits fixed by Italian rules for wastewater reuse

Table 6.4 Main quality parameters issued by Italian legislation for reuse of municipal wastewater

Parameters	Limits	Notes
pH	6.0–9.5	A
SAR	10.0	
Coarse solids	Absent	
TSS (mg/L)	10.0	
BOD_5 (mg/L)	20.0	
COD (mg/L)	100.0	
Total phosphorus (mg/L)	2.0	10.0 mg P/L[a]
TN (mg/L)	15.0	35.0 mg N/L[a]
NH_4 (mg/L)	2.0	A
EC_W (μS/cm)	3000	C
Cl_2 res. (mg/L)	0.2	B
Al (mg/L)	1.0	A
As (mg/L)	0.02	
Ba (mg/L)	10.0	C
Bo (mg/L)	1.0	
Cd (mg/L)	0.005	B
Cr (mg/L)	0.1	B
Cr_6+ (mg/L)	0.005	C
Fe (mg/L)	2.0	A
Mn (mg/L)	0.2	A
Ni (mg/L)	0.2	
Cu (mg/L)	1.0	B
Se (mg/L)	0.01	B
Sn (mg/L)	3.0	A
Zn (mg/L)	0.5	C
H_2S (mg/L)	0.5	C
SO_3 (mg/L)	0.5	C
SO_4 (mg/L)	500	A
Cl_2 res. (mg/L)	0.2	B
Cl- (mg/L)	250	A + B
F (mg/L)	1.5	B
Animal/vegetal oils and fats (mg/L)	10.0	C
Mineral oils (mg/L)	0.05	C
Total phenols (mg/L)	0.1	C
E. coli (CFU/100 mL)	10	D
Salmonella (CFU/100 mL)	Absent	C

(Source: Ministerial Decree 185/2003)
[a]for irrigation use
A: different limits can be fixed by provision of Italian regions, under the supervision of Ministry of Environment and Protection of Land and Sea
B: same limit of drinking water quality criteria
C: limit not mentioned in drinking water quality criteria
D: The limit must be met in 80% of samples, and none of them must exceed 100 CFU/100 mL. Effluents from constructed wetlands or stabilization ponds must not exceed 50 CFU/100 mL in 80% of samples, a maximum limit of 200 CFU/100 mL is allowed

(Ministerial Decree 185/2003) are reported. This approach has led to significant difficulties in promoting water recycling due to its impact on the treatment costs.

Other negative issues (Barbagallo et al. 2012) are: the number of parameters which must be monitored exceeds 50; the monitoring requirements and the sampling frequency can be very high, depending on the regional provisions. Also, it should be pointed out that the regulations do not take into account neither the intended use of crops (e.g. food and non-food crops, crops eaten raw, cooked or processed) nor the irrigation method (e.g. sprinklers or local irrigation methods) and their effects on health risks.

In a study by Barbagallo et al. (2012), the authors showed that, even in cases where *Escherichia coli* counts exceeded the limits set by the Italian regulations, the hygienic quality of the irrigated crops (combining different barriers: e.g. mulching and microirrigation techniques) is preserved accordingly to the revised WHO guidelines (2006). The approach in terms of microbiological quality is very restrictive if it is compared with the microbiological quality of surface water resources generally used in Italy and also compared with the Italian National Standard for bathing (Ministerial Decree 03/30/2010) that allow *E. coli* up to 500 and 1000 CFU/100 mL (EN ISO 9308-1), respectively, in sea and surface waters for recreational uses. The recent proposal of EU regulation on the minimum requirements for water reuse (EC 2018) confirms that this approach of Italian legislation is too restrictive in terms of parameters to be monitored and fixed limits.

Under these conditions, the reuse cost (construction, operation and maintenance), in addition to the costs required for effluent distribution and monitoring, makes this practice sustainable only for large WWTPs, thus reducing the benefit of reuse water and hampering the development of water reuse projects for the smaller facilities. Moreover, several successful reuse activities operating since a few years in small communities of Southern Italy inland areas will face problems difficult to cope with. According to the Italian legislation (article 12, Ministerial Decree, 185/2003), the costs for the RW distribution (power, labour, network maintenance, monitoring, etc.) are covered by the final users; meanwhile the additional costs to achieve the Italian quality limits are covered by the WWTP holders. This means that users of municipal water services must pay this extra cost in the water tariff.

In a set of 54 parameters, which is too large to assure an effective enforcement and monitoring, 11 parameters ask for the same quality of drinking water; 18 parameters are not considered for drinking water (some of them are not justified anyway), and the indication of some other parameters (for instance, biocides and pesticides) is difficult to be explained in an agricultural environment where the use of these substances is a common practice (Cirelli et al. 2008; Salgot et al. 2017).

Some parameters can be modified, under the supervision of the Italian Ministry for Environment, according to the provision of Italian regions taking into account the quality standard of water supply. Specifically, if it allowed fixing limits of water for human consumption less restrictive than reuse standard, the recycled effluents have to match the above limits. The regions can modify also other parameters (see Note A in Table 6.4), with the permission of the Ministry of the Environment; however, the new standard cannot exceed the limits set for the wastewater discharge into the sewage network (Annex 5 – Table 3, Legislative Decree 152/2006).

The cost sharing modality of reclaimed waters has hampered WW reuse. In fact, within the reorganization process of Italian water services in several cases, the "masterplans of urban water management" have not included the implementation of WW reuse systems, with the aim of reducing the total water tariffs. It is foreseen that such an increase of the treatment costs will raise legal controversies, since the costs related to wastewater reuse (charged on municipal users) will produce a direct benefit to farmers, more than to municipal users themselves.

Several Italian regions (e.g. Apulia, Emilia-Romagna, etc.) have issued specific laws or regulations for the reuse of TW. The Regulation 18 April 2012, n.8 established by the Apulia Region (2012) presents a particular interest. This regulation, besides the monitoring parameters of the M.D. 185/2003, indicates the environmental purpose as main objective of the reuse: to restore and improve the water balance of wet areas; to realize an indirect recharge of underground water bodies; to regulate streamflow in intermittent watercourses; and to increase the biodiversity of natural habitats. It requires that, before a public water license withdrawing natural water resources be given or renewed, the possibility to meet water demand by reuse of treated wastewater has to be evaluated (Art. 15). Also it gives specific indications on the sharing of responsibility of water quality control between the managers of the recovery of wastewater and the managers of the distribution network, and a financial contribution to the reuse projects is foreseen.

6.3.2   Reuse Projects for Irrigation and Industrial Purposes

The complexity of the tasks assigned to the regions and the confirmation of severe constraints on the reutilization for limiting the health risk, particularly on microbiological characteristics, have represented the major obstacles to the implementation of the most of the municipal wastewater reuse projects, which have been proposed not only in semiarid regions of Southern Italy, but also in Northern Italy.

It has been estimated that the total TW volume in the country is 5500 hm^3/year, and approximately 50% of it is available for recycling (Barbagallo et al. 2001; Lopez et al. 2006). The interest for the TW use in Italy has begun to grow since the 1980s, when sufficient WW flows became available and reuse projects for agricultural and/or industrial uses began to be considered. One of the first reuse projects was implemented in Emilia-Romagna, where over 450,000 m^3/year of recycled water are used to irrigate more than 250 ha. In Lombardy, since 2006, the effluents of two large WWTPs, San Rocco and Nosedo, which serve the urban area of Milan, have been used by two irrigation consortia with a potential of about 86 hm^3/year. New water reuse projects have been implemented recently in Sicily, Apulia and Sardinia for irrigation of agricultural crops, but notwithstanding the high potential for effluent reuse in southern regions; generally this practice is not yet a strategic role in integrated water management (Aiello et al. 2013; Lopez et al. 2006; Lopez and Vurro 2008; Barbagallo et al. 2012).

However, several experiences of TW reuse have been carried out since the year 1970 in almost all Italian regions. Most of the systems in operation or in progress have been designed for agricultural irrigation purpose, also if several systems are

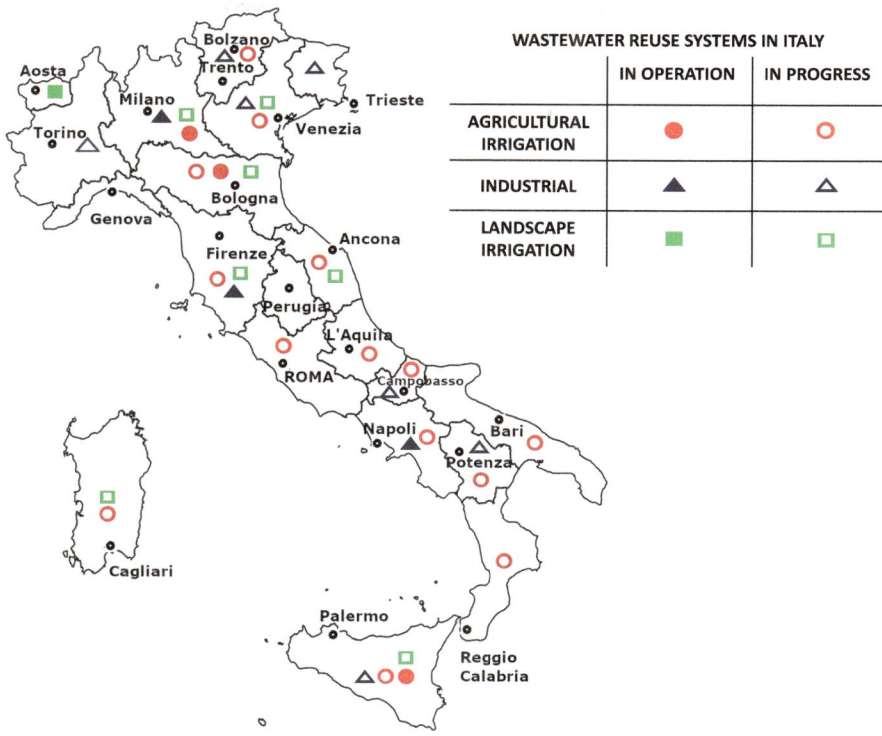

Fig. 6.2 Distribution of wastewater reuse systems in Italy

oriented to industrial reuse or multipurpose reuse, including landscape irrigation and fire prevention (Fig. 6.2). Unfortunately, it is lacking an updated and exhaustive survey of the cases implemented and of the plans in advanced stage of design and/ or of feasibility assessment.

6.4   Perspectives of Wastewater Reuse in Italy

With the severity of Italian wastewater regulations, in the next decade it is foreseen that larger volume of municipal TW will be available for reuse, mainly to meet irrigation demand.

The number of WWTPs distributed in all regions and the amount of the TW support this hypothesis, especially in the regions where a larger gap exists between irrigation requirements and availability of natural water resources.

According to the last ISTAT survey (ISTAT 2015), the total number of WWTPs in operation was 17,897. The WWTPs present very high differences for the percentage of pollution abatement, according to the typology of treatment, which moves from the simple Imhoff tank to primary, secondary or advanced treatment. In detail, the number of plants in operation in each Italian region, distinguishing the treatment level, is presented in Table 6.5.

The greatest number of WWTPs is located in Piedmont (21.7% of the total), followed by Emilia-Romagna (11.4%) and Lombardy (8.4%). The regions which present the higher percentages of plants with secondary (generally activated sludge or trickling filters) and advanced treatments, with reference to the total plants of their territory, are Basilicata (98.8%), Apulia (97.4%) and Sardinia (94.3%). Other regions (e.g. Piedmont, Lombardy and Veneto) show limited percentages of secondary and advanced plants (32.6%, 56.9% and 40.6%, respectively). The plants with advanced treatment, although they cover only 12.9% of the total number of plants, treat a very high percentage of polluting loads since they are generally located in the large urban areas.

The ISTAT survey (2015) gives another interesting information on the source of the polluting loads of wastewaters flowing to the plants. It indicates the estimates of the person equivalent separately for urban wastewater from civil uses and for

Table 6.5 Number of urban wastewater treatment plants (WWTPs) for treatment typology in each Italian region

| Region | Number of WWTPs | | | | |
	Imhoff	Primary treatment	Secondary treatment	Advanced treatment	Total number
Piedmont	2159	460	1177	92	3888
Aosta Valley	272	2	25	4	303
Liguria	600	50	100	26	776
Lombardy	660	65	400	373	1498
Trentino-Alto Adige	113	5	30	87	235
Veneto	664	1	224	259	1148
Friuli-Venezia Giulia	268	130	265	82	745
Emilia-Romagna	1259	82	451	245	2037
Tuscany	520	90	493	200	1303
Umbria	500	11	252	46	809
Marche	171	205	310	119	805
Lazio	32	56	405	142	635
Abruzzo	1009	34	362	30	1435
Molise	5	61	113	23	202
Campania	28	137	219	89	473
Apuly	1	4	8	176	189
Basilicata	–	2	82	88	172
Calabria	46	142	206	48	442
Sicily	63	55	239	57	414
Sardinia	7	15	243	123	388
Italy	8377	1607	5604	2309	17,897

Source: ISTAT (2015)

wastewater from industrial activities. They are reported in Table 6.6, for each region, since the presence of a large percentage of industrial wastewater can hinder the reuse of treated wastewaters.

About half of industrial polluting loads are treated in the plants of the north (48.5%). The amount including also the Central Italy rises to the 74.7%.

However, most of the potential for reuse now discussed refers to the resources deriving from the municipal WWTPs. In the last decades, many research activities have been carried out to investigate the effectiveness of new technologies for treating municipal effluent to be reused in agriculture. For example, a project (Lopez et al. 2006) investigates on four different technologies (membrane filter to improve microbial quality, simplified treatments oriented to save the agronomic potential of organic matter and nutrients, storage reservoirs and constructed wetlands to achieve the efficiency in the removal of various parameters, such TSS, BOD_5, COD, etc., according to the standards fixed by Italian rules). The research developed on wastewater treatment in three Southern regions (Apulia, Sicily and Basilicata) have highlighted the feasibility of conventional and green treatment technologies (in some case integrated each other) to promote the reuse in semiarid areas.

Table 6.6 Estimated person equivalent (PE) for urban and industrial wastewater flowing into the wastewater treatment plants

Region	Person equivalent (10^3)		
	Urban wastewater	Industrial wastewater	Total
Piedmont	5163	1040	6203
Aosta Valley	282	51	333
Liguria	2271	293	2564
Lombardy	9443	1959	11,402
Trentino-Alto Adige	1758	753	2512
Veneto	4094	1367	5461
Friuli-Venezia Giulia	1134	277	1411
Emilia-Romagna	4971	898	5868
Tuscany	3335	2774	6109
Umbria	990	121	1111
Marche	1322	56	1378
Lazio	5758	464	6223
Abruzzo	1716	173	1889
Molise	398	115	513
Campania	5621	1050	6671
Apuly	4678	110	4788
Basilicata	630	32	662
Calabria	2153	107	2260
Sicily	4023	683	4705
Sardinia	1803	1372	3176
Italy	61,544	13,695	75,239

Source: ISTAT (2015)

The DLgs. 152/2006 indicates the possibility of the treatment of municipal wastewater by lagooning or constructed wetlands for less than 2000 PE and had suggested the use of these natural treatment systems up to 15,000 PE as tertiary treatment (Annex 5 – III part).

6.5   Potential Reuse for Irrigation in Sicily

The real possibility of water reuse in an Italian region (Sicily) has been analysed in detail in some researches, aimed to check the quality of effluent of municipal WWTPs and to identify the irrigated areas where the reuse could be more convenient. A study (Barbagallo et al. 2012) has estimated the potential reuse of treated wastewater for Sicily, by using a careful survey of all the urban treatment plants in the island and an updated investigation on the irrigation districts, including the actual irrigation areas served by collective irrigation consortia (excluding the irrigation made by using private water sources). A selection of feasible reuse systems was made, by using the following criteria: (i) urban centres served by WWTPs with person equivalent greater than 5000; (ii) elevation of WWTP higher than the mean elevation of the nearest irrigation district or TW lifting up to 50 m; and (iii) maximum distance between the WWTP and the nearest district of 5, 10 or 15 Km, with reference to the TW volume. The location of urban WWTPs is indicated in Fig. 6.3 and the irrigation district in Fig. 6.4.

The results of a recent investigation (Ventura et al. 2019), that is an up-to-date survey of the previous reported in Barbagallo et al. (2012, 2013) in which a wider

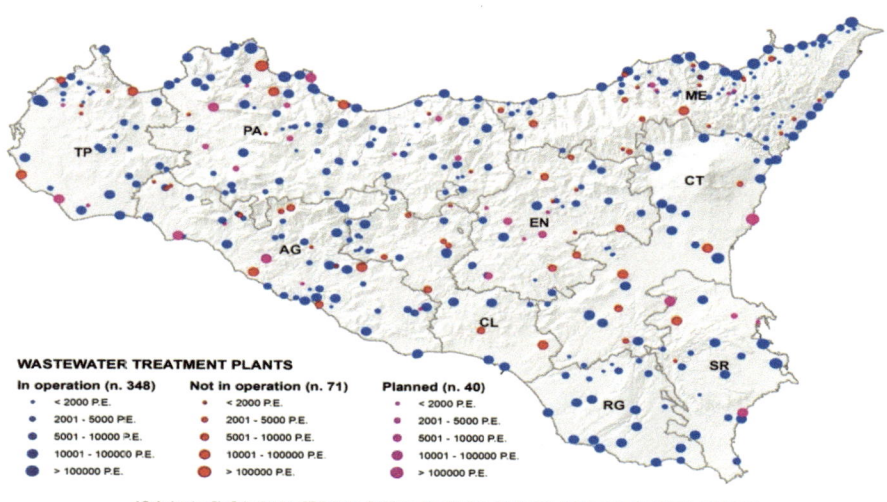

Fig. 6.3 Wastewater treatment plants in Sicily. (Source: Barbagallo et al. 2013)

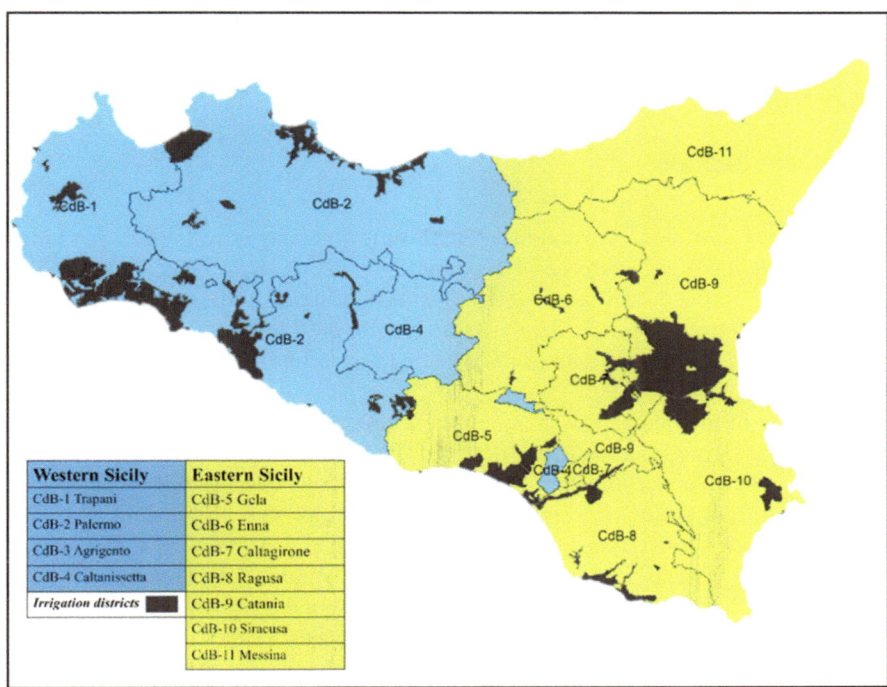

Fig. 6.4 Irrigation districts supplied by collective water distribution networks. (Source: Ventura et al. 2019)

dataset was implemented, showed that the 459 urban WWTPs located in Sicily included 348 plants in operation, 40 planned, while 71 plants not in operation. Also 37 areas irrigated by collective water distribution networks and supplied water volume (most from surface water stored in reservoirs) were identified, and the eventual water deficit was estimated. The comparison between irrigation requirements and water supply gave a deficit of 68 hm³. On the basis of the above cited criteria, 24 irrigation areas resulted eligible out of the 37 areas to receive treated wastewater from 59 plants. The total amount of the WW useable volume was estimated in 87 hm³, greater than the current water deficit of conventional water resources. The distance between selected WWTP and irrigation districts varied from 0 to 14.5 km, and the maximum elevation difference was 45 m. Irrigation districts with surplus resources could extend the irrigated area or save high-quality water for other uses or avoid aquifer overexploitation.

Besides, the cited studies (Barbagallo et al. 2012, 2013; Ventura et al. 2019), by using the experimental results of quality monitoring of wastewater in three treatment plants, proposed to overcome the constraints fixed by M.D.185/2003 on *Escherichia coli* contamination through a post-treatment health protection measure. Also it was suggested to adopt a procedure for microbial risk assessment (quantitative microbial risk analysis QMRA), which has been used by the WHO guidelines (WHO 2006).

6.6   Conclusions

Although the use of non-conventional water resources today is very limited in Italy, it is expected that it will increase in the future.

A growth of the desalinated water to match drinking requirements of small islands with a strong touristic demand occurred in the last decades. Likely it will continue due to the innovations of the desalination technology which has reduced the production costs.

The reuse of municipal TW in agricultural, urban, industrial and environmental sectors has been proposed since the 1970s as an opportunity to cover water short-age, save freshwater resources of high quality and contribute to the protection of environment from pollution. However, the growth of positive experiences has been very slow. Two main factors will contribute to a significant increase of the reuse, particularly for irrigation: the evolution of the techniques for chemical-physical treatment and disinfection and the increase of the cost of the conventional resources, due to depletion of the cheaper surface and groundwater sources. However, uncer-tainty about the future of WW reuse feasibility derives from the proposed regulation on minimum requirements for water reuse, in particular in agricultural irrigation, recently issued by European Union (EC 2018). Anyway, two strong conditions should occur: (i) the revision of current Italian legislation on the WW reuse, in order to obtain a wise balance between the primary health protection objective and the necessity of increasing water resources (also requiring a more careful monitoring in operation stage) and (ii) a clear rule for sharing the total costs of treatment between the users of municipal service, who pay the part of the treatment required by dis-charge standards, and the users in agriculture, who should support the supplemen-tary costs related to the treatment for achieve the quality standards for reuse.

References

Aiello R, Cirelli GL, Consoli S, Licciardello F, Toscano A (2013) Risk assessment of treated municipal wastewater reuse in Sicily. Water Sci Technol 67(1):89–98

Apulia Regione (2012) Regolamento 18 Aprile 2012, n.8 - Norme e misure per il riutilizzo delle acque reflue depurate [Rules for the reuse of treated wastewater], Bollettino Ufficiale della Regione Puglia - n. 58 del 20-04-2012

Barbagallo S, Cirelli GL, Indelicato S (2001) Wastewater reuse in Italy. Water Sci Technol 43(10):43–50

Barbagallo S, Cirelli GL, Consoli S, Licciardello F, Marzo A, Toscano A (2012) Analysis of treated wastewater reuse potential for irrigation in Sicily. Water Sci Technol 65(11):2024–2033

Barbagallo S, Cirelli GL, Consoli S, Licciardello F, Marzo A, Toscano A (2013) The use of reclaimed water for agriculture in Sicily, Italy. Abstract In: Proceedings of IWA International conference on water reuse, Windhoek, Namibia, 27–31 Oct 2013

Cirelli GL, Consoli S, Di Grande V (2008) Long-term storage of reclaimed water: the case studies in Sicily (Italy). Desalination 218(1–3):62–73. https://doi.org/10.1016/j.desal.2006.09.030

CITAI Comitato Interministeriale per la Tutela delle Acque dall'Inquinamento (1977) Smaltimento dei liquami sul suolo e nel sottosuolo Allegato 5 [Disposal of wastewater on the soil and in the subsoil, Annex 5]. Delibera 4.2.1977, G.U.R.I. 48, 21 February 1977

CNA Conferenza Nazionale delle Acque (1972) I problemi delle acque in Italia. Relazioni e documenti, [The water problems in Italy: reports and documents], (in Italian). Tipografia del Senato, Roma, p 815

EC European Commission 2018 Proposal for a regulation of the European Parliament and of the council on minimum requirements for water reuse. COM (2018) 337 final 2018/0169 (COD), Brussels, 28 May 2018

ISTAT (2012) Report "Censimento delle acque per uso civile - anno 2012" [Urban water census-year 2012], 30 June 2014. https://www.istat.it/it/archivio/127380

ISTAT (2015) Report "Censimento delle acque per uso civile-anno 2015" [Urban water census – year 2015], 14 Dec 2017. http://dati.istat.it/Index.aspx?lang=en&SubSessionId=12 65d969-d042-4e97-81e9-feda20189913

L.D. Legislative Decree 152/2006 - Decreto Legislativo 3 aprile 2006, n. 152, "Norme in materia ambientale" ["Standards on the environment"], G.U.R.I. n. 88, 14 April 2006 - Supplemento Ordinario n. 96

Lopez A, Vurro M (2008) Planning agricultural wastewater reuse in Southern Italy: the case of Apulia region. Desalination 218(1–3):164–169. https://doi.org/10.1016/j.desal.2006.08.027

Lopez A, Pollice A, Lonigro A, Masi S, Palese AM, Cirelli GL, Toscano A, Passino R (2006) Agricultural wastewater reuse in Southern Italy. Desalination 187:323–344

MAF Ministero Agricoltura e Foreste (1990) I problemi delle acque in Italia. Aggiornamento al 1989 dei risultati della CNA. [The water problems in Italy. Updating of the results of the National conference for Water], (in Italian) Edizioni Agricole, Bologna, p 398

Ministerial Decree 185/2003 - Decreto Ministeriale 12 giugno 2003, n. 185. "Norme tecniche per il riutilizzo delle acque reflue" ["Technical standards for wastewater reuse"], G.U.R.I. n.169, 23 July 2003

Salgot M, Oron G, Cirelli GL, Dalezios NR, Diaz A, Angelakis AN (2017) Criteria for wastewater treatment and reuse under water scarcity. In: Eslamian S, Eslamian F (eds) Chapter 14 handbook of drought and water scarcity: environmental impacts and analysis of drought and water scarcity. CRC Press/Taylor & Francis Group, Boca Raton, pp 263–282

Tran QK, Schwabe KA, Jassby D (2016) Wastewater reuse for agriculture: development of a regional water reuse decision-support model (RWRM) for cost-effective irrigation sources. Environ Sci Technol 50(17):9390–9399. https://doi.org/10.1021/acs.est.6b02073

Ventura D, Consoli S, Barbagallo S, Marzo A, Vanella D, Licciardello F, Cirelli GL (2019) How to overcome barriers for wastewater agricultural reuse in Sicily (Italy)? Water 11(335):1–12. https://doi.org/10.3390/w11020335

WHO World Health Organization (2006) Guidelines for the safe use of wastewater, excreta and greywater, vol.2: wastewater use in agriculture, WHO Handbook. Geneva

Chapter 7
Assessment of Water Requirements

Marcello Benedini

Abstract The evaluation of water demand, described in this chapter, is based on the data available by the various institutions and authorities responsible of water resources management. Particular attention is to the withdrawal for an increasing population and for the best economic development. The available resources in the country meet principally the urban and domestic demand, but other fundamental sectors, like agriculture, industry and electric energy generation, request large amount of water, often giving rise to undesirable competition and conflict. Other uses, normally considered less important, cannot be neglected, because they can interfere to the achievement of the prospected goals for the principal exploitations. The chapter reflects the difficulty due to the lack of reliable data in some specific cases, but the recourse to accurate estimates can help to identify reliable figures, useful for characterizing the actual and future management of Italian water resources.

7.1 Introduction

Ascertaining the water demand for the various activities is a difficult task in a complex country like Italy. Beside the complexities due to its geographic and climatic conditions, Italy still denotes a fragmentation in sharing the responsibility of water management. This persistent situation can affect the search of significant and homogeneous data able to give a comprehensive picture at the country level.

An increased sensitivity about the water problems characterizes several sectors of the country life. It has motivated a new legislation, while new administrative structures are now working, as shown in the preceding chapters. At the same time, specific problems have attracted the scientific and technical communities to develop up-to-date reliable estimates, which, although limited to special cases, can be useful to draw reliable considerations extended to all the country area (AMBIENTEITALIA 1997; EU 1997).

M. Benedini (✉)
Italian Water Research Institute (Retired), Rome, Italy
e-mail: benedini.m@iol.it

© Springer Nature Switzerland AG 2020   143
G. Rossi, M. Benedini (eds.), *Water Resources of Italy*, World Water Resources 5,
https://doi.org/10.1007/978-3-030-36460-1_7

The evaluation of the actual and future water demand must take into account some important aspects that affect all the uses and attract the attention of the responsible institutions, in particular:

- The climatic change, already experienced in terms of drought and shortage of available water, only apparently balanced by a higher frequency of very intensive precipitation and floods
- An augmented demand of water for all the uses, due to an increasing population and to the improvement of living conditions
- An increased environment deterioration, effect of the discharge of solid, liquid and gaseous wastes
- The constriction imposed by the limited amount of necessary financial means

These aspects, already in the concern of the responsible authorities, should be brought to the attention of all the interested people, the participation of which is essential to achieve the best management procedures. In this context, the collection of data must be done in view of a better future for all the vital aspects of the country.

The water demand estimated in the following paragraphs is an attempt to work out the information necessary for a reliable picture of the present and foreseeable situation of Italian life, relevant to the way it depends on the available water resources.

7.2   Urban Water Services

Great attention is in Italy for the urban and domestic use, considered with the highest priority in the management of national resources. The Italian governing structures have always underlined the importance of this use. In early 1963, the Ministry of Public Works promoted a National Masterplan for Aqueducts, aiming at a rational settling of the potable water problems. Based on the data available at that time and prospecting a great increase of the Italian population, the foreseeable situation of 2015 prospected a maximum demand of 11 km^3 for an estimated population of 74 million inhabitants. This amount was not recognized after suitable considerations on the population growth and the available information acquired in the meantime. The National Water Conference (CNA 1972) estimated an annual requirement between 6 and 8 km^3 for the year 1980, also confirmed 10 years later by estimates that are more accurate. The primary role of the urban and domestic use has been officially recognized in 1994 with Law 36/1994.

Now, following frequent interventions of the responsible authorities, in a large portion of the national territory, the demand is satisfactorily met, even though some situations of scarce availability persist. Proper regulation rules have been promulgated for the existing plants, promoting new facilities in areas of critical situation, also in accordance with the Directives of the European Union (Benedini et al. 1999).

The actual structures for withdrawing, conveying and delivering water to the users can meet the needs of an increasing population that has reached a high standard of life, characterized not only by the domestic demand but also by the request of essential services in the urban agglomeration, including shops and frequently small industrial plants, as well as gardens and amenities.

The majority of the Italian urban and domestic schemes consist of structures connected to single urban settlements, but some extended networks supply the aggregation of several dwelling places. The resulting dissemination and disparity of the management structures hampers to draw up a complete picture of the Italian situation, and then only some estimates are feasible in order to assess the amount of water withdrawn from the natural bodies and that effectively delivered to the users.

The water requested at national level has been estimated in 2015 in a census of the Central Institute of Statistics (ISTAT 2015a). The situation at regional level, shown in Table 7.1, refers to the total amount of water inside the whole of conveying ducts in the country. Such an amount is the overall demand that impends over the national water resources, in competition with other existing and foreseeable utilizations. The figures in the table give some considerations relevant to the various regions.

The largest demand is for Lombardy, where the high number of users can benefit from the extensive natural sources of lakes, rivers, and underground. The overall demand takes in due account also the non-potable uses sharing the works of the

Table 7.1 Actual urban water in Italy

Region	Population (10³)	Withdrawn (m³·10⁶/y)	Pro-cap. (l/d·inh.)	Delivered (m³·10⁶/y)
Piedmont	4393	584	*364*	378
Aosta Valley	127	26	*563*	21
Liguria	1565	239	*418*	160
Lombardy	10,019	1392	*381*	993
Trentino-Alto Adige	1063	160	*412*	112
Veneto	4908	648	*362*	388
Friuli Venezia Giulia	1218	196	*440*	102
Emilia-Romagna	4449	471	*290*	326
Tuscany	3742	427	*312*	241
Umbria	889	102	*314*	54
Marche	1538	167	*298*	110
Lazio	5898	973	*452*	458
Abruzzo	1322	231	*478*	120
Molise	310	53	*472*	28
Campania	5839	820	*385*	437
Apulia	4064	427	*288*	231
Basilicata	570	98	*470*	43
Calabria	1965	350	*488*	206
Sicily	5057	683	*370*	342
Sardinia	1653	275	*456*	122
Italy	**60,589**	**8320**	*399*	**4875**

potable water, like intakes from natural bodies and storage reservoirs. The actual withdrawal satisfactorily meets the demand, with a pro-capita close to the national average estimated around 399 l/d. Satisfactory is also the situation of other regions of the north. In less populated areas, like Trentino-Alto Adige, where the pro-capita is higher, great amount' of water remains in the natural bodies to the benefit of other users.

Acceptable is the situation in the centre, where the actual withdrawal can fulfil aggregated and spread users, while in Lazio the demand of Rome is remarkable, where the conveying and delivering structures, designed for a total flowrate of more than 16 m³/s, supply the town and the surrounding places for more than three million inhabitants. Figure 7.1 shows the actual works, drawing from remote springs in the Apennine and with a total 400 km of mains.

In the South, scarce local resources characterize the situation of Apulia, partially balanced with a conspicuous amount of water conveyed from contiguous regions. A worrisome scarcity of natural resources has motivated the governments, since the first decade of the twentieth century, to devise and construct a masterpiece of hydraulic engineering, the Apulian Aqueduct. After some decades of works, great part of an intensively populated region could eventually overcome a long traditional gap. After the Second World War, the governing institutions, supported by the Southern Italy Development Fund, carried out some remarkable works for improving the municipal water supply in all the southern regions.

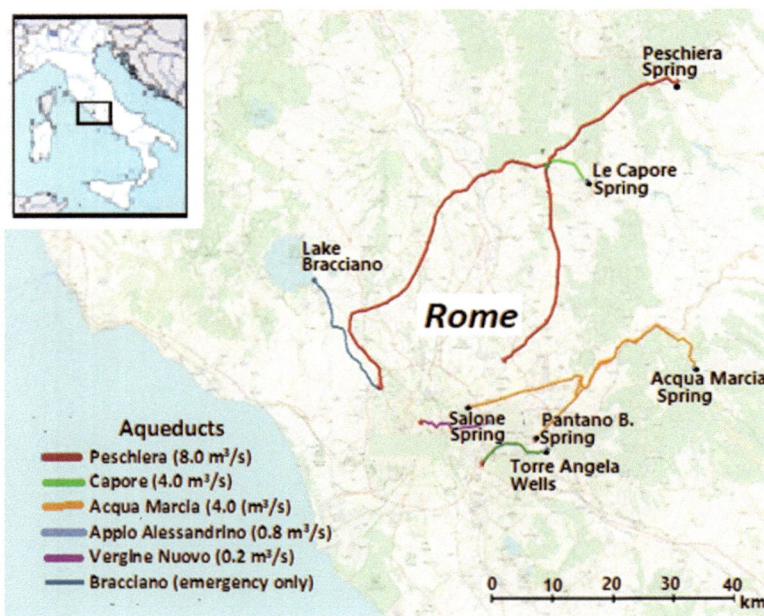

Fig. 7.1 The urban supply scheme of Rome. (Source: ACEA 2018)

The last column of Table 7.1 shows an estimate of the amount delivered to the final users. The difference from the withdrawal is due principally to the losses in the mains and in the delivery networks, as well as to the difficulty of measuring the quantity supplied in the urban context but outside the domestic use.

The Central Institute of Statistics (ISTAT 2015b) has extended the census to the natural sources from which water is withdrawn. At regional level, Fig. 7.2 shows the percent of the various sources in respect to the total amount withdrawn in the regional territories.

The major sources are springs and wells, connected to the natural groundwater normally recharged through the annual rainfall. Direct intakes from rivers are frequent in northern regions, where also the great lakes can provide considerable amount of water. Artificial reservoirs of medium and large size are located in all the regions, especially in the South, where the domestic and potable supply share frequently the works for irrigation and electricity generation. New works are in progress in the Apennine valleys.

Marginal is still the desalination of sea and brackish water, especially in coastal dwelling places and the small isles of the Mediterranean Sea. The reverse osmosis is the predominant desalination process, with plants able to produce up to 15,000 m^3/d.

Normally, the quality of water supplied for the urban and domestic use complies naturally with the potable standard imposed by the European Union and is frequently

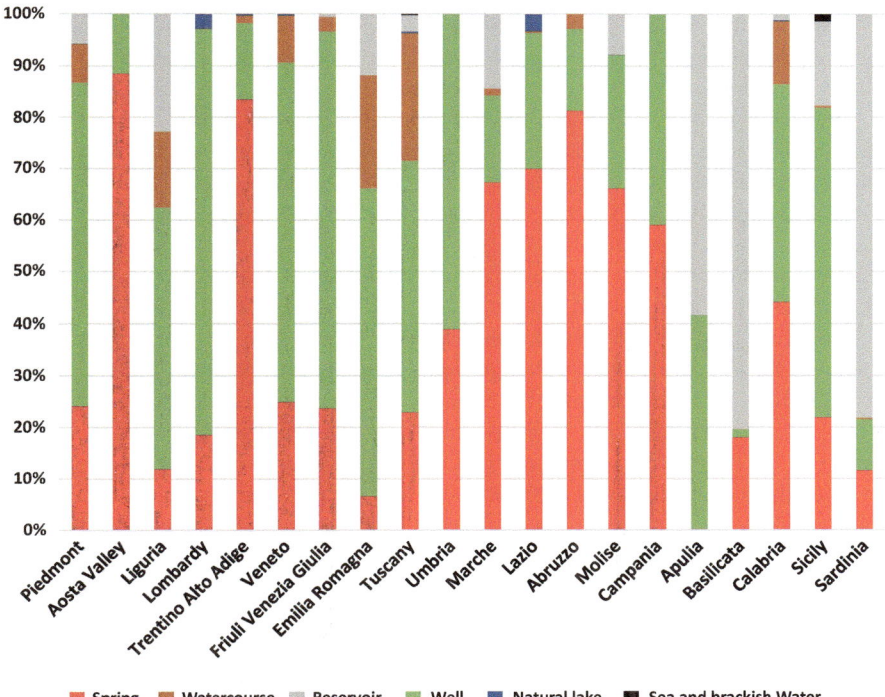

Fig. 7.2 Intakes of water withdrawal for the urban and domestic use. (ISTAT 2015b)

controlled by sanitary authorities. Suitable treatment plants have been necessary for the original source affected by unacceptable pollutant concentration. Remarkable is the new plant for the urban agglomeration of Florence, with a total output of 4.00 m^3/s, directly supplied from River Arno.

Beside the works connected to networks for large urban centres, a more detailed evaluation should take into account the great number of users that provide direct withdrawal from near water bodies, preferably from aquifers. This is a frequent case of scattered dwelling houses in the countryside, mostly without any control of the responsible authority.

Unfortunately, there are still some situations not properly working, due to inefficient control or lack of the financial support necessary for the correct running and maintenance. This induce the users to rely on bottled water, purchased in a flourishing market, the total annual amount of which is around 8,000,000 m^3 (see Chap. 8).

The amount of water requested for the urban use has been tailored according to several factors. Determinant are the level and the efficiency of the public services, as well as the pattern and habit of the population served, now in accordance with socio-economic concerns. The climate conditions, variable with time, affect the delivering of the water to the urban agglomeration.

Moreover, following an overall glance to the national situation, urban water is not yet evenly distributed in time and space for all the country, where households and services tend to demand more, particularly in hot and dry periods. Consequently, several measures have been adopted in order to overcome unpleasant situations, now and in the future.

Rationing the amount delivered to the users is a frequent escape for temporary occurrences, but more efficient and reliable interventions are devised and recommended (Masciopinto et al. 1999; Benedini and Giulianelli 2003; MAP 2007).

An efficient way for reducing the water demand is inducing the users to avoid unnecessary withdrawals by means of suitable tariffs proportional to the used amount. A promising solution, already successfully tested in some restricted cases, is the "dual network", consisting of two independent schemes, one for pure drinkable water and another for the "grey water", safe under the sanitary view but undrinkable because of the content of unacceptable substances. In an urban context, this solution can save the high-quality water for the benefit of potable needs.

The future of urban and domestic use is conditioned by the amount of population to be served, which grows although with a decreasing annual growing rate. Consequently, also the demand at the country level is expected to increase. Anyhow, taking into consideration the measures based on the rational use of the available water (UTILITAS 2017), combined with a widespread consensus of the users, the actual estimated amounts will probably hold also for the years to come.

7.3 Agricultural Water Requirements

7.3.1 Agriculture in Italy

Agriculture is traditionally a primary activity in Italy and greatly affects the national economy. As shown in Table 7.2, about 74% of the territory is dedicated to agriculture, almost equally distributed among the various regions.

These figures refer to a census done at the end of last century, but today the area pertinent to agriculture has to face the effect of a negative trend, experienced also in other EU countries, that aims at enhancing other economic sectors.

Several factors condition the agricultural activities, which are greatly variable from one region to the other, while the confirmed climatic change has a remarkable impact on the cultivable area and the way the ordinary cultivation practice is conducted.

Up today, several institutions deal with the main aspects of Italian agriculture, also in relation to the water resources. Some estimates, at national and regional level, revise the actual situation in various locations of the country, focusing all the connections with physical and economic considerations. Figure 7.3 emphasizes the various cultivation modes. Sown land and permanent crops are the most tied to the use of available water.

The partition is in line with the Mediterranean characteristics of Italy and its climate. In a future perspective, the Italian agriculture is undergoing a remarkable

Table 7.2 Agriculture development in Italy (IRSA 1998)

Region	Total regional area	Area pertinent to agriculture		Region	Total regional area	Area pertinent to agriculture	
	(ha·10³)	(ha·10³)	(%)		(ha·10³)	(ha·10³)	(%)
Piedmont	2539	1776	70	Marche	940	793	84
Aosta Valley	326	202	62	Lazio	1723	1244	72
Lombardy	2386	1591	67	Abruzzo	1083	802	74
Trentino-Alto Adige	1361	1108	81	Molise	45	34	77
Veneto	1841	1298	71	Campania	1367	990	72
Friuli-Venezia Giulia	786	491	62	Apulia	1954	1597	82
Liguria	541	333	61	Basilicata	1007	846	84
Emilia-Romagna	2245	1722	77	Calabria	1522	1141	75
Tuscany	2299	1776	77	Sicily	2583	1913	74
Umbria	846	689	81	Sardinia	2410	2049	85
				Italy	**30,207**	**22,394**	**74**

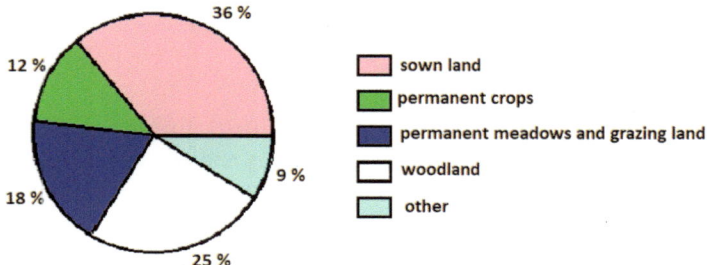

Fig. 7.3 Typical crops and their extension in Italy (IRSA 1998)

transition in favour of crops and cultivation techniques less dependent on water. The uncultivated land tends to increase, and the impact to the agricultural water management will certainly change.

7.3.2 Irrigation Requirements

Due to the predominance of cultivated land, and in line with peculiar climatic conditions, the first impact of agriculture on water resources is appreciable on the demand for irrigation, which still accounts for the largest portion of the total water used in the country (Bazzani et al. 2004). Other water uses are included in the agriculture, also determinant for the specific economic development, which depend on water although with minor demand.

Following what synthetically shown in Fig. 7.3, the zones in which the irrigation can be considered belong principally to the sectors cultivated with sown species and, secondarily, to those with permanent crops. These sectors include extensive fields of herbaceous species, like vegetables, corn, soybean, sugar beet and fodder, as well as arborous species like fruit, vines and olive trees. The economic value of these species is now especially appreciated. The actual trend in all the Italian regions is towards large fields equipped with advanced irrigation systems, possibly with automated devices.

The sector of permanent meadows and grazing land includes almost entirely the hilly and mountainous grounds, practically with null or very little need of irrigation. Small plots, generally with family conduction, are both in the sectors of sown land and "others", according to their extension or the size of watering equipment.

The above considerations lead to the identification of the territory that is potentially suitable to be irrigated, namely, the "irrigable area". This term follows scientific considerations used principally in a general water resources planning, but normally it does not give any real assessment on the amount of requested water. The current practice, both in designing and running an irrigation work, requests the evaluation of the "irrigated area", necessary to quantify the amount of usable water.

Anyhow, the first step in assessing the irrigation characteristics is the identification of the irrigable area, which, in a territory having a great complexity like that of Italy, is conditioned by many factors and depends on the particular zone in consideration. Physical factors concern the geological and pedologic characteristics, since the irrigation requests land without rocky formations and with a limited slope. Therefore, the most efficient systems are located in the flat lands of Northern Italy, while in many hilly zones, costly interventions have been necessary in order to improve the running and the effect of delivered water.

A second condition concerns the natural availability of sufficient quantity of water that can be withdrawn from a water body, without interfering with other existing or planned utilizations. The threat to the surrounding environment has become now the main risk to avoid. Several works have been realized, useful to capture the necessary quantity of water, to store in proper reservoirs and convey to the delivery places.

A third experienced condition refers to the cultivated species, which should have a promising economic return able to justify the cost of developing and running the relevant facilities. During the past decades, the cultivation pattern has frequently changed to follow the request of the market. This happened especially for vegetable and sometime for the vine, with costly transformations of the irrigable land. All these conditions persist today and probably will increase in the future, varying not only from one region to the other but also with time.

During the last century, the National Water Conference (CNA 1972) made an estimate in aggregated form for the outstanding partitions of the Italian territory and for the years having reliable meteorological information. The estimate started from a detailed evaluation of the potential evapotranspiration, following reliable theories able also to point out some approximated values of the necessary volume of water, as shown in Table 7.3.

These values, also underlined by several contemporary authors (MAF 1990; Faostat 2018), did not take into consideration several aspects determinant for the identification of the irrigated area and the real volume of water given to the crops.

Since such values were generally greater than those expected in the practical use, they must be considered just an attempt to an order of magnitude, in comparison with other forms of water utilization. The values confirmed the different situation in the various parts of the country. The greater extension refers to the zones in which

Table 7.3 Irrigable area estimated by the National Water Conference

Aggregated zones	1968		1980	
	Area	Volume	Area	Volume
	$(ha \cdot 10^3)$	$(m^3 \cdot 10^6)$	$(ha \cdot 10^3)$	$(m^3 \cdot 10^6)$
North	2329	20,000	2752	22,500
Centre	299	1400	502	2100
South and islands	718	4200	1435	7600
Italy	3345	25,600	4693	32,200

water is naturally abundant and the irrigation was already active in accordance with current traditional practices. Moreover, there was also a hint to the national policy aiming at increasing the irrigation in southern regions.

More recently, detailed information is available in the report of the national census conducted in 2010 by the Central Institute of Statistics. By means of questionnaires sent to a large number of farmers, many data have been collected and analysed, although the results refer only to 1 year and do not reflect the contribution of the farmers that were not able to give satisfactory answer to the high number of requested items (ISTAT 2016a, b). Therefore, the census cannot show entirely the real situation that could be appreciated only by means of gross figures. The reported values are the outcome of an exceptional investigation carried out with the best tools available at the time. Anyhow, these values can be now accepted and their order of magnitude allows reliable evaluations to be done.

The Central Institute of Statistics has conducted a census also for year 2013 (ISTAT), adjourning the existing values of irrigable and irrigated areas, but without any direct evaluation on the amount of used water. This amount should be then of the same order as that measured in the preceding investigations. The most significant terms, summarized in Table 7.4, can focus the points to which the attention of water policy makers should be called, for the present and future.

Table 7.4 Annual values of irrigation in Italy

Region	Area pertinent ($ha \cdot 10^3$)	Irrigated area ($ha \cdot 10^3$)	Specific volume (m^3/ha)	Water volume ($m^3 \cdot 10^6$)
Piedmont	1776	366	5047	1849
Aosta Valley	202	15	693	11
Lombardy	1591	582	8085	4703
Trentino-Alto Adige	1108	61	1226	75
Veneto	1298	242	2657	643
Friuli-Venezia Giulia	491	63	3318	209
Liguria	333	5	1749	9
Emilia-Romagna	1722	257	2988	769
Tuscany	1776	33	3121	102
Umbria	689	20	3355	67
Marche	793	16	2551	41
Lazio	1244	76	4212	321
Abruzzo	802	29	2380	69
Molise	34	11	3151	34
Campania	990	85	4519	384
Apulia	1597	239	2793	666
Basilicata	846	34	3868	131
Calabria	1141	75	3604	269
Sicily	1913	147	5182	763
Sardinia	2049	63	4888	308
Italy	22,394	2419	4666	11,287

The table specifies how Piedmont and Lombardy predominate in terms of irrigated area and amount of used water. Remarkable is also the situation of Emilia-Romagna, while Lazio, Apulia, Sicily and Sardinia give evidence to the efforts aiming at increasing the productivity of central and southern regions. Significant variations can be expected in the future, due to the foreseeable trend of the European climate.

Acceptable can be therefore the total annual amount of water destined to irrigation in the country, with an order of 11 km^3. This value is lesser than that of previous estimates, which referred to assumptions unlikely justifiable, motivated also by the difficulty of distinguishing between the amount of water withdrawn at the intake from rivers and lakes and the amount really delivered to the crops.

The development of irrigation fostered also the numerous works for conveying and delivering to the users the water withdrawn from the natural bodies. This happens mostly in northern regions, where large channels are still in operation. It is worthwhile to notice that some of them are now conveying much more water than necessary and dispose downstream the unused quantity without any appreciable benefit, with the only exception of their contribution to recharge the underground aquifers (Giordano et al. 2015).

On the other side, more recently, especially in southern regions, new irrigation schemes have been developed, with long lines of pipes replacing very often the original free surface ducts in which considerable amount of water went lost due to leakages and evaporation.

Like the other uses of water, also the actual pattern of Italian irrigation, in all the national territory, is oriented to reduce the amount of water withdrawn from the natural bodies and destined to the crops.

As emphasized by the census and shown in Fig. 7.4, the sprinkler prevails in a large part of the country. In several areas of the South, several sprinkler schemes are now progressively changing in favour of the drip, which entails an appreciable saving of water. Traditional furrow persists in northern areas for infiltration and flowing, essential in the cultivation of rice.

In the best equipped farms, water reaches now the crop following advanced procedural schemes, taking in due consideration the phytological aspect that governs the growing of the plant. Specific conceptual and simulation models, often connected to automatic devices, help to deliver the water to the cultivated fields.

During the last decades, advanced mobile sprinkle devices, able to spray water on a large cultivated area, help to speed up the irrigation procedure.

In the overall framework of the water resources management, the above-mentioned quantities, which refer to annual values, are in practice restricted within the irrigation season, normally lasting few months during the summer. Watering activity is split down a sequence of short periods, in accordance with the particular growing step of the crop and the availability of usable water.

The development of irrigation schemes has to face also the current use of fertilizers and pesticides, during the various steps of crop growing. A conspicuous spreading of synthetic fertilizers containing nitrogen occurs particularly in those areas having high natural fertility, in which the competition for high-quality products

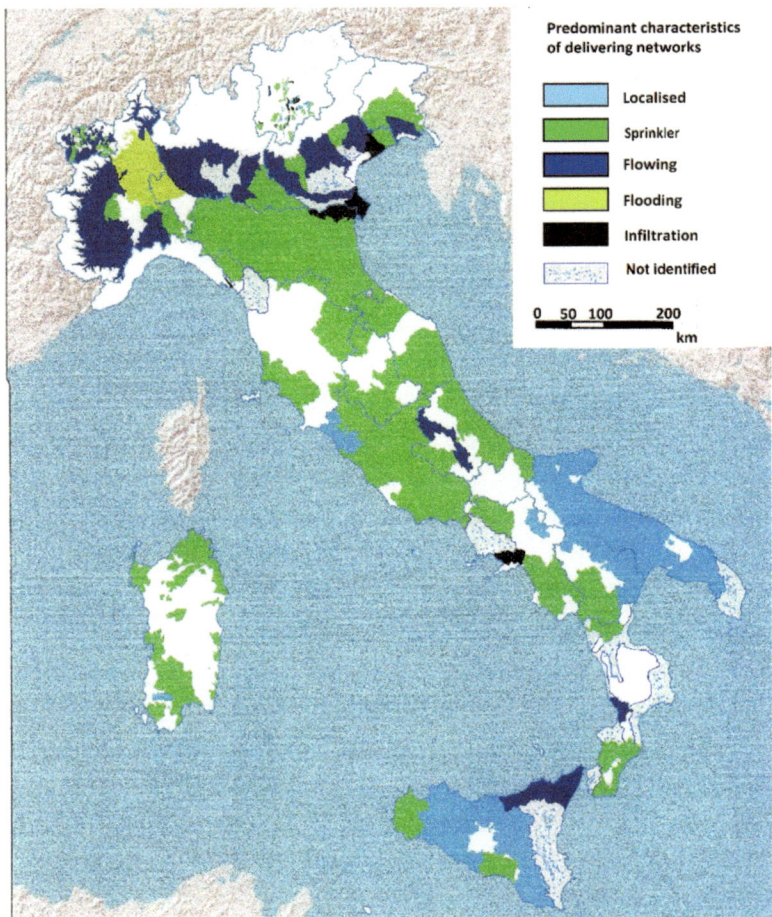

Fig. 7.4 Typical irrigation procedures in Italy. (Source: Zucaro 2014)

is great. Phosphorous is predominant in areas having specialized crops, like Liguria, where flower and vegetables are main cultivation (IRSA 1998). This practice has normally a remarkable effect on the amount of water for the crops. Actually, the constraints for abating the pollution of water body have enhanced a progressive reduction of chemicals in agriculture, with possible effects also on the amount of water for irrigation.

Various are the natural sources, from which the water is withdrawn. In a glance at the national territory, the majority of farms can rely on large aqueducts (56% of the total irrigated area), connected to rivers and lakes, mostly realized and run by local farmers associations ("consortia"). Direct withdrawals from groundwater (27%) and from surface bodies (17%) are also frequent for scattered plots.

The need of sufficient amount of water during the irrigation season has motivated the responsible institutions to realize many reservoirs able to store the precipitation in rainy periods. Today, more than 200 large structures are in operation, often sharing their water with urban supply and electricity generation. Typical of the little extension farms in the hilly territory are the ponds with a capacity lesser than 100,000 m^3 (Benedini 1995), while in many parts of the country, the family plots still avail themselves of cisterns and sumps to store rainfall water.

To reduce the amount of water withdrawn from nature and used for irrigations, a feasible solution can be the change of crops, adopting species that request less water, but this can involve unfavourable effects to farming economy.

All the above considerations presume the use of freshwater, with acceptable quality, as available in the natural rivers, streams and lakes, which, as mentioned above, have often satisfied other valuable uses.

Another promising alternative for reducing the amount of irrigation water could be the use of urban and industrial wastewater, after a low-cost treatment that abates the pollutant concentration down to values acceptable for the environmental and human sanitary protection. Several realizations are already working in the various parts of the country, with encouraging results. It should be remembered that the use of urban wastewater was already experienced centuries ago in Lombardy with the so-called "marcite", meadows cultivated with grass. This practice, even though abandoned decades ago, could be now another low-cost solution.

7.3.3 Irrigation for Frost Prevention

Frost conditions in winter or early spring can cause serious damage to the extensive cultivations in the hilly lands. In particular, the fruit trees face every year the risk of losing their product, very often with irreparable effects. Following advanced research, the farmers have now developed the practice of sprinkling water on the plants during the frost period, with encouraging results.

Special irrigation systems are working when a low air temperature is expected, for a treatment that can last several hours, mostly during the night. Normally up to 40 mm are sprinkled for 5–8 h, with a seasonal amount of more than 300 m^3/ha of water.

According to some estimates (Zinoni et al. 2005), at the present time such treatment concerns large part of total area cultivated with fruit trees in Lombardy, Piedmont, Emilia-Romagna and Trentino-Alto Adige, evaluated by Central Institute of Statistics (ISTAT 2014). Such area is destined to increase, given the high commercial value of the fruit.

Table 7.5 shows two different hypotheses for an estimate of the area requesting frost prevention, prospected after direct contacts with farmers adopting the treatment. The specific amount of requested water is assumed equal to 300 m^3/ha in both the hypothesis, which could take in due count the "overhead irrigation", requesting

Table 7.5 Frost prevention

	Total area of fruit trees	High hypothesis			Low hypothesis		
		Treated area		Water volume	Treated area		Water volume
	(ha)	(%)	(ha)	$(m^3 \cdot 10^3)$	(%)	(ha)	$(m^3 \cdot 10^3)$
Northwest	9188	30	2756	827	20	1838	551
Northeast	28,712	30	8614	2584	20	5742	1723
Centre	6168	10	617	185	8	493	148
South	18,170	10	1817	545	8	1454	436
Islands	7361	5	368	110	4	294	88
Total	69,599		11,832	3550		8352	2506

higher quantity of water. In both the hypotheses, the overall result, in terms of treated area and amount of requested water, is comparable with that of the large values of total agricultural demand, keeping in mind that this application is concentrated only in few months.

7.3.4 New Irrigation Possibilities

In many parts of the country, particularly during the summer season, the recent years of drought and the constant increase of freshwater demand for the urban and domestic sector have made problematic the supply of water for irrigation. In this occurrence, the use of reclaimed wastewater can be an acceptable solution (Bonomo et al. 1999; Lopez et al. 2006; Barbagallo et al. 2001; Portoghese et al. 2013).

A survey of the national treatment plants has estimated the total annual of treated effluent at 2,400,000 m^3, potentially usable for irrigation (Libutti et al. 2018). Nowadays, treated wastewater used for agricultural irrigation covers more than 4000 ha. However, reuse of municipal wastewater in agriculture is not yet active in most Italian regions and has recently decreased due to the high cost of installing new treatment plants and relevant facilities for water conveyance and storage.

Anyhow, several cases are already confirming this validity, even though specific regulations are not yet officially in force, because the existing legislation for environmental protection is particularly oriented to the treatment of municipal and industrial discharge for the benefit of the receiving bodies and the reuse of the treated wastewater for agriculture is controlled principally within local management systems. The attention of the farmers already practicing this kind of irrigation is particularly on the quality of the treated water, which must comply not only with the limits imposed by the national and regional laws but also with the specific requests of the crops, during their various stages of growth.

Following specific research and the development of strategic projects starting from year 2002, the wastewater treatment is achieved by means of simplified processes, mostly membrane filtration. In the meantime, proper reservoirs are available, to store the temporarily unused water. Wetlands technologies are also under investigation, at field scale, to evaluate the effectiveness of treating processes in relation to the quality requested by crops. Very often, treated effluents, with an acceptable microbial concentration, can contribute to save high-quality local groundwater (Giordano et al. 2015); likewise, a content of organic matter and nutrients can be useful for particular crops. Some species, like olive trees, do not request biological processes, with a remarkable reduction of the operation costs.

A planned exploitation of municipal wastewater could be useful particularly in southern regions, where farmers have been practising uncontrolled reuse for a long time (Raso 2013). In Northern and Central Italy, where available water resources generally meet the demand for different purposes, wastewater could play also an important role in controlling the pollution of water bodies.

In Aosta Valley some municipal wastewater treatment plants, in operation since the end of year 2000, can provide great amount of reclaimed water, in part usable for irrigation purposes.

Applications are in progress in Veneto, where some irrigation schemes have been designed for up to 70% of the available reclaimed wastewater, and in Emilia-Romagna, mainly in coastal areas, with many utilizations of treated municipal wastewater in conjunction with environmental protection purposes. In this region, a large wastewater irrigation plant was already in operation since the late twentieth century for a daily 1250 m^3 flow, covering about 400 hectares of orchards.

The regional administrations of Abruzzo and Basilicata have recently included norms concerning the wastewater reuse in their regulations for water resources management and reclamation water planning, in which the reuse for irrigation is recommended. In Campania new municipal and industrial (agro-food) wastewater plants are in operation mainly for tomato fields, while in Apulia about 16,000 m^3/ day of treated wastewater are available for irrigation in several zones of the regional territory. In Sardinia the wastewater reuse is considered a key action to face frequent emergencies of scanty water for irrigation.

In Sicily, where an uncontrolled wastewater reuse is traditional, several important projects are developed, among which that for the town of Palermo, where, in a first stage of working, about 28,000 m^3/day of treated wastewater, integrated with storage reservoirs, is available for the irrigation of several hectares.

A technological improvement of the devices to convey and trickle the irrigation water has favoured the development of the drip systems, now frequently applied also for the fertigation (Incrocci et al. 2017), the advanced system releasing water with dissolved fertilizers. Such system is now frequent in the orchards of the South and in some vegetable crops of the River Po Valley, for a total national area of 500,000 hectares. A further improvement of this system is the possibility of trickling the slurry of animal farming, with high fertilizing effect. Some promising experimental applications are in progress (Bortolini 2009; Capra 2016; Capra and Sicolone 2016).

7.3.5   Water for Animal Farming

Livestock in Italy is also a heritage of hold practices, which are characteristic of every region. So far, the relevant water demand has rarely caused particular concern, especially because most animal farming is carried out following the old tradition of grazing that directly benefits from the water naturally available on the Alpine and Apennine slopes. This happens, in particular, for sheep and cattle during the hot summer.

Only recently, the development of intensive farming, conducted following industrial criteria, has introduced some problems of water supply and wastewater disposal. This activity has become prominent in some restricted areas, located in the flat land of the North and in some Apennine valleys, where economic and environmental aspects have allowed, in particular, swine farms with thousands of animals to be developed.

The identification of a countrywide figure of water demand for animal farming is a difficult task but cannot be avoided, as it can help to estimate how an intensive farming affects the overall water resources management in the areas mentioned above. Moreover, the problem becomes significant for the biggest animals, like cattle and horses, as well as for the flock of minor ruminants (sheep and goat) during the stabling period. The growing of poultries and other small animals is commonly included in the general practice of the farm, where water is supplied principally for the needs of the farmers (FAO 2018a).

A remarkable effect of the animal farming is the resulting wastewater, which gives rise to unavoidable problems of environmental protection.

In spite of all that, wastewater can be a promising alternative resource, usable principally for irrigation, provided an efficient removal of all the contaminants harmful to human health and the surrounding environment.

Table 7.6 gives an adjourned synthetic view of the livestock present in the main regional aggregation in which the national territory can be analysed (ISTAT 2015a).

An estimate of the water demand can be done in terms of the daily amount per animal, according to the values of Table 7.7, recognized in the ordinary practice (OMAFRA 2019).

In cooler season, the animals require little quantity of water beyond that they receive with grazing, but in dry weather they need a greater quantity, especially when retained in stable, where water is necessary also for keeping the farm in a

Table 7.6 Animal farming in Italy (1000 animals)

	Cattle	Equines	Minor ruminants	Pigs
Northwest	2361	53	326	5872
Northeast	1586	47	224	2272
Centre	488	48	1411	587
South	928	38	1564	384
Islands	590	32	4120	216
Italy	5953	219	7644	9331

Table 7.7 Daily water demand for livestock (l/animal)

	Range	Average
Cattle	5–155	115
Equines	13–60	33
Minor ruminants	4–11	8
Pigs	0.1–35	9

good condition. Normally, considerable amount of water is necessary for washing, subtracted from other important agricultural activities. Breeding 5000 pigs, now a frequent case in Lombardy and Umbria, the daily water request can be comparable with that normally necessary for other purposes in the farm.

7.3.6   Water for Aquaculture

Aquaculture in Italy, based on a long tradition, denotes now a high level of specialization and large-scale production. In 2014, several species have been bred, both in marine and inland farms, with a total amount of 149,000 ton. The annual output of freshwater fish amounted to 40,700 ton, of which rainbow trout has the major share, followed by sturgeon and eel. In spite of a large import from nearby countries, these farms are becoming an interesting factor of economy, and many plants are now active, especially in northern and central regions as shown in Fig. 7.5, to meet an increasing demand of this kind of food.

Aquaculture occurs normally beside the natural bodies, where the necessary water is available at acceptable cost. In 2008, more than 750 production farms operated in Italy (FAO 2018b; Spagnolo 2012).

Various is the way of breeding, according to the species and their level of growth, and consequently various is the quantity of requested water.

In restricted areas, at the foot of the mounts, an old tradition still survives, for which the fishes are left free in restricted reaches of the streams having naturally high-quality water. This way of breeding imposes obviously severe environmental constraints, particularly in the upstream reaches of the river, in which any contaminant discharge must be banned, with restrictions for other possible uses. This way of farming can give normally a high-quality product but with limited quantity. The increasing demand for the consumers has now favoured the development of an intensive farming practice to produce every day large quantity of fish. The breeding occurs normally in specific ponds of more than 5000 m^3, able to contain several thousands of animals. Large amount of water is then necessary to keep the best vital conditions for the fish, as well as to save the sanitary aspects of all the farm premises, which undergo frequent health controls. Water runs almost continuously through the ponds and returns eventually to the natural bodies. Only a small quantity is consumed through evaporation, losses and secondary services in the farm.

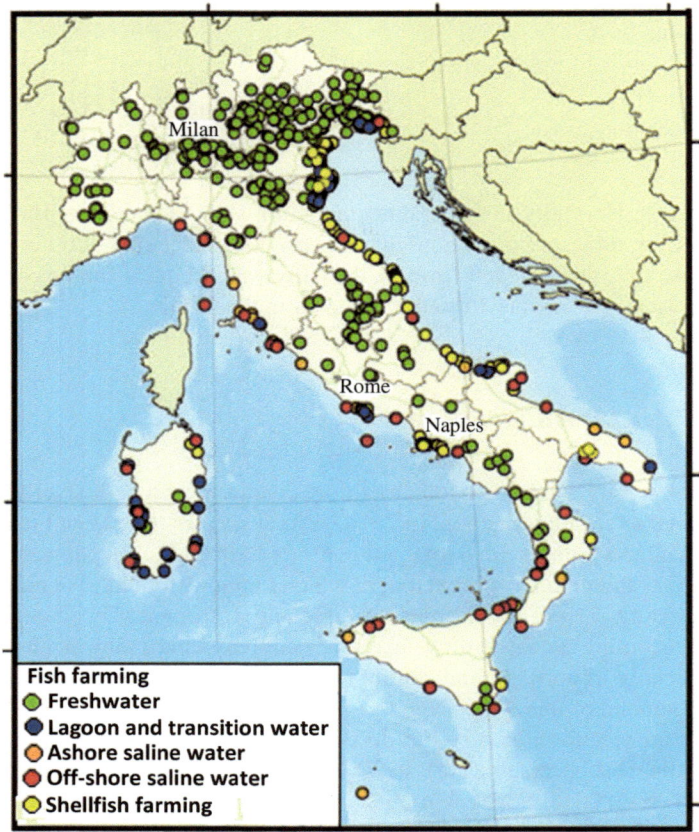

Fig. 7.5 Location of aquaculture farms. (Source: FAO 2018)

7.4 Water for Industry

Industry is an important factor of Italian economy and greatly impacts on the national management of water resources. In the industrial activity, the use of water is for process, washing and cooling. The use of water in thermal power generation is in a separate section of this chapter. The considerations of these paragraphs concern only the freshwater withdrawn from rivers, lakes and underground, in a view of a possible competition with other utilizations. The use of saline water, withdrawn from sea or brackish aquifers, frequent in the plants located along the coast, is not considered.

Industrial water in Italy concerns a few large production plants accompanied by a very great number of small enterprises, often conducted at family level, located in all the country, mainly in northern and central regions.

The resulting fragmentation, worsened by the great variability of the productive structures in terms of size, work force and technology level, makes it difficult to obtain a reliable knowledge of the amount of water used in this sector.

An attempt that can be still valid is the survey carried out by the Water Research Institute, in cooperation with the Central Institute of Statistics, in the early 1970s (IRSA 1973; Giuliano and Spaziani 1985). The collected data available at that time allowed to identify a first criterion of analysis, based on the amount of water required for the unit-specific output of the industrial plant, in terms of cubic metre of water per one ton of produced goods, as summarized in Table 7.8.

The survey concerned only the amount of water withdrawn in some significant samples of the various industrial settlements that responded to specific questions. Many small plants withdraw water directly from surface and underground sources, very often without any control of the responsible authority, and their amount of used water is unknown.

The resulting annual quantity at country level for year 1971, taking in due count the significance of the samples, was 7.8 km^3 (Benedini and Spaziani 1994). Therefore, this value refers only to the largest plants that installed measuring devices for the water withdrawn and used in the various steps of the productive process. Moreover, the collected data did not tell much about the way of using water inside the plant, without any chance to focus on the existing technology and on the possibility of saving water.

Table 7.8 Industrial water demand for unit product

Sector of activity	Water withdrawal (m^3/ton)	Sector of activity	Water withdrawal (m^3/ton)
Preserved foodstuffs	10.5	Hides and leather	443.5
Frozen foodstuffs	161.8	Oil	1.4
Beer	23.8	Metal industry	21.7
Chemicals	84.3	Glass	21.2
Paper	184.6	Mining	3.4
Rubber	148.3	Textiles	147.8
Steel	10.9	Textiles-dyeing	333.2
Cast iron	90.2	Ceramics and refractories	9.3
Metal sheet (iron alloys)	41.3	Abrasives	25.0
Coke	3.3	Cement	0.5

Table 7.9 Industrial water demand for unit employee

Sector of activity	Annual water withdrawal (m³/employee)	Sector of activity	Annual water withdrawal (m³/employee)
Cereals and "pasta"	3500	Wood	1100
Confectionery	500	Metalworking	3900
Food preserves	2200	Metal industry	550
Cheese dairy	1100	Vehicle construction	600
Vegetable and animal fats	6600	Non-metallic minerals	1700
Sugar refineries	4000	Chemicals	5500
Wines and spirits	3500	Rubber	1700
Non-alcoholic beverages	1800	Synthetic fibres	5000
Tobacco	350	Paper	16,000
Textiles	1500	Photography	280
Leather	1200	Plastic products	1100

Only rough estimates are then possible at the national level. Various attempts, carried out later, could give only values restricted to particular cases and locations.

Anyhow, in spite of these negative aspects, the census mentioned above allowed also an evaluation on the amount of water required for one employee working in the productive line, for all the produced goods. The outcome, in terms of annual withdrawal per unit employee, is in Table 7.9.

Late in the 1980s, the National Water Conference adopted the criterion of water per unit employee for a new estimate relevant to year 1971, taking into account the basic information acquired in the meantime (CNA 1972). Eventually, the obtained value was 7.2 km³, which is lesser than that obtained with the criterion of the unit product, as described above.

The discrepancy between the two values can be due to the scarce significance of the available information. Anyhow, the order of magnitude of both the estimates underlined the role of industry for its economic level achieved by Italy in that period.

The available data allowed also an evaluation of the 1981 situation, giving the annual amount of 7.5 km³.

In more recent time, the considerations and the results proposed by the Water Research Institute, accepted in in a wide international concern, gave rise to evaluations that confirm the order of magnitude of 7 km³, for a long period till the beginning of the twenty-first century. Some projections for the future are also available, with a decreasing value around 5.1 km³ for year 2025 (Margat and Vallée 1999).

A more recent survey of the Central Institute of Statistics (ISTAT 2016a, b) has also confirmed this trend, prospecting an amount of 5.5 km³ for year 2012.

Several aspects can justify a decreasing value expected for the incoming years. Very important can be a scaling down of some productive plants, due to an increasing import of goods from other counties at a very low cost, responsible also of enhanced unemployment with serious social and economic impact. For some particular productive line, remarkable can be the effort for reducing the withdrawal by expanding tools and technologies suitable to work efficiently with less amount of water, including a low-cost recycle and the multiple reuse of the discharge from upstream utilizations.

Among the various productive sectors, the annual water demand for chemicals (0.7 km^3) and for rubber and plastic products (0.6 km^3) is predominant, followed by that for metalworking (0.5 km^3).

7.5   Water for Electricity

7.5.1   Electricity Demand in Italy

Italy ranks among the major European electric energy producers. A strong progress in the energy policy is the notable improvement and implementation of a comprehensive long-term energy strategy. The medium- and long-term objectives of the government are clearly oriented to the reduction of costs, meeting the environmental targets and fostering a sustainable economic growth. In line with a worldwide trend, the future perspectives take into account an increase of the renewable sources and the replacement of natural fuel combustion, to which the country should rely, after a referendum has banned the nuclear plants. Today, the national electricity production is still insufficient to match an increasing development of the most vital economic sectors, and a considerable number of kilowatt-hour is imported from the nearby countries, at a cost that affects the delivery to the final users.

An evaluation of the energy produced in Italy, adjourned to year 2013, is in Table 7.10 (ENEL 2018). Other recent estimates (Celentani 2017) give values of the same order of magnitude.

Table 7.10 Annual electricity production from various sources (TWh)

Hydroelectric	52.77
Thermal	198.65
Wind	14.90
Solar	21.59
Total	**287.91**

7.5.2 The Hydropower

Water has been the primary source of electricity for many decades, starting from the beginning of the twentieth century, when the "white coal" was the main chance for energy production, as mentioned in Chap. 2. During the years following the Second World War, the demand of electricity for an increasing economic development drove to the construction of numerous high-power plants. From 1900 to 1960 hydro-electricity contributed for about 87.5% of the total energy produced. Since the 1960s the share of hydroelectricity decreased following a greater increase of thermal power and to a progressive exhausting of usable water. By 1980 hydroelectricity fell below the 25% of total annual production.

The great interest for hydropower has motivated in the past an investigation of the annual gross hydroelectric potential of the national territory, estimated about 200 TWh (Lehner et al. 2005), of which the limit of possible exploitation has been reached (WEC 2010; Garegnani et al. 2018).

The consistency of the Italian hydropower relevant to year 2010 is summarized in Table 7.11. Such values are acceptable also today (TERNA 2018).

The table regards all the existing plants with more than 10 MW installed and the mini and micro plants with a few kW, officially recognized by the governing institutions.

The largest plants belong mostly to the Italian National Electricity Board (ENEL), a company partially under public control, while the smallest installations are generally owned and run by private enterprises. The partition of the values described in the table can help to appreciate the way of using water for electricity generation. A consistent and reliable value of the amount of water for this use is not feasible, because customarily no direct flowrate measurement is available for the utilized water, while the produced electric energy is carefully gauged at the exit of the generation groups. Moreover, the water for hydroelectric generation is not consumed, since it almost entirely returns to the natural bodies. Such use is therefore of scarce importance in an overall balance of resources management, and an attention to the hydropower plants concerns mainly their location and the impact on other possible utilization of the same water body.

Table 7.11 Actual hydroelectric power in Italy

	Plants with reservoir		Of which pumped-storage		Run-of-flow plants		Total	
	n.	(MW·10³)	n.	(MW·10³)	n.	(MW·10³)	n.	(MW·10³)
North	237	12.9	15	5.4	1953	4.3	2190	17.2
Centre	49	1.2	–	–	290	0.5	339	1.7
South and islands	57	4.4	7	2.8	150	0.5	207	5.0
Italy	**343**	**18.6**	**22**	**8,2**	**2393**	**5,3**	**2736**	**23.9**

Numerous plants for very great power are especially in the Alps and Apennines, with large reservoirs and tall dams (ENEL 2012), built during the twentieth century with a renowned expertise in hydraulic engineering. A high usable operating head allows a discharge up to some tenths of cubic metres per second (Fig. 7.6). The more up-to-date plants are equipped with pumped-storage facilities, which are also a promising tool to adjourn some old power station and increase the amount of usable water.

Normally, the large reservoirs contribute to the landscape, favouring tourism and recreation activities. The artificial dams hinder the sediment transport in rivers and streams, and, consequently, several large reservoirs have undergone a massive siltation, with considerable reduction of their storage capacity (Di Silvio 2004). Due this occurrence, some plants have been dismissed.

Large run-of-river plants are in the main rivers, like Po (Bizzi et al. 2015; Basso and Botter 2012), Adige and Tiber, where they contribute also to control the natural flow and the riverbed incision (Fig. 7.7).

The high exploitation of the national hydroelectric potential will not allow a further realization of large plants; therefore the actual policy aims at improving the existing structures by means of a more efficient hydraulic and electric machinery.

During the last decades, there was a particular attention to the plants with a power less than 1 MW, particularly in Abruzzo, Marche, Molise and Sicily. The development of mini and micro plants is also favoured, in many cases accompanied by public financial support and fiscal incentives. These plants are connected to the country's electrical network to which they can transfer the amount of energy not

Fig 7.6 The Chiotas arch-gravity dam in Piedmont, 88 m tall, is a part of the Entracque hydropower plant, actually the largest in Italy with 1318 MW installed (Source: ENEL 2012)

Fig. 7.7 The Isola Serafini power plant on River Po, with four 300 m³/s generation groups, is one of the largest run-of-river plants in Europe (Source: Bizzi et al. 2015)

used for their own purposes, contributing to the availability of the national power. This is also in accordance with the directives of the European Union (Bodis et al. 2014).

The enhancement of mini and micro plants, which are scattered over the total national territory, has increased the water withdrawal from the natural bodies, which, in many cases, have lost their original hydrologic characteristics and undergone worrisome drought conditions. The effects of water abstraction from rivers and streams alter the conditions for the original aquatic life, giving rise to serious environmental problems. The 28.7.2004 Decree of the Ministry of Environment (as discussed in Chap. 11) prescribes the "ecological flow" in rivers and streams, which is another restriction for the possibility of developing new plants.

In the plants with large reservoirs, another constraint is due to needs of leaving sufficient empty volume for storing the foreseeable floods.

The future of Italian hydropower is therefore destined to face a compromise that will eventually stop the development of new plants and perhaps imposing a reduction of the actual production possibilities. For meeting an increasing demand of energy, other sources will be necessarily exploited (ISTAT 2018).

7.5.3   Water and Thermal Power

In past decades, the constraints imposed to hydropower have favoured the development of thermal power, which now provides about 70% of the total electric energy produced in the country. Numerous plants have risen, with very great power. They work with the steam produced from the primary water in a boiler through the combustion of coal, mineral oil or gas.

In the majority of plants, the dissipation of the residual heat occurs by means of cooling water, in exchangers where the steam is condensed and put again into the boiler, for a new productive cycle. The cooling process requests a large amount of water, withdrawn from a natural body, in which it eventually discharges with increased temperature. In terms of total produced energy, according to the Italian National Electricity Board (TERNA 2018), the water demand for cooling is of the order of 0.30 m³/MWh for a plant fed with mineral oil or gas and 1.05 m³/MWh for a plant fed with coal.

The Italian legislation imposes that the maximum temperature increase of cooling water, from its inlet to the outlet after the cooling process, should not exceed 4 °C.

High quality is essential for the primary water to be transformed into steam and sometime needs a preliminary treatment.

The plants are located almost in all the Italian regions, mostly in the North, where the energy demand is higher. The actual consistence of the thermal power is shown in Table 7.12, for large aggregations of the Italian territory and in terms of nominal values of installed power. The values refer to year 2010, but they give an order of magnitude acceptable also for the present time.

The geographic partition in the table includes in the North the predominant situation of Piedmont and Lombardy, where the cooling freshwater is withdrawn from the large rivers and lakes, sometimes from underground. In other northern regions, such as Liguria, Veneto and Friuli-Venezia Giulia, some large plants are located on the coast, where they can benefit from using seawater. In the other aggregations of centre, south and islands, the coastal location and the use of seawater predominate.

Today, an evaluation of the amount of cooling water is not yet available, because there are very few direct measures in the various plants. Only some estimates can be inferred from the situation described above. According to a survey of the Central Institute of Statistics (ISTAT 2016a, b), the estimated demand of the thermal plants for year 2012 is shown in Table 7.13.

Table 7.12 Actual thermal power in Italy

	Number of plants	(MW·10³)
North	730	34.3
Centre	195	13.4
South and islands	172	26.3
Italy	1097	74.0

Table 7.13 Freshwater for thermal plants (m$^3\cdot10^6$)

	Primary water	Cooling water	Total
North	37.52	1724.47	1761.99
Centre	8.11	82.70	90.81
South and islands	29.75	314.83	344.58
Italy	**75.38**	**2122.00**	**2197.38**

The above estimates consider only the freshwater, with its impact on other possible uses. The predominance of thermal plants along the coast underlines the advantage of seawater.

The discharge of cooling water with high temperature hinders the original conditions of the receiving body. Problems of environmental protection can arise, with the introduction of limits to the discharge that eventually obliges to reduce the electric production.

To overcome these problems, several plants have adopted the cooling towers, instead of the heat exchangers. Such structures request only a small amount of water, of the order of 1.0 m^3/s, necessary to replace the amount lost through evaporation. Consequently, the cooling towers can save considerable resources, but now only few of them are working in Italy.

In a more general context of environmental protection, besides the dissipation of residual heat, the plants working with fuel combustion are responsible of releasing smoke and gas into the atmosphere. This imposes a severe constraint in the present and future perspective of the national thermal power, in a way that the actual situation should be considered a maximum reachable limit, with a possible reduction of the number of existing plants.

An overall glance on Italian thermoelectric power includes the heat source of some geothermal springs; the principal ones are in Tuscany, for a 769 MW generation. The impact on these plants on natural water is practically null.

7.6   Secondary Uses of Water

7.6.1   Inland Navigation

Inland navigation was a prosperous activity in Italy during the past centuries, to connect large rivers, lakes and lagoons for low-cost transportation of passengers and bulky materials. Now, the development of more efficient transportation means, especially on road, has greatly reduced this activity, which survives only in restricted zones, also for the benefit of touristic resorts. Nevertheless, the convenience of low-cost transportation in places of great economic interest has rediscovered the opportunity of the inland navigation, in particular areas of the country, to freight large quantities of rough material and liquid fuel necessary for civil constructions and for an efficient running in the industrial settlements. Consequently, great concern arises on the relevant impact expected on the general management of Italian water resources.

An efficient inland navigation for massive goods can be restored only in Northern Italy, particularly in the lakes and rivers of Piedmont, Lombardy and Veneto. In the Venice lagoon, the navigation has been always the only way of transportation, also related to the connection with the open sea. In the lower reaches of large rivers in Central Italy, like Tiber and Arno, freight transport has disappeared and only local touristic chances persist, in conjunction with sport and recreation facilities.

Proper installations could make easier the activity for the best accomplishment of navigation, especially locks, ports and mooring facilities. Artificial channels connect now the natural streams, in a wide intersection of ducts open to high tonnage vessels. An evaluation of the present situation estimates an annual capacity of 29 million tons of freight for the whole of Italian networks, with a large increase in the future. The development of the Northern Italy inland navigation is conceived in view of an extended interaction with all the transportation systems existing or designed in neighbouring countries (Mantua 2011), as part of a great political action of the European Union.

Conflicts of navigation with the other uses of water are expected particularly during the abnormal hydrological events that now occur with an increased frequency. In case of drought, when the water level can be very low but at the same time the withdrawal for primary uses is high, the manoeuvre of ships is hampered, with serious impact for some important economic activity. Considerable amount of water must remain in the river in order to avoid siltation dangerous for ships mobility and mooring. Moreover, the water velocity in the river must be as low as possible, in order to allow safe operation for the ships. In an overall framework of the water management, this means that suitable amount of water must be left in the water body, subtracted from important in-stream and abstraction uses. The prospected use of vessels of more than 1000 tons requests at least 6 m of water depth.

Beside the water quantity constraints, the inland navigation, especially for tourism and passenger transportation, requests proper high-quality level, necessary to avoid undesirable contamination, bad smell and unbearable looks. Great attention is therefore necessary on the discharge of used wastewater and the capability of the receiving body for reducing the pollutant concentration.

Therefore, a new set of constraints arises in the management of the water resources systems. In spite of the long tradition of Italian inland navigation, a nationwide official regulation is not yet in force, but the responsible institutions, controlled by the Ministry of Transport, are now committed to set up suitable rules, compatible with the overall policies of water resources management.

7.6.2 Preservation of Monuments, Amenities and Touristic Resorts

Although less important than the principal uses described in the preceding sections, the demand of water can cause management problems in some unexpected situations. A typical and more frequent case is the presence of historic witness. All over Italy, monuments and buildings give evidence of a high cultural level developed in the sequence of centuries. Art accompanies the engineering aspects giving rise to marvellous realities that characterize all the places in the country, and, in many situations, it is not difficult to perceive a connection with water. This occurs especially where water is the principal component of a masterpiece or when the work is explicitly dedicated to a water body.

One of the most significant examples of such tie are the monumental fountains, which are present almost in all the urban establishments, where they contribute to make an attractive look. Very often, in a wide context of their existence and current working aspects, their presence cannot be neglected, especially where the water resources have to meet several requirements, particularly during the period of drought. It should be remembered also that the good condition of artefacts and monuments is a principal incentive for tourism, drawing the attentions of the governing institutions, which must provide for the necessary maintenance. There is also the need to consider the economic aspect.

The fountains characterize in particular the city of Rome, where more than 50 monuments attract visitors from all over the world. The most significant one is the Trevi Fountain, erected in the seventeenth century (Pinto 1986). Originally designed for a continuous supply every day of more than 80 m^3, it was connected directly to one of the oldest aqueducts of the Imperial Period dating from more than twenty-one centuries ago. Now, an increased demand for urban uses has motivated the municipal authority to install a pump for recycling the amount of water necessary for the normal working of the monument. Many situations in all the country are now adopting a similar solution.

Likely for the historical monuments, Italy attracts visitors also for its natural amenities and mild climate, in a way that tourism has been always a remarkable source of economy.

Among the attractions for tourism, the winter sport has become now very important in many parts of the country. The favourable meteorological conditions in Alpine and Apennine has fostered the development of skiing resorts equipped with highly efficient facilities. Winter sport requests a proper layer of snow on the ground, which normally is the effect of frequent abundant snowfall during the season. Unfortunately, low precipitations, due to the recent climate change, cannot guaranty the right performance of all the sporting activities. To overcome such undesirable situation (Damm et al. 2014), which can give rise to serious negative impacts on the economic aspects, an efficient practice is now developed, consisting of spreading "artificial snow" on the ground. Special equipment, the "snowmaking gun", is now in operation, which requests a conspicuous amount of water. It consists of a mobile device along the ski slope, able to transform into snow 3000–4000 m³ of water per hectare of covered ground. About 0.5 m³ of water is necessary for a machine working at more than 0.5 m³/min. A significant amount of this water is lost through evaporation, while not all the snow on the ground, after melting, returns in form of water to the original body.

The practice of artificial snow concerns now more than 70% of the total skiing resorts in Italy, requesting every season about 75,000,000 m³ of water for all the country, expected to further increase in the future. This gives rise to a worrisome situation for the communities living in the mountain, especially in the Alpine zone, where the natural availability of usable water is normally poor during the winter. Another problem arises for the planning of local water resources.

The need of preserving some touristic attraction and enhancing the economy of particular places can give rise to more undesirable conflicts. Frequent is the case of natural amenities already exploited for high productive uses. An example is in Central Italy, relevant to the beautiful Marmore Waterfalls, transformed at the beginning of the third century BC. The biggest waterfall in Europe, with a flow of 15 m³/s falling for 165 m, attracts always numerous visitors with the development of a fruitful tourism in nearby area. More recently, the water diverts to a large hydroelectric plant of 654 MW. The conflict remains in spite of adopting two alternate periods, for which the total amount of water is for electricity during the working days, while the fall remains active during the day in which the touristic demand is predominant.

The water demand for tourism is already included in that of the domestic use for the municipality in which the tourist resorts are located. A remarkable peculiarity is the seasonal fluctuation, which restricts the demand within a few months, mostly in summer, when the number of users to be served can increase several times over that of resident population. Since the majority of touristic resorts are located in zones where the natural resources are limited, such demand gives frequently rise to difficult problems of water supply. In the small islands, the conveyance by means of tanker is necessary. An estimate at the country level, in comparison with the principal countries interested in tourism (Gössling et al. 2012), gives for Italy an annual demand of 118,000,000 m³, a figure to be valid also for the future.

The above examples show once more how the water management in Italy can comply with the needs of uses that are fundamental for the daily life of the country.

7.7 Conclusions

The above paragraphs describe how the available resources, as pointed out in the preceding chapters, can meet the overall demand necessary for the actual and foreseeable development of the country. As shown in Fig. 7.8, agriculture historically characterizes the main economic activity of Italy, while the urban and domestic use underlines the efforts in the whole territory for maintaining the living conditions at a level now recognized for the social and economic progress. All these activities use freshwater abstracted from the natural bodies, which request the greatest attention

Fig. 7.8 The water use in
Italy

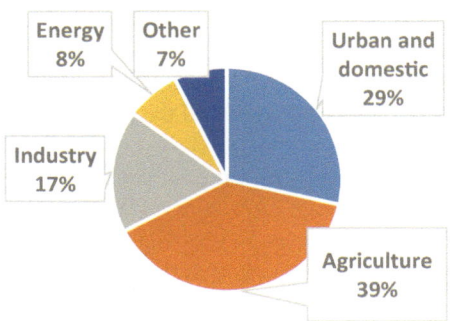

for water quantity and quality control. The demand for industry and thermoelectric energy takes into consideration not only the possible withdrawal from sea and brackish resources available in the long coastal line, but also the application, in several uses, of advanced technologies aiming at reducing the requested amount of freshwater.

Beside the primary uses of urban and domestic, agriculture, industry and energy generation, the water demand cannot neglect some uses normally retained less important, but always determinant for the water resources management. Very often, they can request considerable amount of water giving rise to worrisome constraints and conflicts. Moreover, the overwhelming goal of an acceptable environmental condition already imposes to revise how to withdraw water from the natural bodies.

The peculiarities of the various uses are considered in the frame of the existing and available resources, stressing the way each use can affect the others. These considerations have induced the responsible authorities and the scientific community to search for reliable data on the hydrologic aspects of the natural water bodies, also in view of a possible climatic change, as well as in accordance with an expected development of the national and European economic conditions.

These considerations have led to reliable evaluations, also in comparison with past estimates that proposed values not anymore acceptable. In fact, the most recent investigations, supported by improved information tools, have put into evidence that some estimates proposed 20 years ago are not confirmed today and lesser values have to be considered, also in formulating future trends and perspectives.

Comparing Italy water demand with that of the other European countries facing the Mediterranean (including Portugal), as shown in Fig. 7.9, some characteristic aspects can be recognized, like the predominance of agriculture, which can be principally attributed to the particular geographic and climatic conditions. The same conditions in the other countries of central Europe can favour the development of other fundamental uses.

Some particular aspects of the actual Italian water demand can relate to the efforts of the governing authorities, which, especially in recent years, induced the search for suitable tools able to overcome an existing discrepancy among the various regions. As already mentioned, the traditional lagging behind of places where

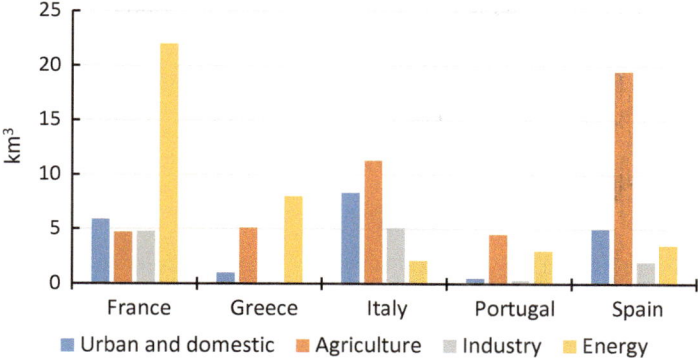

Fig. 7.9 Water use in some European countries

water is naturally scarce was responsible also of worrisome cases of unemployment and emigration towards more economically developed locations, in the country and abroad.

The scientific community has now faced the challenge of the future demand. Not all the actual existing resources could be able to meet the needs of a population potentially growing. Moreover, an increasing pollution of surface and underground water bodies may reduce the naturally available water, and more advanced treatment processes must accompany the demand in some particular areas of the country. The rational management of the existing water resources, besides the way for finding suitable economic and political expedients, requires primarily an active participation of the users, which, in turn, need a proper commitment of education and information.

References

ACEA (2018) Rome water utility, ACEA ATO 2, Rome, Italy

AMBIENTEITALIA (1997) Collection of data concerning water resources use and management in Italy. Final Report, Contract n. 12515-F1PC I with the Joint Research Centre of the European Commission. Ispra, Italy

Barbagallo S, Cirelli GL, Indelicato S (2001) Wastewater reuse in Italy. Water Sci Technol 43(10): 43–50 © IWA Publishing

Basso S, Botter G (2012) Streamflow variability and optimal capacity of run-of-river hydropower plants. Water Res Res. 48(10):W10527. https://doi.org/10.1029/2012WR012017 ISSN 1944-7973

Bazzani GM, Di Pasquale S, Gallerani V, Viaggi D (2004) Irrigated agriculture in Italy and water regulation under the European Union water framework directive. Water Res Res 40:w07s04. https://doi.org/10.1029/2003wr002201

Benedini M (1995) Improving the practice of small ponds in the Italian hilly land as a tool to mitigate the effect of droughts. Water Research Institute, Rome

Benedini M, Giulianelli M (2003) Urban water and climate change. In: Proceedings of international N.A.T.O. research workshop enhancing urban environment by environmental upgrading and restoration. Rome, pp 311–320

Benedini M, Spaziani FM (1994) An attempt to estimate the industrial water demand. In: Proceedings of the international congress on water resources. Cairo

Benedini M, Giulianelli M, Paoletti A, Passino R (1999) Urban water management in Italy, Country paper of Italy, EURAQUA Sixth scientific and technical review, Lisbon, Portugal

Bizzi S, Dinh Q, Bernardi D, Denaro S, Schippa L, Soncini-Sessa R (2015) On the control of riverbed incision induced by run-of-river power plant. Water Res Res. https://doi.org/10.1002/2014WR016237

Bodis K, Monforti F, Szabó S (2014) Could Europe have more mini hydro sites? A suitability analysis based on continentally harmonized geographical and hydrological data. Renew Sustain Energ Rev 37:794–808

Bonomo L, Nurizzo C, Rolle E (1999) Advanced wastewater treatment and reuse: related problems and perspectives in Italy. Water Sci Technol 40(4/5):21–28

Bortolini L (2009) Under canopy distribution of cattle manure on maize using a traveller boom equipped with drop tubes: first results. XXXIII CIOSTA CIGR V Conference Artemis, Reggio Calabria, pp 953–956

Capra A (2016) Role of micro irrigation filters and farm wastewater treatment to improve the performans of emitters. In: Goyal MG, Tripathi VK (eds) Research advances in sustainable micro irrigation. Wastewater management for irrigation. Principles and practices. Apple Academic Press, Oakvill, pp 97–143

Capra A, Sicolone B (2016) Management of emitter clogging with municipal wastewater. In: Goyal MG, Tripathi VK (eds) Research advances in sustainable micro irrigation. Wastewater management for irrigation. Principles and practices. Apple Academic Press, Oakvill, pp 81–95

Celentani G (2017) Impianti idroelettrici [Hydroelectric Plants] (in Italian), L'Acqua, Italian Association for Hydrotechnics, n. 4/2017:67–95

CNA (1972) National Water Coference. I Problemi delle Acque in Italia" [Water problems in Italy] (in Italian), Tipografia del Senato, Rome, Italy p 812

Damm A, Koberl J, Prettenthale F (2014) Does artificial snow production pay under future climate conditions? A case study for a vulnerable ski area in Austria. Tour Manag 43:8–21

Di Silvio (2004) Modelling long-term reservoir sedimentation for optimal management strategies, 6th conference on hydro-science and engineering, Brisbane (Australia)

ENEL (2012) The upper Gesso plant. Active Communications International, Rome

ENEL (2018) Wikipedia electricity sector in Italy. https://en.wikipedia.org/wiki/Electricity_sector_in_Italy#cite_note-7

EU (1997) (European Commission) Forward studies unit, long-range study on water supply and demand in Europe, Level A: studies at country level. Italy Water Research Institute, Rome

FAO (2018a) Water use of livestock production systems and supply chains – guidelines for assessment (Draft for public review). Livestock Environmental Assessment and Performance (LEAP) Partnership. FAO, Rome

FAO (2018b) Fishery and aquaculture country profiles. Italy country profile fact sheets. In: FAO fisheries and aquaculture department Rome. http://www.fao.org/fishery/

Faostat (2018) Food and Agriculture Organization of the United Nations, Statistics Division. http://www.fao.org/faostat/en/#data. Accessed on 12/07, 17/07, 04/09

Garegnani G, Sacchelli S, Balest J, Zambelli P (2018) GIS-based approach for assessing the energy potential and the financial feasibility of run-of-river hydro-power in Alpine valleys. https://doi.org/10.1016/j.apenergy

Giordano R, D'Agostino O, Apollonio C, Scardigno A, Pagano A, Portoghese I, Vurro M (2015) Evaluating acceptability of groundwater protection measures under different agricultural policies. Agric Water Manag 147:54–66

Giuliano G, Spaziani FM (1985) Water use statistics in industry. Experiences from regional surveys and planning in Italy". Stat J U N, ECE 3:229–245, North-Holland

Gössling S, Peeters P, Hall CM, Ceron JP, Dubois G, Lehmann L, Scott D (2012) Tourism and water use: Supply, demand, and security. An international review. Tour Manag 33:1–15

Incrocci L, Massa D, Pardossi A (2017) New trends in the fertigation management. Horticulturae 3:37. https://doi.org/10.3390/Horticulturae3020037

IRSA (1973) Water Research Institute. L'impiego dell'acqua nell'industria [Industrial use of water] (in Italian). Quad Ist Ric Acque, 15, p 73

IRSA (1998) The quality of receiving water bodies in agriculture-dominated areas. Country Paper presented by the Water Research Institute. In: Proceedings of the 5th EURAQUA Scientific and technical review "Farming without harming". Oslo

ISTAT (2014) Utilizzo della risorsa idrica a fini irrigui in agricoltura [Use of water resources for irrigation in agriculture] (in Italian), 6th General Census of Agriculture, ISBN: 978-88-458-1805-9 https://www.istat.it/it/files/2014/11

ISTAT (2015a) Public water supply use. http://dati.istat.it/Index.aspx?DataSetCode=DCCV_CONSACQUA

ISTAT (2015b) Water abstraction for drinkable use http://dati.istat.it/Index.aspx?DataSetCode=DCCVPRELACQ&Lang=en

ISTAT (2016a) Annuario Statistico. Agricoltura [Yearbook of Statistics. Agriculture] (in Italian). https://www.istat.it/it/files/2016/12/C13.pdf

ISTAT (2016b) Giornata Mondiale dell'Acqua [World Water Day] (in Italian). https://www.istat.it/it/files//2016/03/Focus_acqua

ISTAT (2018) Production of gross electricity from renewable energy sources. http://dati.istat.it/Index.aspx?lang=en&SubSessionId=ea39d752-37c9-4cf3-ae9e-be24310db082

Lehner B, Czisch G, Vassolo S (2005) Europe's hydropower potential today and in the future. Center for environmental systems research, University of Kassel, Institut für Solare Energieversorgungstechnik (ISET), Kassel, Germany, lehner@usf.uni-kassel.de

Libutti A, Gatta G, Gagliardi A, Vergine P, Pollice A, Beneduce L, Disciglio G, Tarantino E (2018) Industrial wastewater reuse for Irrigation of a vegetable crop succession under Mediterranean conditions. Agricultural Water Management 196-1

Lopez A, Pollice A, Lonigro A, Masi S, Palese AM, Cirelli GL, Toscano A, Passino R (2006). Agricultural wastewater reuse in southern Italy. Desalination, 187 (1–3): 323-334

MAF (1990) I problemi delle acque in Italia [Water Problems in Italy] (in Italian). Ministry of Agriculture and Forestry. Edizioni Agricole, Bologna, p 398

Mantua (2011) Master Plan of the Northern Italy Waterway System, Province of Mantua Sector Planning. http://www.provincia.mantova.it

MAP (2007) Mediterranean Action Plan, Water demand management, progress and policies. Proceedings of the 3rd regional workshop on water and sustainable development in the Mediterranean, Zaragoza, Spain

Margat J, Vallée D (1999) Mediterranean vision on water, population and the environment for the XXI century. MEDTAC Plan Bleu

Masciopinto C, Barbiero G, Benedini M (1999) A large scale study of drinking water requirements in the Po Basin (Italy). Water Int 24(3):211

OMAFRA (2019) Factsheet. Water requirements of livestock, order no. 86-053 http://www.omafra.gov.on.ca/english/engineer/facts/07-023.htm

Pinto JA (1986) The Trevi fountain. Yale University Press, New Haven

Portoghese I, D'Agostino D, Giordano R, Scardigno A, Apollonia C, Vurro M (2013) An integrated modelling tool to evaluate the acceptability of irrigation constraint measures for groundwater protection. Environ Model Softw 46:90–103

Raso J (2013) Final report on wastewater reuse in the European Union. EC-reference:. TYPSA-7452-IE-ST03. WReuse_Report-Ed1.doc

Spagnolo M (2012). What kind of management for Mediterranean Fisheries, Note to the European Parliament. DG for International Policies. Policy Dept B, PE

TERNA (2018) Impianti di generazione [Electric generation plants] (in Italian). www.terna.it

UTILITAS (2017) Il Rapporto Generale sulle Acque [2nd General Report on Water] (in Italian), Premedia http://asvis.it/goal2/home/353-1949/

WEC (2010) https://www.WorldEnergyCouncil.org/data/resources/country/Italy

Zinoni F, Antolini G, Palara U, Rossi F, Reggidori G (2005) Physical and eco-physiological aspects in forecasting and crop protection of fruit trees from late frost. Review n. 2- Italus Hortus 12(4):63–78

Zucaro R (ed) (2014) Atlas of Italian irrigation systems. INEA, Rome

Part III
Problems

Chapter 8
Management of Municipal Water Services

Michele Di Natale and Giuseppe Rossi

Abstract Municipal water services represent a key element for an effective sustainable and equitable water resources management. Although drinking water supply is guaranteed to almost all the Italian population and sewerage and wastewater treatment serve a large part of the population, several problems strike the management of municipal water services and are perceived by the society as severe troubles. This chapter describes the evolution of regulatory framework, driven by the Italian rules on all public municipal services and by the European water directives. The features of the infrastructures for water supply, drainage and treatment of wastewater are analysed. Obsolescence of water supply networks and sewer systems, leakage in the networks, inadequate fulfilment of the mandatory quality standards and difficulty in sludge disposal are identified as the main critical issues. Some proposals to face the challenges of municipal water services in the next decades are highlighted. They include improvement of the governance of the integrated water service, a plan of investments for completing infrastructures, advanced criteria to compute the service tariff according to the perspectives adopted by the Authority for Regulation of Energy, Networks and Environment (ARERA) and a wider use of new technologies for monitoring and control.

8.1 Introduction

Municipal water supply, drainage and wastewater treatment are probably the most relevant water resources issues in Italy. According to the recent census estimation (ISTAT 2015), although the supply of drinking water is guaranteed to almost the whole Italian population (98%), and also sewage systems and wastewater treatment serve a large part of the population (93% and 88%, respectively), several problems

M. Di Natale (✉)
Department of Engineering, University of Campania "L. Vanvitelli", Aversa (CE), Italy

G. Rossi
Department of Civil Engineering and Architecture, University of Catania, Catania, Italy

© Springer Nature Switzerland AG 2020
G. Rossi, M. Benedini (eds.), *Water Resources of Italy*, World Water Resources 5,
https://doi.org/10.1007/978-3-030-36460-1_8

179

concern the management of municipal water services. The main technical problems perceived by public opinion include the reduction of water availability during frequent severe droughts which hit various regions and produce water emergencies or the poor functioning of treatment plants and the connected increase of pollution risk in the receiving streams, or coastal areas, where outfalls are located. However, other problems, often not perceived by the community, can be more serious, such as the progressive pollution of sources of supply, particularly groundwater, due to the industrial outfalls or to the use of pesticides in agriculture (diffusive pollution), as well as the inadequate treatment of new biological contamination (e.g. *Giardia* cysts). The more widely known issue about management of the municipal water services refers to the controversial choice between private and public company. This issue arose during the referendum of 2011, which abrogated the priority of private companies to entrust the service and eliminated the portion of tariff allocated to cover the costs of capital.

This chapter presents the evolution of the regulatory framework of municipal water services and a state-of-the-art view of the main technical and management problems in Italy. Several perspectives for facing the main technical financial and environmental challenges of the future are discussed. The chapter is organized as follows. First, an analysis of the development of legislation on municipal water services is carried out (Sect. 8.2), highlighting that most of the rules derive from the general legislation acts regulating municipal public services and from the European water directives. Afterwards, the technical features and the management characteristics of the infrastructures for water supply, sewerage systems and wastewater treatment are described (Sects. 8.3 and 8.4). Main problems identified in the structure and/or in the operation of each component of the urban water service are discussed, and several measures aimed at facing the main challenges are proposed (Sect. 8.5). Lastly, some concluding remarks are drawn (Sect. 8.6).

8.2 Evolution of the Regulatory Framework in Urban Water Services

In the second half of the nineteenth century, throughout Europe (starting from England, France and then Germany), there was the phenomenon of "municipalization" of public services, including that of the distribution of drinking water. Therefore, such services had been managed chiefly by public enterprises. This phenomenon developed as a result of new social needs and pressures linked, first and foremost, to urbanization processes in the big cities.

In Italy the process of municipalization of public services took off with Law 103/1903 (the so-called Giolitti Law). The municipalities were granted the possibility of providing, through a special company, the creation and management of public services plants, as well as "the construction of reservoirs and fountains and the distribution of drinking water". Subsequently, Royal Decree 2578/1925 established that the management of public service could be carried out again directly by the municipal and provincial authorities, on a "time and materials basis", that is, through an internal office. Since that time the municipalities have had a central role in water resources management for urban uses. However, though it had resisted for a long time, the model of direct management of public services by municipalities and provincial administrations ("municipalization") evinced various critical aspects. In particular, there was excessive fragmentation in service management due to the high number of management bodies (municipalities and provincial administrations) plus the absence of entrepreneurial know-how.

Based on the impulse given by European law on competition and free circulation of services, and on the basis of the progressive emergence of environmental and social interests, starting from the 1990s onwards, the Italian legislator has intervened several times. Law 142/1990 introduced a uniform discipline of the "local" services – i.e. public services including water – with the purpose of the production of goods and activities aimed at achieving social ends and promoting economic and civic development of local communities. For these services, it was established that they could also be managed by a public-private partnership (so-called mixed company) or by private management. In any case planning and control of the services remained the competence of the local institutions (municipalities and provincial administrations). Subsequently Legislative Decree 267/2000 subdivided local public services into those "of economic relevance" and "not of economic relevance" (social services, excluded from the rules of competitiveness and the market). This division was recently superseded and replaced with Legislative Decree 175/2016, which reproduced the category of European law on services of general interest, i.e. "services of general interest supplied for payment on a market".

With specific reference to water services, only with Law 319/1976 (so-called Merli Law), water was considered in an environmental vision, disciplining discharge of wastewater. The regional administrations were charged with drawing up the rehabilitation plans, as well as the task of financing drainage system and treatment plants. Water services discipline was deeply innovated by Law 36/1994 (so-called Galli Law) which introduced into Italy the "Integrated Water Service" ("SII"). This should be understood as the grouping of public services of collection, adduction and distribution of water for civil use, of drainage and treatment of wastewaters. In particular, the "Galli Law" intervened in an organic manner, modernizing the water sector in an industrial key and establishing that:

- The Integrated Water Service be managed in accordance with the principles of effectiveness, efficiency and cost-saving in observance of national and community regulations.
- The water services should be reorganized on the basis of Optimal Territorial Unit (OTU whose territorial definition – entrusted to the regional administrations – should respect the unity of the catchment area).
- A tariff system be introduced based on the principle of "full cost recovery" including investment and operation cost.

In the following years, Europe dictated mandatory rules for member states in the environmental sector and in waste disposal which were then acknowledged by

Legislative Decree 152/2006. This decree and subsequent regulations redefined the water services sector, envisaging:

- New obligations on water quality and the discipline of wastewater discharge.
- The establishment between municipalities of each OTU of the governing bodies with the task of planning interventions to be carried out by the management bodies and of checking the management.
- The regulation of the system through the establishment of Committee Supervising the Use of Water Resources, subsequently National Commission Supervising the Water Resources, the latter's tasks then passed (according to L.D.214/ 2011) on to Authority for Electric Energy and Gas and Water Systems (AEEGSI) which recently took the name of Authority for Regulation of Energy, Networks and Environment (ARERA); this authority is tasked with defining the new tariff mechanisms in accordance with the community principle of "full cost recovery".
- Fixing of qualitative standards.
- Checking the plans of territorial unit and providing agreements for assignment of the service by the governing bodies.

In particular, the tariff constitutes payment of the integrated water service and is determined by taking into account the quality of resource and service supplied, the necessary works and updates and the extent of management costs of the protected areas, this in such a way that full cover of investment and running costs is assured. Certain components of the tariff must take into account also environmental load, in application of the community "polluter pays principle". The tariff is based on the "price cap method" and applied to cover "standardized costs" in accordance with the principle of the dynamic revenue cap. Moreover, it is pointed out that the current tariff system excludes remuneration of invested capital, in line with the people's referendum held in 2011.

The evolution that one may grasp is therefore also of an ideological nature, since the different approach the legislator reserves for the safeguarding of water resources, which are no longer considered as limitless but rather a resource to safeguard and manage in accordance with the principles of effectiveness, efficiency and cost saving.

8.3 Infrastructures for Water Supply

Supplying water to an urban centre is a primary need which the various civilizations since antiquity have sought to meet. As regards Italy, the first significant contribution to the creation of aqueducts came from the Romans who, as early as the third century BC, encouraged the building of huge works for collecting and delivering water. For example, we may recall the eleven aqueducts of the city of Rome, which, even in the early centuries, had extensive water resources (estimated as twice the current supply). Among the main Roman aqueducts, we may mention Acqua Marcia (91 km), Aqua Anio Novus (87 km) and Acqua Claudia (68 km). Moving to more recent times, we should remember that in the twentieth century, important water supply systems were built in different parts of the country which today are still mainstays of the national infrastructure system, for example, the Acquedotto Pugliese, the largest aqueduct in Europe, which serves 340 municipalities in Apulia with around 4 million inhabitants; the Acquedotto Istriano, built in the 1930s and formed by the 3 systems of the Quieto, the Risano and the Arsa, serving the Istrian peninsula; the Acquedotto Campano, built in the 1960s by the former Southern Development Fund (Cassa per il Mezzogiorno), an institution set up in 1950 to favour economic development in the south, to serve 42 municipalities of the Campania region with 580 km of waterways and a maximum capacity of 5.3 m³/sec; the Acquedotto della Romagna, fed mainly by the storage backwater of Ridracoli and developed through the region for an overall length of more than 600 km with an average discharge of around 3.5 m³/sec.

As things stand, the overall volume of water abstraction for municipal use in our country (2015) amounts to roughly 9.49 billion cubic meters (ISTAT 2018). This water is drawn by more than 1800 utilities that manage the water sources. 76.3% of this volume, just over seven billion cubic metres, was measured with the appropriate instruments, while the remaining 23.7% was estimated by utilities. In the absence of measurers, the maximum flow rate of water concession is often used as an estimate parameter, but this does not correspond to the real value of the delivered volume.

The volumes of water delivered into the network (and therefore inclusive of the rate of losses) in 2015 (Utilitatis 2017b) were on average around 145 m³/inh year, corresponding to a water abstraction of around 396 l/inh day (with considerable differences between the various cities, shifting from 180 l/inh day measured in Lanusei (Sardinia) to around 750 l/inh day in Frosinone). It is clear that these data are heavily influenced by the problem of water leaks, which will be discussed below.

International comparison of unitary freshwater abstraction for public water supply from groundwater or surface water bodies in the 28 countries of the European Union (Fig. 8.1) shows that Italy, with 156 m³/inh year, is the country with the great-

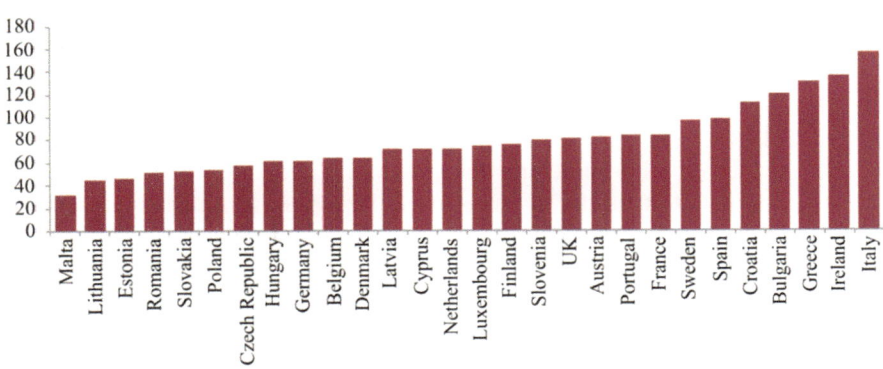

Fig. 8.1 Freshwater abstraction (cubic metres per inhabitant) for public water supply in 28 EU countries. (Source ISTAT 2018)

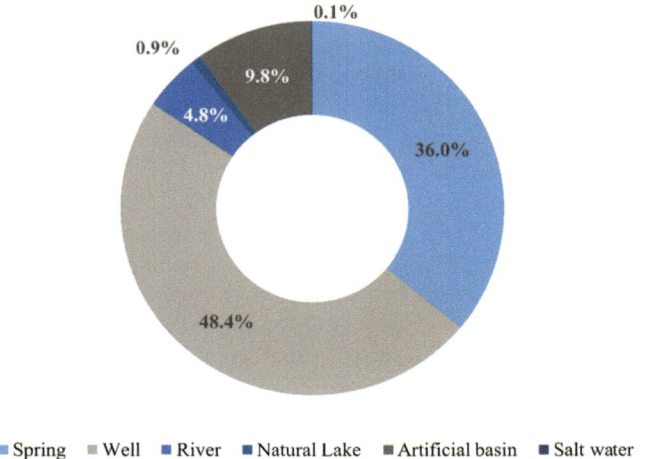

■ Spring   ■ Well   ■ River   ■ Natural Lake   ■ Artificial basin   ■ Salt water

Fig. 8.2 Withdrawal of drinking water (2015) by source typology. (Source: ISTAT 2017)

est volume, followed by Ireland (135 m³/inh year) and Greece (131 m³/inh year) (ISTAT 2018). Lower figures have been recorded in some Eastern European countries.

As the type of supply sources is concerned (ISTAT 2017), Fig. 8.2 shows that 84.3% of national municipal water comes from groundwater (48% from wells and 36.3% from springs) and 15.6% from surface waters (9.9% from artificial reservoirs, 4.8% from surface waters and 0.9% from natural lakes); the remaining 0.1% comes from sea or brackish waters. Previous numbers show that the main resource to satisfy the country's drinking water needs is groundwater. Another important datum concerns water that undergoes a purification process, a figure equal to roughly 3.1 billion cubic metre per annum.

In the context of drinking water use, the consumption of bottled mineral water should be highlighted. According to recent surveys (IRi 2016), sales of mineral water in Italy reached peaks of 8–9 billion litres, with a growing increase in recent years. Italy is Europe's highest consumer of bottled mineral water with around 206 l/inh year, which means 29 l more than the Germany, 84 l more than the France and 85 l more than the Spain. The average monthly family expenditure for mineral water in 2016 was estimated in 10.75 euros, with growing increases (ISTAT 2018).

With regard to the infrastructures for transporting drinking water, the national aqueduct system (Utilitatis 2017a) has an overall extension of roughly 425,000 km (measured over a management sample of 40 million of inhabitants out of a total of roughly 61 millions), differentiated in supply networks (15%), distribution networks (69%) and user connections (16%). Figure 8.3 shows the percentage subdivision among the different types of network at national level, for the northern, southern and central areas (data referring to a 2014 sample of 54 management bodies for 31 million inhabitants).

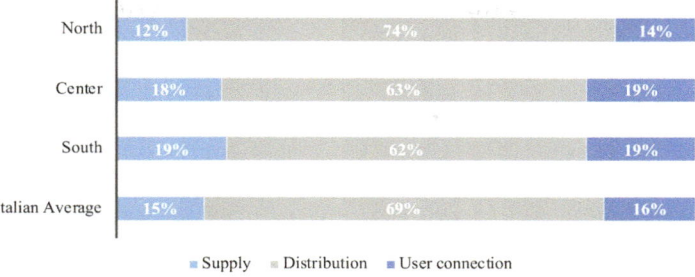

Fig. 8.3 Length percentage of different types of network. (Source: Utilitatis 2017a)

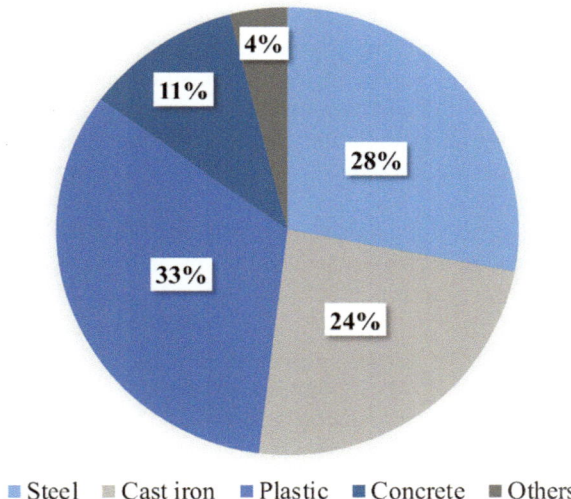

Fig. 8.4 Division of main aqueduct network analysed by type of material. (Source: Utilitatis 2017a)

Another item of technical interest regards the materials employed for aqueduct piping. According to recent survey (Utilitatis 2017a), about half of piping is in metal (24% cast iron, 28% iron-steel), approximately 33% in plastic (mainly PVC and PeaD) and 11% in cement-based material (above all the great conveying pipes created in the 1950s/1960s) (see Fig. 8.4).

Setting out from this general technical information on the Italian aqueduct systems, in Sect. 8.5 we shall analyse the main critical aspects thereof and the possible short- and medium-long-term actions to be taken in order to overcome them.

8.4   Infrastructures for Drainage and Treatment of Wastewater

8.4.1   Sewage Network

The most efficient sewage networks of Italian ancient times were – once again – those built by the Romans. By way of example we point out that, in the sixth century BC, a massive infrastructure was built, known as the *Cloaca Maxima*, in the form of a great open channel. Used for many centuries, some remains still exist today. Recently, there have been built some sewage systems, localized in the most central areas of the biggest cities. We may mention, for example, the Cuma outfall, a very large sewer tunnel about 15 km long, created at the end of eighteenth century for the collecting and removal of much of the sewage of the city of Naples.

Only in the mid-twentieth century, stimulated by the great urbanization of the city centre, the construction of systems for the collection and drainage of wastewater and rainwater along with the construction of treatment plants resumed throughout the national territory. In Southern Italy the process was implemented chiefly due to the extraordinary intervention of the Southern Italy Development Fund to narrow the gap with Northern Italy.

Currently, on the basis of the data sample examined in 2016 (Utilitatis 2017a), the main drainage system in Italy is characterized by an average length per inhabitant of 3.5 m and changing in function of the extent of the served catchment area (Table 8.1).

For sewer networks, 60% consists of a mixed system, 33% of a separate system for black water and 7% a separate system for white water (ARERA 2017a). These average values depend a lot on population density of the served area, and, where this exceeds 600 inh/km², data show a percentage growth of up to 74% in mixed networks (Fig. 8.5).

The still very high percentage of mixed type sewers creates significant difficulties in the treatment plants downstream of the drainage networks, as it will be discussed in Sect. 8.5.

Table 8.1 Indicators of network length by dimensional class of catchment area served (2014)

Class of inhabitants	Main network length/inhabitant [m/inh.]	Main network length/sewer system surface [km/km²]
≤150,000 inh.	4.91	0.87
150,001–350,000 inh.	5.07	1.18
350,001–500,000 inh.	5.44	1.00
500.001–1,000,000 inh.	4.31	1.23
≥1,000,000 ab.	2.32	1.14
Italian average	**3.54**	**1.12**

Source: Utilitatis (2017a)

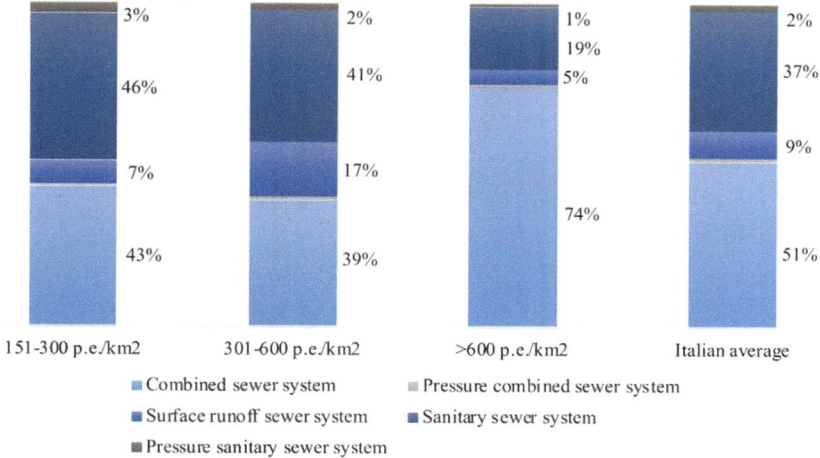

Fig. 8.5 Division of sewer network typology by population density (2014). (Source: Utilitatis 2017a)

8.4.2   Treatment Plants

A general illustration of the state of treatment services in Italy is outlined by using data from ISTAT (2018) and Utilitatis (2017a). The biodegradable pollutant load generated in territories where the sewer service exists is equal to around $33 \cdot 10^6$ p.e. (1 p.e. = 60 gr/day of BOD_5). About 80% of this load is collected by the network. With reference to industrial wastewater, an estimate of the volumes of contaminated water is about 150 million per year. Regarding the numerical consistency of the treatment plants, it is pointed out that those actually operative are about 18,000; those not in service are about 545, while there are about 80 plants under construction. The wastewaters treated are composed as follows: around 86% as civil waste, 12% as industrial waste and 2% from other sources. Figure 8.6 shows distribution of the polluting load by plant typology in the various areas of the country. With reference to the type of the plants, 56% is of primary type, 31% of secondary type and 13% of tertiary plant. Average distribution of the total load shows 78% is treated by tertiary-type plants (51% usual and 27% advanced tertiary), 19% by secondary plant, 2% by primary plant and around 1% by simple Imhoff tanks.

A singular factor is the low number of advanced tertiary plants in Central Italy where the percentage decreases to 6% as against that of other areas which is 30% greater.

In relation to plant size, it emerges that about 10% treat more than 60% of the pollutant loads since they are at the service of large cities. Also interesting is the datum concerning the potentiality of the plants, which is always greater than the collected pollutant load (average ratio around 1.53, with a great difference between north, which reaches 1.76, and south which stands at around 1.13). This is on the one hand an encouraging result, since it demonstrates good flexibility of the plant

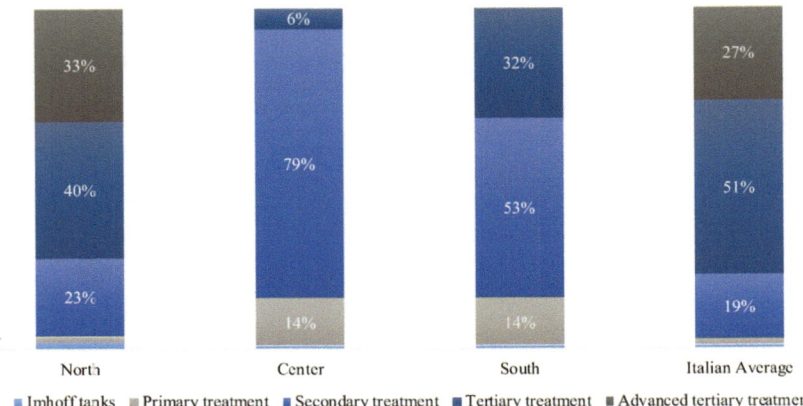

Fig. 8.6 Distribution of purified load by plant typology (2014). (Source: Utilitatis 2017a)

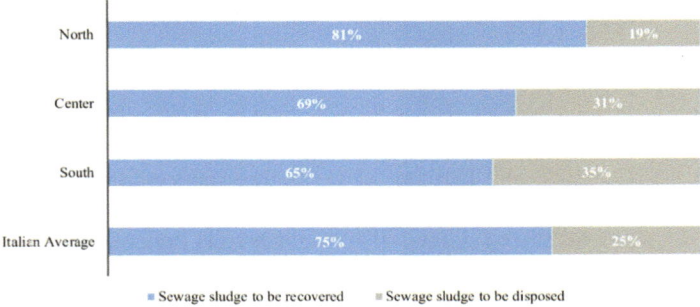

Fig. 8.7 Destination of sludge produced in treatment plant (2014). (Source: Utilitatis 2017a)

systems, but on the other hand it is worrying if we take into account that – especially in the south – the collection system is not yet completed for the carrying wastewaters to the plant. Therefore it cannot be taken for granted that when the sewer collector network will be completed the plants will still be capable of treating the whole inlet pollutant load. With regard to the sludge produced by the treatment system, the amount is about 13 kg for each cubic metre of water treated. On average, about 75% of sludge upon leaving the plants is sent for reutilization, whereas about 25% is disposed of in the landfill. Figure 8.7 also shows the distribution for the individual areas of the country. It may be seen that the percentage of dried sludge sent to the landfill is increasing from north to south, where the values become almost double those of the north, with a series of problems associated to the general system of sludge disposal, which will be dealt with below.

Reutilized sludge is used mainly in agriculture and for the production of compost and in chemical, physics-related or biological-type plant. A small percentage goes to incinerators. Figure 8.8 shows the percentage for the various end uses.

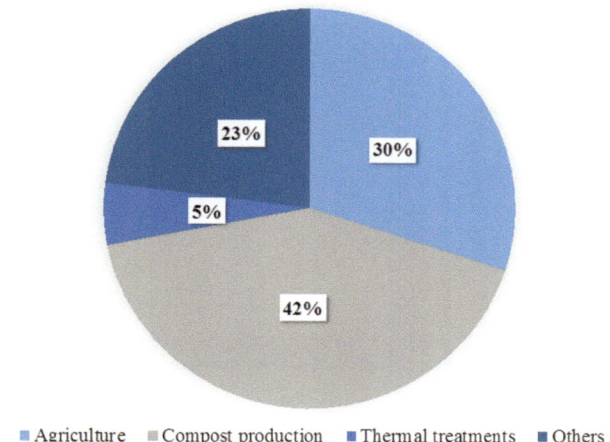

Fig. 8.8 Destination of sludge produced in plant and set to reutilization (2014). (Source: Utilitatis 2017a)

In Sect. 8.5, there will be an analysis of the critical aspects of the infrastructural system of aqueducts, sewers and treatment plant and a preliminary description of actions to be taken to eliminate shortcomings.

8.5 Problems and Challenges of Urban Water Services

Numerous problems, both administrative-managerial and technical-economic, prevent the proper functioning of urban water services. With reference to the first aspect, the inadequacy of governance is attributable, on the one hand, to a confused legislative framework and an overlap of powers and, on the other hand, to the excessive fragmentation of water service management. According to official data (Utilitatis 2017a), there are over 700 water utilities divided into different types of legal entities and 72 concessions granted by about 90 territorial units. Besides these data we must take into account a high number of micro-managements, on a municipal scale, which are still in existence and make the entire system even more fragmented and of poor efficiency. It should be, again, highlighted that at regional level, there is considerable lack of homogeneity in the choices and in the management models employed by the exceedingly numerous territorial units. Additionally, a fundamental problem in many cases is that the planning tool (plan of territorial unit) is often obsolete and in need of at least an updating.

As for the technical-economic-type aspects, it should be recalled first and foremost that the services' degree of cover (Table 8.2) is not yet up to scratch, especially for what concerns sewer networks and the overall treated pollutant load.

Table 8.2 Coverage of services by geographic area. (Source: Utilitatis 2014)

Geographic area	Aqueduct		Sewerage system		Treatment (capacity)		Treatment (treated load)	
	Coverage	Deficit	Coverage	Deficit	Coverage	Deficit	Coverage	Deficit
North	95.1%	4.9%	94.8%	5.2%	93.2%	6.8%	84.9%	15.1%
Centre	94.2%	5.8%	92.6%	7.4%	87.2%	12.8%	81.1%	18.9%
South	98%	2.0%	90.9%	9.1%	71.1%	28.9%	68.6%	31.4%
Italy	**95.6%**	**4.4%**	**93.1%**	**6.9%**	**85%**	**15%**	**78.5%**	**21.5%**

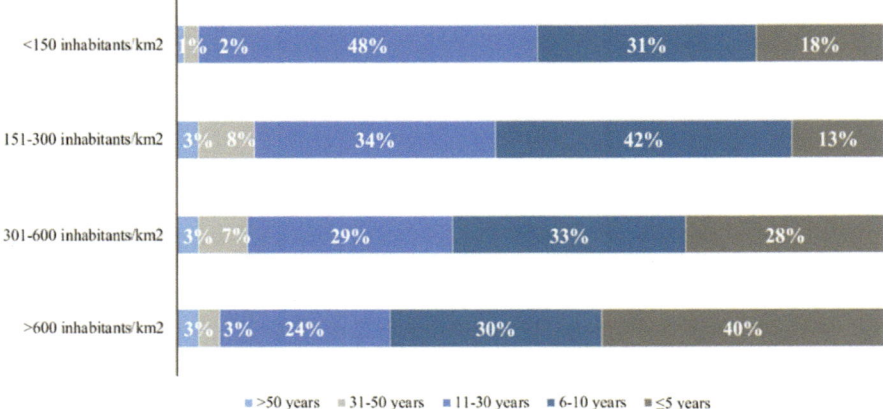

Fig. 8.9 Distribution of the main aqueduct network by date of laying and density class of population served (2014). (Source: Utilitatis 2017a)

A further critical aspect concerns obsolescence of networks and plants. Figure 8.9 shows average values of the age of national networks. One may note that most of the piping (60–70%) has an average lifetime of 30+ years, with a high percentage (40%) of 50+ years. There also emerges a tangible difference between the northern and southern areas of the country, the latter having even greater problems of superannuated networks.

Directly linked to the structural age of the networks is the problem of the amount of leakage of drinking water as against the volumes fed into the network. These losses are on average equal to 41.4%, of which 38.3% is estimated as actual leaks (due to breakages in pipes, defective joints, etc.) and 3.1% apparent water leakage (unauthorized consumption and errors of measurement). The significant differentiation on national territory in the 2015 indices should be noted: 30% in the northwest, 32.6% in the northeast, 41.4% in Central Italy, 40.9 % in the south and 48.3% in the islands. In absolute terms, these percentages result in a loss of roughly 46 m³/inh year of drinking water, for a total of about 3.45 km³ per annum. Table 8.3 shows total water losses for the various Italian regions, as reported in the 2015 ISTAT census.

Table 8.3 Total losses in the municipal water distribution networks-percentage values of the volumes released. (Source: ISTAT 2017)

	2012	2015	Difference 2015–2012
Geographical Distributions			
Northwest	30.0	30.7	0.7
Northeast	32.6	37.0	4.4
Centre	41.4	48.2	6.8
South	40.9	46.2	5.3
Islands	48.3	51.6	3.3
Regions			
Piedmont	38.0	35.2	−2.8
Aosta valley	21.9	18.7	−3.2
Liguria	31.2	32.8	1.6
Lombardy	26.5	28.7	2.2
South Tyrol-Trentino	25.6	29.8	4.2
Veneto	35.6	40.0	4.4
Friuli-Venezia Giulia	44.9	47.8	2.9
Emilia-Romagna	25.6	30.7	5.1
Tuscany	38.6	43.4	4.8
Umbria	38.5	46.8	8.3
Marche	28.9	34.1	5.2
Lazio	45.1	52.9	7.8
Abruzzo	42.3	47.9	5.6
Molise	47.2	47.4	0.2
Campania	45.8	46.7	0.9
Apulia	34.6	45.9	11.3
Basilicata	38.5	56.3	17.8
Calabria	35.4	41.1	5.7
Sicily	45.6	50.0	4.4
Sardinia	54.8	55.6	0.8

To the great environmental impact, deriving from enormous waste of water resources, the greater burden – environmental and economic – must be added, which derives from the energy consumption necessary for lifting the dispersed outflows.

On this subject we recall that on the basis of data gathered by TERNA for the year 2015 (Utilitalia 2018; TERNA 2016), "aqueducts" alone (data supplied by water service managements) consumed 6092.7 GWh of electricity, more than 2% of the overall national energy requirement. Electricity is one of the main cost items in water service management, standing at between 10% and 30% of the total costs of the service (TERNA 2016), but specific cases that exceed this interval are frequent. The high percentage of water lost – set against the previous datum on costs – gives a more precise idea of the extent of the economic damage caused by water that is dispersed in the networks.

The structural deficit of integrated water systems emerges also downstream of the resource management cycle (sewer networks and treatment plant). Indeed not all wastewaters are collected in suitable drainage networks, and the treatment plants

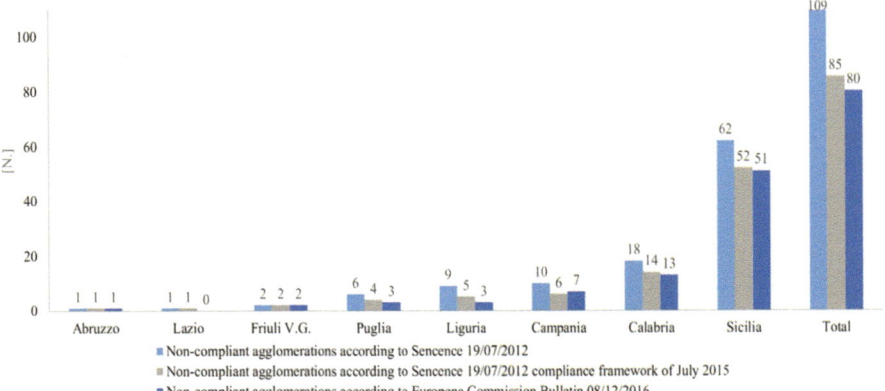

Fig. 8.10 Number of conurbations subject to Sentence C-565-2010. (Source: Utilitatis 2017a)

often have a capacity lower than that required for the polluting load produced on the served territory. As a consequence, the systems are unable to guarantee the compliance with European community directives (Directive 91/271/EEC), causing administrative and monetary sanctions. It is worth noticing that Italy is currently undergoing three sets of trials, already at sentencing stage, for breaking the rules of the above-mentioned directive. Figure 8.10 shows the number of conurbations, grouped for Italian regions, which are interested by each of these three sentences: the most critical aspects lie with regional administrations in the south.

The 2015 ISTAT census data inquiry into the provincial capital municipalities (Utilitatis 2017b) highlights how, as against an aqueduct service coverage of 98% of resident citizens, the sewer service touches 93% of citizens, whereas the purification service reaches no farther than 88%. The sewer networks, prevalently of mixed type, are often characterized by structural defects due to obsolescence but recently also by inadequacy of water conveyors due to the increase in runoff linked both to the increase of impervious areas in the catchment basins and to the intense rainfalls caused by the climatic changes under way. In mixed-type sewers, an important hydraulic aspect is the inadequacy of the structures for the derivation of first rainwaters and the overflow of excess storm capacities. In many cases we find a dimensional or maintenance-related deficiency of the said structures that compromises correct separation of wastewaters from overflows, with consequences for the observance of quality standards in the receiving bodies of water.

A problem that remains unresolved is that of effective responsibility for the drainage system for rainwater in city centres, but also in the external drainage networks. In fact, it is known that, according to current regulations, these hydraulic systems are under the responsibility of organizations other than those for water service management (municipalities, reclamation consortiums, regional administrations, etc.). This notwithstanding, due to the dominance of mixed sewer systems, all the functional inefficiencies of rainwater drainage systems are, very often and improperly, placed on the shoulders of the water service managements. This leads to frequent administrative squabbles between bodies and above all disservices to citizens.

Analysis of purifying systems (Utilitatis 2017a) reveals, from the viewpoint of management shortcomings that there is a great parcelling out of plant, with 67% of plants having a potentiality under 2000 p.e. and only 2% over 100,000 p.e.

Today, one of the biggest problems associated to correct purifying plant management is the treatment and disposal of produced sludges. With reference to reutilization in agriculture (today around 75%), it should be highlighted that European and national discipline classify urban sludge as "special non-dangerous waste" which may therefore be reutilized. However, the possibility of its recovery in agriculture may be conditioned by the presence of industrial discharges into sewer systems and by the delivery of wastewater from septic tanks into the plant. This latter aspect has been receiving special attention, and in some parts of the country (Lombardy), the first prohibitions against reutilizing sludge in agriculture have been applied. Since we do not yet have mature technological solutions or plant capable of producing solutions alternative to the reuse of sludge in agriculture, a great number of problems are being created for service managements.

Infrastructural and management lacunas in the sewer networks and treatment plant bring with them, as consequence, the fact that around 10 million citizens cannot have an adequate purifying system or still cannot have one at all. The Italian regions most affected by this criticality are in the south, and the consequential environmental damages are huge, not only for the environment itself (quality of sea and land surface) but for the tourism economy which is highly penalized by the non-use of water recreation purpose.

The perspective lines for overcoming these critical aspects that we have briefly sketched out are to be sought in joint action of certain technical-political levers represented by (i) a programme of infrastructural investments; (ii) a new perspective on the tariff structure of the integrated water service; and (iii) the use of new technologies in the world of water services.

8.5.1 Infrastructural Investments

With reference to the investment issue there are two orders of problem. The first is that the levels of economic programming do not widely cover the investments required; the second is the fact that interventions effectively implemented are too low referring to programmed investments (on average only 75% of the investment gets spent). Figure 8.11 shows the trend of investments made in the period 2007–2015 (Utilitatis 2017a).

An estimate of the real amount of investment needed, carried out in 2013 by ARERA, indicates a unit value of 80 €/inh year (ARERA 2017a). This estimate is equal to almost double the value programmed by the integrated water service operators for the 4-year period 2014–2017, just equal to roughly 41 €/inh year (inclusive of financial resources from the tariff and from public funds and contributions). The tariff contribution alone is therefore far from being adequate if we take into account that, at national level, in one year an average of 3.8 metres of pipeline is replaced for every kilometre thereof, and with this rate of replacement, it would take an average of 250 plus years to replace the whole network existing today.

Fig. 8.11 Trend and composition of investments made in the period 2007–2015, in millions of euros. (Source: Utilitatis 2017a)

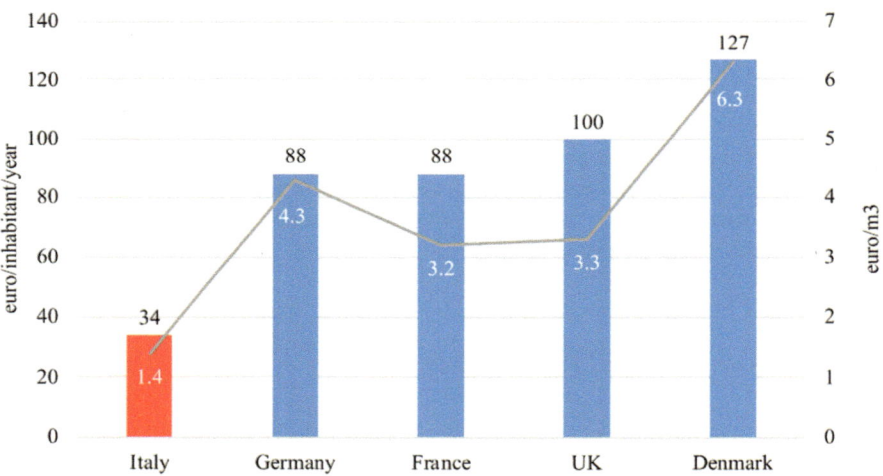

Fig. 8.12 Average investments and tariffs in some European countries. (Source: Utilitatis 2017a)

Comparison with what happens in other countries is quite singular, as may be seen from Fig. 8.12 which shows how the current level of Italian investments pro capita is far lower than the levels in other European countries.

It is important to note that the political choice of investing in the water services field would represent growth in GDP and employment. Recent studies (Ref Ricerche 2015) predict that one billion euros invested in the water sector would produce 2 million euros of GDP and 34.000 new jobs, just as 5 billion would lead to a 0.7 increase in overall GDP, with 183,000 jobs.

8.5.2 Tariff Charging for the Integrated Water Service

On the subject of tariff regulation, it should be remembered that ARERA has been developing (since 2012, year of constitution in accordance with the DPCM of 20 July 2012) an intense activity for regulation of the water systems based on the principles of efficiency and low cost of the service. During the first 3-year period (2012–2014), a regulatory project was carried out, characterized by an "asymmetrical" and "innovative" regulation which, taking in account the different territorial situations, has adopted differentiated regulatory schemes. In the second triennial, the authority set forth its activity on the stability of the regulatory framework to favour infrastructural investments, the sustainability of water resources, safeguarding of users and the reduction of bad debts. The activity of regulation envisaged among other things the introduction of an accounting unbundling, the definition of standard costs and the analysis of benchmarking for assessment of management efficiency. The resolution no 917 of 2017 (ARERA 2017b) is noteworthy inasmuch as it deals with the regulatory problem with reference to technical quality. More specifically, quality standards are introduced with macro-indicators specified by numerical parameters. In Table 8.4 the macro-indicators set with the description of the contents are shown.

Regulation of technical quality obligates the management to monitor, record and communicate technical data. Moreover, there is an incentive mechanism (prize-penalty) for the management and an automatic indemnity to users in cases of failure to meet standards.

The great stimulations set by ARERA activities have, as a direct consequence, impact on the calculation of tariffs. From this mechanism we expect efficacious stimuli for tackling and resolving – also from the economic-financial viewpoint – the complicated load of criticalities in water systems which have slowly accumulated over the years and now reached levels of warning can no longer be postponed.

Table 8.4 Macro-indicators set introduced by ARERA on 2017

Macro-indicator		Contents
M1	Water losses	The containment of dispersions with efficient surveillance of the aqueduct infrastructures
M2	Interruptions of the service	The rules for service maintenance also through appropriate configuration of the supply sources
M3	Quality of water delivered	Identifies the parameters for a suitable quality of the resource destined for human consumption
M4	Adequacy of sewage system	The environmental impact deriving from conveying wastewaters and to the storm drains discharging into mixed sewers
M5	Sludge disposal in landfill	Shows the parameters regulating the production and disposal of sludge
M6	Quality of purified water	Regards the classification related to the rate of exceeding the limits in the samples of discharged wastewater

8.5.3 Use of New Technologies

The third incentivizing lever for overcoming the criticalities of the water system is the development, implementation and employing of new technologies. This aspect is well to the fore in international and community programming.

Agenda 2030 for Sustainable Development is a programme of action for people, the planet and prosperity, signed in September 2015 by the governments of the 193 countries that are members of the UNO. It incorporates 17 objectives for Sustainable Development. For the water sector, strongly supported by the 2030 Agenda, there are some possible and prospective turning points not only to guarantee everyone the availability of water and the sustainable management of sanitation. The new objectives are in fact also to make the water sector sustainable and resilient in the face of climate change. In a European community framework, the Horizon 2020 Programme, which is financing projects for research and innovation in Europe from 2014 to 2020, has plenty of space for water sector. The European Water Supply and Sanitation Technology Platform (WssTP) has described its vision of attaining to a "water-smart" society which reduces by at least 50% of its own impact on natural water resources in accordance with four key components that research and innovation must possess: (a) manage the "true" value of water; (b) adopt digital technologies and solutions (information and communication technology (ICT)) for the entire water cycle; (c) integrate "green" and "grey" infrastructures and develop ecosystem services; and (d) adopt inclusive and multi-stakeholder systems of governance.

The themes of innovation in the context of the integrated water system are numerous; some concern individual components such as innovative technologies in the field of water treatment or sludge treatment, while others are transversal and concern, for example, the adoption of IoT (Internet of Things) infrastructures and solutions to guarantee control, safety, efficiency and water/energy saving. Below, in a manner anything but exhaustive, we look at certain innovations that regard the future of urban water services. A few examples are given.

8.5.4 Smart Metering

The development of "smart metres", which identify consumption data and buildup user profiles based on behaviours and habits (Cominola et al. 2015), is one of the main innovations, especially in the aqueduct field. For example, with one single domestic smart metre, it would be possible to trace out the consumption of each single water source (tap, bath, washing machine, shower, etc.) (Cominola et al. 2018). On the one hand, this information could be used to stimulate users to carry out water-saving actions and, on the other, to optimize the distribution system based on user behaviours that are not otherwise predictable. In Italy this new technologies are rapidly spreading for its application to remote measuring and control of water consumption. Interesting examples of existing and developing applications are those of the aqueducts of CAP and MM companies in Lombardy and by the Acquedotto Pugliese in Apulia.

8.5.5   Big Data Management

Profitable use of the huge quantity of information that technology puts at our disposal today has enormous potential for public services. Network interconnection, amplified by the introduction of information and communications technologies (ICT, IoT and smart technology) spur us towards an integrated vision of infrastructure and water service, which allows us to use system connections and the Big Data, linked thereto, for a sustainable and efficient use of the resource. Setting out from mathematical models of identification and data mining (based, e.g. on the neural and neuro-fuzzy networks), it is possible to achieve intelligent infrastructures that are innovative from the viewpoint of both company and end user. Numerous national and European projects (European Commission 2018) are working in this direction with the aim of supplying water utilities with innovative methods for the interpretation and exploitation of the enormous amount of data produced by the new technologies but no longer analysable by traditional methods. The result would be innovative management of the service.

8.5.6   New Wireless Technologies for Data Transmission

The recent development of sensors of measurement and control in real time opens up new challenges in the field of data transmission protocols in the era of smart cities and the Internet of Things (IoT). In fact, one of the fundamental elements for implementation of the IoT is a network to which the various sensors are connected and which responds to the requisites of high penetration, broad coverage and low consumption. Contrariwise, if the IoT does not always necessitate real-time monitoring but only in the case of variations in the magnitudes measured, it does not require high bit rate. Implementation of new data communications technologies is also essential in the context of water services, at domestic, district and network levels. One of the most promising technologies is ZigBee, a grouping of high-level communications protocols which use low power digital antennas and are based on the standard IEEE 802.15.4 for WPAN (wireless personal area network) and LoRaWAN (Long Range wide-area network), based on new specifications and protocols for low-consumption wide-area networks that use a wireless spectrum without need of any licence. This technology is capable of linking up sensors over long distances, while at the same time offering optimal battery duration and requiring a minimum of infrastructure (Pule et al. 2017).

8.5.7   Partitioning of the Water Networks

The paradigm of "divide and conquer" in water supply systems was introduced in the UK in 1980 (WAA and WRC 1985; WIR 1999; Wrc/WSA/WCA 1994) and then adopted in Italy through Decree of the Ministry of Public Works N° 99 of 1997. It consists of identifying permanent districts called DMAs (district metred areas) in which the original network is subdivided by inserting measurement systems of the incoming and outgoing flows, this with view to favouring different actions for improving management of the water distribution networks: improvement of management; water balances for localization of leaks; pressure control; and protection against accidental or intentional contamination. In Italy just over 10% of the water networks is districted, also due to the complexity of applying this technique to networks already functioning and modifying their performances. Recently, the development of automatic techniques of partitioning (Di Nardo et al. 2013, 2017, 2018a) might turn out to be an important opportunity for the planning of optimal districts which, coupled with the use of intelligent sensors and valves appropriately placed within the network, would permit the achievement of dynamic water distribution network configurations, thus improving management performances and quali-quantitative monitoring of the resource. Applications of this technique are underway for the aqueduct networks managed by the GORI Inc. in the Campania region for 76 municipalities and for a total of about 5000 km of piping.

8.5.8   Reducing Losses of Water and Energy

As we have seen, about 2% of the energy consumed in Italy is used in the aqueduct systems. In fact water systems include numerous pumping plants and, unfortunately, a considerable portion of energy required for the pumps is wasted in compensating for the deficit in flow and pressure caused by aged piping, tanks, valves, control devices, etc. So, energy is destined for the distribution of water that is dispersed in vain. The management of district pressures (Alonso et al. 2000), the adoption of better performing pumping systems and of incentivizing policies for the reduction of water and energy waste and, above all else, the use of advanced simulation and optimization software could lead to notable results in terms of water and energy saving.

8.5.9   Water Safety Plan

Recent adoption of the procedures prescribed by the water safety plans (WSP), which envisage the control of water not only on a simple analytical basis but also through monitoring and territorial risk mapping, opens up new challenges for the Integrated Water Service. Within this framework, the possibility of having smart water networks, based on accurate modelling and controlled in real time by measuring devices, is a fundamental element in responding to regulatory requirements and ensuring public safety, also with considerable advantages in terms of system management, as well as being a guarantee for end users. Moreover, the use of new technologies for carrying out measurements and transporting them rapidly to the control centres – e.g. by means of drones – is an innovative theme on which both the scientific and technical communities are working on.

In particular, downstream of recent world events involving terroristic phenomena, the WSPs might also represent an effective model of protecting networks from intentional contamination. In this sense, partitioning and sectioning of water networks could be an efficacious methodology for the protection of water distribution networks against accidental and intentional contamination, observing the criterion of the dual-use value (Grayman et al. 2009; Di Nardo et al. 2018b).

8.5.10   Purification of Drinking Water

New water service challenges have also opened up in the field of drinking water purification. In fact applied research is seeking advanced chemical treatments to replace traditional ones (oxidation, disinfection, softening, etc.) such as the so-called AOPs (advanced oxidation process) which are based on the combined application of various oxidation/disinfection systems (e.g. membranes and new adsorbent materials) which evidence greater efficacy with regard to the contamination of drinking water caused, for example, by protozoan cysts (*Giardia* and *Cryptosporidium*) (Utilitatis 2017b).

8.5.11   Urban Meteoric Water Reuse

Water saving using suitably accumulated meteoric water is one way of saving resources of drinking water. The study of systems of rainwater harvesting for recovery and recycling of rainwater already has various applications for one-family homes and small-size dwelling complexes (Di Natale and Di Nardo 2016). Development of this technique could be a highly interesting economic and environmental prospect for the near future.

8.5.12   Treatment of Wastewater

One ambitious aim in the international scientific community is to find a way of making wastewater not only perfectly compatible with the environmental characteristics of its receiving bodies (to avoid pollution of rivers, lakes and seas) but also capable of extracting material resources therefrom (organic carbon, fibres, nutrients, etc.) and energy (thermal, chemical, kinetic energy, etc.). Some applications are based on the use of pre-treatments of anaerobic digestion of sludge to maximize the reduction of volatile solids and therefore the production of energy. An interesting research frontier regards the possibility of exploiting the substratum in wastewater for cultivation, simultaneous with the purification process, of selected biomasses capable of accumulating special organic compounds within the cells (typically, polyhydroxy-alkanoate, PHA) which, once extracted from the cells themselves, may be used for the production of bioplastics (biodegradable polymers) (Utilitatis 2017b).

8.5.13   Energy Production

Thermal and chemical energy can be recovered from sewage. Heat, hard to recover at source due to practical difficulties, may be recovered at the wastewater plant with the use of heat pumps. Experiments (Utilitatis 2017b) have demonstrated the feasibility of this approach, the ratio of energy recovered/electricity expended being greater than 3 (the threshold determining the economic convenience of the process). Chemical energy may be recovered in part from sludge, essentially in two ways: (a) by producing biogas (through anaerobic digestion) and thence electricity and heat and (b) by using sludge as fuel in incinerators.

8.5.14   Online Sensors

The impact of ICT has become essential also in the water sector. In fact, the integration and implementation of new monitoring and control devices in the Integrated Water Service is the main challenge, offering new and formerly unthinkable management scenarios. Over and above improvement in quantitative measuring (pressures, capacities, temperatures, etc.), the main novelty is market availability of online sensors for the measurement of water quality. Traditionally, the commonest method (for drinking water and wastewater) involves sampling and laboratory analysis. The main novelty is the introduction of new methods of analysis which supply online measurements, monitor bacteria present in the watery matrix analysed and include automatic sampling plus acquisition of the results in a few minutes without the addition of chemical reagents or compounds, mainly for the identification of pathogenic agents. A recent investigation (Di Nardo et al. 2014, 2018b) describes most of the innovative technologies in this field.

8.5.15   No-Dig Techniques

No-dig technologies permit underground pipe or cable laying, or the functional, partial or total recovery, or the replacement of existing buried conduits without having to resort to open trench/open cut, thus avoiding tampering with surfaces (roads, railways, airports, woodlands, rivers and canals, areas of high environmental value, historic piazzas, etc.). In the aqueduct and sewer sector, these techniques, which subdivide into various typologies (e.g. horizontal directional drilling, rod pusher, micro-tunnelling, tunnel boring machine, raise borer, mole, pipe jacking, etc.), may open new scenarios for the maintenance of piping systems. Also, they can increase the possibility of simplifying the installation of innovative technologies in a considerable way (Ariaratnam et al. 2008).

In Italy, these technologies are in rapid development, with their most relevant implementation to the relining of urban sewer system. Significant applications have been implemented recently to drainage canals of the urban sewer of Trento, Pisa and Altamura (Bari) cities.

8.6 Conclusions

The analysis of Italian water systems highlights certain significant aspects regarding the criticalities of the urban water management. In the first place, we note the presence of a regulatory framework still in evolution. In fact, even if certain concepts like the vision of an integrated water service and the principles of effectiveness, efficiency and cost-saving are, now, consolidated at conceptual level (even if often disregarded in reality), there are still many open spaces in the discussion regarding management models in entrusting the service. This aspect is sharpened by the great fragmentation of a system that sees around 1800 management bodies most of which are characterized by micro-company systems or concessions on a municipal scale. Moreover managements do not always regard the integrated cycle but limit themselves to only a part of it. ARERA, in the sector of regulation, is developing a series of actions aimed at eliminating these shortcomings with an approach to managerial and technical improvement of the management activities that has direct repercussions on the tariff provisions.

Among the merely technical aspects of the network systems (aqueducts and sewers), we find that the level of territorial cover is not yet adequate (especially for sewers where the average values are around 93.1%) for wastewater treatment (85%). Moreover, another critical infrastructural lacuna emerges in the network systems due to their obsolescence, with high average working life of pipes and equipment (more than 50 years in 40% of cases and 30 years in 30% of cases), and to the chronic absence of suitable maintenance. The critical aspects described are systematically more marked in the south of the country.

Unfortunately, the aqueduct networks are affected, though in a differentiated manner, by the serious problem of leaks (physical and apparent) which on average reach around 40% of water delivered to the network, with percentages around 30% in the north, 40% in the centre and 50% in the south of Italy.

The sewer networks, mainly of the mixed type (60%), do not appear adequate for collecting all the wastewater produced and suffer from hydraulic problems, regarding both carrying capacity, often no longer sufficient to contain the meteoric flows of the drains, and the inadequacy of the structures for the derivation of first rainwaters and the overflow of excess storm capacities.

Wastewater treatment plants still have an insufficient degree of territorial coverage to treat all the wastewaters produced by the catchment areas in question. There is still a high number of small plants (less than 2000 inhabitants served) with difficulty in finding efficient managements, and many are still primary and secondary treatment schemes. Lastly, an important critical aspect is the quantity of sludge produced and the problems of treating and dumping it. It is increasingly less likely to end up in agriculture due to the presence of unacceptable contaminants.

A great obstacle in the resolution of infrastructural problems is the lack of funds. This gap is highlighted by the fact that the annual financial resource available for infrastructures was on average 41 €/inh year in 2014 (inclusive of tariff proceeds and public funds and contributions), as against a requirement of almost double. A broad economic-political discussion opens up on this aspect that has yet to find a concrete solution in defining how to obtain the necessary financial resources to tackle such a serious problem (tariff, public/private funds, European funding).

Promising research is under way for the fine-tuning of innovative monitoring instruments and new technologies capable of mitigating the critical aspects highlighted herein.

References

Alonso JM, Alvarruiz F, Guerrero D, Hernandez V, Ruiz PA, Vidal AM, Martìnez F, Vercher J, Ulanicki B (2000) Parallel computing in water network analysis and leakage minimization. J Hydr Eng ASCE 126:251–260

ARERA, Authority for Electric Energy and Gas and Water Systems (2017a) Relazione annuale sullo stato dei servizi [Annual Report on the status of services] vol. I https://www.arera.it. Accessed Mar 2018

ARERA, Authority for Electric Energy and Gas and Water Systems (2017b) Regolazione della qualità tecnica del servizio idrico integrato [Regulation of the technical quality of the integrated water service] Deliberation N.917/2017. https://www.arera.it. Accessed Mar 2018

Ariaratnam ST, Koo DH, Sihabuddin S (2008) Sustainability in trenchless construction. 26th No-Dig International Conference and Exhibition ISTT, No-Dig 2008 Moscow, 144–150

Cominola A, Giuliani M, Piga D, Castelletti A, Rizzoli AE (2015) Benefits and challenges of using smart meters for advancing residential water demand modeling and management: a review. Environ Model Soft 72:198–214. ISSN 1364-8152

Cominola A, Spang ES, Giuliani M, Castelletti A, Lund JR, Loge FJ (2018) Segmentation analysis of residential water-electricity demand for customized demand-side management programs. J Clean Prod 172:s1607–s1619

Di Nardo A, Di Natale M, Santonastaso GF, Venticinque S (2013) An automated tool for smart water network partitioning. Water Res Manag 27(13):4493–4508

Di Nardo A, Di Natale M, Musmarra D, Santonastaso GF, Tzatchkov V, Alcocer-Yamanaka VH (2014) Dual-use value of network partitioning for water system management and protection from malicious contamination. J Hydroinform 17(3):361–376

Di Nardo A, Di Natale M, Giudicianni C, Santonastaso GF, Tzatchkov V, Rodriguez Varela JM (2017) Economic and energy criteria for district meter areas design of water distribution networks. WATER 9:463–472

Di Nardo A, Di Natale M, Gargano R, Giudicianni C, Greco R, Santonastaso GF (2018a) Performance of partitioned water distribution networks under spatial-temporal variability of water demand. Environ Model Soft 101:128–136

Di Nardo A, Baquero González D, Baur T, Bernini R, Bodini S, Capasso, et al (2018b) On-line measuring sensors for smart water network monitoring. Proceedings of HIC 2018, Palermo.

Di Natale M, Di Nardo A (2016) Urban meteoric water reuse and management. In: Urban water reuse. Taylor & Francis Group, Boca Raton, pp 533–546. ISSN: 978-1-4822-2915-8

European Commission (2018) ICT4 water, roadmaps & action plan. http://www.ict4water.eu. Accessed Feb 2018

Grayman WM, Murray R, Savic DA (2009) Effects of redesign of water systems for security and water quality actors. Proceedings of the world environmental and water resources congress, 49 Kansas City (S. Starrett, ed.). Kansas City, MO

IRi-Information Resources (2016) Il clima impatta i consumi: il mercato dell'acqua minerale [The climate impacts consumption: the mineral water marke] https://www.iriworldwide.com /it-IT/ insights

ISTAT-National Institute of Statistics (2015) Focus Giornata mondiale dell'acqua 20 marzo 2015 [Focus World Water Day 20, March2015]. https://www.istat.it/it/archivio/153580

ISTAT-National Institute of Statistics (2018) Focus Giornata mondiale dell'acqua 20 marzo 2018 [Focus World Water Day 20 mar. 2018]. https://www.istat.it/it/files/2018. Accessed Jan 2018

ISTAT-National Institute of Statistics (2017) Censimento delle acque per uso civile [Census of water for civil use]. Statistical Report https://www.istat.it/it/files/2017. Accessed Jan 2018

Pule M, Yahya A, Chuma J (2017) Wireless sensor networks: A survey on monitoring water quality. J Appl Res Technol 15:562–570

Ref Ricerche (2015) Le infrastrutture idriche: un patrimonio comune [Water infrastructure: a common asset] Acqua n. 47 pp 3–14

TERNA (2016) Dati Statistici sull'energia elettrica in Italia [Statistical data on Electricity in Italy]. http://www.terna.it

Utilitatis (2014) Blue Book:I dati del Servizio Idrico in Italia [Blue Book: Data from the Italian Water Service]. https://www.utilitatis.org. Accessed Jan 2018

Utilitatis (2017a) Blue Book:I dati del Servizio Idrico in Italia [Blue Book: Data from the Italian Water Service] https://www.utilitatis.org. Accessed Feb 2017

Utilitatis (2017b) II Rapporto generale sulle acque: Obiettivo 2030[II General Report on Water: 2030 target]. Pubblimedia Ed.

Utilitalia (2018) Linea Guida per l'esecuzione della diagnosi energetica [Guideline for the execution of energy diagnosis]. http://www.utilitalia.it/focuson, series "Focus on" June 2018. Accessed Apr 2018

WAA -Water Authorities Association- and WRC -Water Research Centre (1985) Leakage control policy and practice, Technical working group on waste of water. WRC Group, London

WIR -Water Industry Research Ltd (1999). A manual of DMA practice. UK -WIR

Wrc/WSA/WCA Engineering and Operations Committee (1994) Managing leakage. UK Water Industry Managing Leakage, Report A-J. London

Chapter 9
Water Pollution Control

Marcello Benedini

Abstract Italian water resources face a high risk of contamination, due to the discharge of solid and liquid waste coming from the various activities connected with the use of water. Efficient tools are available to tackle the water quality problems. Beside the methods for the analysis of water quality, some criteria for a first-hand evaluation of the pollution level in surface and underground water are available to identify some "classes", according to the concentration limits of the most significant pollutants. Important are the biological indicators necessary for the ecological status of the water bodies. The existing monitoring system is able to work out an acceptable evaluation of the actual water quality in Italy, even though an improvement would be necessary, in terms of number and efficiency of gauging stations. The present evaluations denote remarkable differences of pollution level among the various regions, also in relation with the existing wastewater treatment plants and in view of future developments. Some remarkable cases, relevant to specific locations in the country, are described, considered significant for characterizing the Italian water quality problems.

9.1 Water Quality Problems in Italy

Like in the other industrialized countries, the Italian water resources undergo now a high risk of contamination, due principally to a massive discharge of solid and liquid waste coming from the various activities directly or indirectly connected with the use of water. Such contamination is recognized as "pollution level", which makes the water resources unusable, with a serious impact on the interested population, not only as concerns the economic aspect but very often also the living conditions. The water pollution is thus the main concern in today's water resources

M. Benedini (✉)
Italian Water Research Institute (Retired), Rome, Italy
e-mail: benedini.m@iol.it

© Springer Nature Switzerland AG 2020
G. Rossi, M. Benedini (eds.), *Water Resources of Italy*, World Water Resources 5,
https://doi.org/10.1007/978-3-030-36460-1_9

management, and the responsible institutions are facing a consistent challenge, on the success of which our future is seriously conditioned. The Italian water resources, particularly during the more recent decades, have undergone a progressive deterioration, and now suitable measures are necessary and cannot be deferrred.

For several centuries, the main object of water management was the control of the rivers crossing the urban agglomerations, in a view of reducing the effects of the frequent floods. In the largest rivers, the intake for irrigation and potable uses affected the water bodies, as did the interior navigation, flourishing in the lowest reaches of the rivers in the Venetian Plain, as well as in River Po and River Tiber. At the end of the nineteenth century, the need to suppress endemic diseases like malaria in several wet zones, among which the Pontine marshes, the delta of River Po, and some coastal areas of Sardinia, implied huge hydraulic works, often diverting the flow of natural rivers. At the same time, in many cases the river harnessing for hydropower made long stretches completely dry, with a very high environmental impact. Such a tendency lasted till the early 1960s, and in practice several rivers were barred with dams, which caused a complete alteration not only of the relevant ecosystem but also of the whole hydrological regime. Moreover, very often the water of a river was mixed with that diverted from another water body, causing a further irreversible damage. A significant example in the early 1960s was a plan for a conspicuous hydropower scheme in Central Italy to divert the flow of the middle stretches of River Tiber into the volcanic lakes of Northern Lazio, for supplying several power stations. If realized, such a scheme would have mixed different kinds of water, destroying the original ecosystems. The plan was eventually rejected under high public pressure, as people became increasingly environmentally conscious. Since then, the water management authorities began to be concerned with the need of quality preservation in water bodies, also threatened by the discharge of polluted wastewater of urban and industrial origin.

The effects of chemicals used in agriculture further underlined the need to intervene, and, following the new tendency, all the hydraulic works aimed at water use were submitted to severe constraints, imposed by the need of preserving the aquatic ecosystem, prescribing the maintenance of suitable conditions, in terms of flow, level and usable volume.

It is important to mention a problem pertaining to the water quantity and the increasing request of water withdrawals for human activities. Since the conservation of aquatic life depends on the quantity of water available in the water body, the "ecological use" shows the highest flow requirement. Consequently, in the last decades, the concept of minimum acceptable flow has been developed and introduced in the law in order to protect the aquatic species. The size of this problem in Italy can be appreciated considering, for example, that the protection of aquatic life from the effects of hydropower diversions would cause a 6–10% decrease in energy production and an additional cost of hundreds of millions Euro per year for a surrogate source of energy. On the other hand, there is an increasing awareness that a management approach focusing on a single water use (hydroelectric, agricultural) is no longer acceptable. Imposing by law the "ecological flow" in a river introduces severe constraints on the management of water resources, especially in running existing structures. Worrisome is the case of the reservoirs built for irrigation that must release a conspicuous part of the stored water in order to maintain the minimum flow in downstream reaches.

Water pollution problems in Italy were officially considered early the 1970s, when the Parliament promulgated Law 319/1976, which recognized the real situation all over the country and prospected some solutions in political and administrative terms. Today, in spite of several interventions, done with appreciable results, several cases still survive in many parts of the country, in which the pollution persists, sometime also aggravated by uncontrolled discharges. Several laws have been proposed in the meantime, and new structures are in operation, which have been able to draw a reliable picture of the situation, focusing also the points requesting major attention.

The principal authority in duty for these problems is the Ministry of Environment, Land and Sea, which avails itself of operation structures at the level of the central government and detached in form of regional agencies. Fundamental is the role of the Institute for Environmental Protection and Research (ISPRA), a state agency that promotes the monitoring activity and controls how the national rules and the directives of the European Union are applied.

9.2   Operation Tools

The responsible structures and authorities, as described in Chapter 3, with the support of the scientific community, have been able to identify some of the more essential tools necessary to tackle the water quality problems. One of the first products was the identification of the chemical methods for the analysis of water quality, promoted by the Italian Water Research Institute in the early 1970s and currently adjourned in order to include the pollutants that could be progressively detected in wastewater and in the natural bodies (IRSA/CNR 2004). The scientific and technical literature contains now a long list of physical, chemical and biological indicators, regarding particularly the pollutants at a very little concentration.

The Water Research Institute, following examples in other countries, proposed some criteria for a first-hand evaluation of the pollution level in surface water (Benedini et al. 1998), as shown in Table 9.1. It identifies four "classes", according to well-defined concentration limits of the most significant indicators.

Table 9.1 Quality classification for surface water proposed by the Water Research Institute

Quality indicator	Unit of measure	Class 1 From	To	Class 2 From	To	Class 3 From	To	Class 4 From	To
DO	(mg O2/l)	7.01	10.00	3.01	7.00	1.01	3.00	0.00	1.00
BOD	(mg O2/l)	0.00	3.00	3.01	7.00	7.01	10.00	>10.01	
COD	(mg O2/l)	0.00	10.00	10.01	20.00	20.01	30.00	>30.01	
N-NH4+	(mg N/l)	0.00	0.030	0.031	0.50	0.501	1.00	>1.01	
P-PO4	(mg P/l)	0.00	0.05	0.051	0.10	0.101	0.20	>0.201	
Faecal col.	(n/100 ml)	0	100	101	2001	2001	20,000	>20,001	
N-NO3	(mg N/l)	0.000	0.050	0.051	1.000	1.010	10.000	>10.01	

By means of such indicators and their prospected range, it is possible to identify:

- Class 1: water of good quality that does not need treatment
- Class 2: water with an acceptable pollution level
- Class 3: polluted water
- Class 4: water with high pollution level

This classification was largely adopted by the responsible authorities, above all because it gave a first-glance evaluation of the status of water bodies. The selected indicators are in fact able to characterize a warning situation for the utilization of an available resource.

The Water Research Institute proposed also a similar classification for groundwater, for which Table 9.2 recalls the basic scheme and points out a way of application. Three levels of judgment include the most significant indicators that can affect the use for drinking purpose, which is the most frequent way of exploiting an aquifer.

The classes in the table have the following meaning:

- A suitable for potable use with no treatment, acceptable for the majority of industrial and agricultural uses
- B suitable for potable use with no treatment, with some limitation for industrial and agricultural uses
- C unsuitable for potable use with some limitation for other uses

    - C1 requiring some specific treatment
    - C2 requiring simple or advanced oxidation

The classifications and the monitoring procedures mentioned above encompass a very large set of indicators, able to characterize the quality of a receiving body in respect to all kinds of pollution sources. The main objective was to guarantee the availability of potable water, always considered the ultimate requirement for any water management policy. Some "guide values" have been proposed for the classification.

More accurate and detailed monitoring procedures have been subsequently developed. During the last decades, the worsening of environmental conditions has moti-

Table 9.2 Classification scheme of groundwater for physical and chemical undesirable substances

Judgment	Class	F (mg/l)	Electr. cond. (μS/ cm)	SO_4 (mg/l)	Cl (mg/l)	NO_3 (mg/l)	Fe (mg/l)	Mn (mg/l)	NH (mg/l)
Optimal	A	15°–30[a]	<1000[a]	<50[b]	<50	<10[a]	<0.05	<0.02	<0.05
Acceptable	B	30°–50	1000[a]–2000	50[b]–250	50–200	10[a]–50[a]	0,05–0,2	0,02–0,05	0.05–0.5
Poor	C	>50	>2000	>200	>200	>50	>0.2	>0.05	>0.5

°Minimum recommended value
[a]Indicative value between the maximum acceptable concentration (MAC) and the guide value (GV)
[b]Double value with respect to GV

vated the governing structures towards more efficient tools for the water quality control, starting from a more refined evaluation of the water body pollution. In line with the Water Framework Directive (WFD) 2000/60 of the European Union, an intensive work has been done in Italy in order to identify the source and the behaviour of the various pollutants in water, also in view of the most appropriate measure for their abatement. A great complexity characterizes the problem, conditioned by the very high number of indicators. Following the directive, the Italian action was centralized in the Decree 260/2010 issued by the Ministry of Environment, Land and Sea. It contained the main directions for defining two significant kinds of qualifying classes according to the specific nature of the pollutants, namely, respectively, the chemical and the ecological class. The application of the proposed directions is mandatory of the regional agencies for environmental protection (ARPAs) and concerns a different procedure for rivers, lakes and groundwater. In practice, every regional administration, with the supervision of the Institute for Environmental Protection and Research, has worked out its specific rules, according to the local consistence of water pollution.

The fundamental aspect of the chemical status is determined according to the limit imposed for the organic and inorganic compounds relevant to the main uses of the water resource, in terms of GOOD and NO-GOOD classes. They can be applied to the surface water bodies, as rivers, streams, natural lakes and artificial reservoirs, as well as to groundwater.

More complex is the assessment of the ecological status, for which six classes are to be considered, as summarized in Table 9.3.

This classification is essential for surface water bodies, while some middle class could be not necessary for groundwater resources.

Following the European Directive, any deterioration or improvement of the ecological status in surface water should mainly refer to the response of the biota, rather than to the changes in usual parameters. The "Biological quality elements" (BQEs) are therefore compulsory. They include the composition and abundance of aquatic flora and benthic macroinvertebrates, as well as the abundance and age structure of fish fauna. Currently, most of the European Union member states have established methods for assessing the ecological status, and these have been intercalibrated at European level (Bennett et al. 2011; Buffagni et al. 2007; Erba and Buffagni (2009)).

In Italy, for the benthic macroinvertebrates in rivers, a specific monitoring system is used, according to the national legislation. Such system combines information related to some basic elements relevant to the main aspects required by the

Table 9.3 Ecological classification of water resources. (ISPRA 2015)

High	Good	Sufficient	Scarce	Bad	Not classifiable
No alteration of the natural aspect due to human intervention	Low alteration with no serious effects	Significant alteration due to human intervention but minor effects on the living organisms	Alteration of living organisms	Heavy alteration of living organisms	Poor availability of significant data

European Water Framework Directive. Four quality classes are identified (high, good, moderate, poor and bad) on which management actions should be based (Erba and Buffagni 2009).

9.3 Pollution Sources

9.3.1 Urban Pollution

The first cause of pollution is relevant to the urban and domestic use, as outlined in Chapter 7. Large amount of water characterizes the life in every house, responsible of the discharge of wastewater with high values of oxygen demand and ammonia concentration. The discharge from secondary uses, particularly washing and cleaning, through the presence of phosphorous, contributes to the eutrophication in the receiving natural bodies. Substances that are more sophisticated, like chemicals in medical compounds and pesticides, are more and more frequent, which, even though in very small concentration, can alter the vital conditions of aquatic fauna. The discharge from domestic use can convey also bacteria and viruses, with the risk of disseminating worrisome diseases. At the same time, considerable amounts of hydrocarbons, due to vehicle engines, often associated with appreciable concentration of lead, characterize the urban discharge, which already contains heavy metals and chemicals due to the handicraft and industrial activities very common in many Italian urban agglomerations.

Quite worrisome is becoming now the presence of small particles of plastic materials, coming from the smashing of ware that ordinarily should be disposed together with the solid litters. Such particles, conveyed through the sludge, enter the food chain of aquatic species in a complex environmental, chemical and biological alteration (Barbiero and Cicioni 2000; Benedini and Giulianelli 2003).

The actual policy of pollution abatement imposed by law, also in line with the European Union Directives, is concentrated on the wastewater treating process, fostering the construction of plants at the end of collecting sewerage, before the final disposal into the natural body. The majority of plants are designed to abate the organic pollutants identified by oxygen demand, which can be directly assessed in relation to the real and equivalent population, but little they can do for other numerous pollutants, especially metals and synthetic wares with complex molecular composition.

The actual situation of treatment plants for urban and domestic wastewater is assessed in a survey of the Central Institute of Statistics (ISTAT 2017). At the end of 2015, Italy could rely on more than 17,000 plants. The survey underlines the predominance of the Imhoff process, which still covers about 47% of the total treating capacity in the country. This very simple process, able to abate large part of the polluting charge expressed in term of oxygen demand, is currently adopted in the small urban communities that are not in a position to install more efficient and expensive equipments. The Imhoff process can release to the water body a considerable amount of residual pollutants.

More efficient are the plants with primary level of treatment (9% of the total number in the country) and those with secondary level (30% in number), while more than 2000 plants for tertiary level (13% in number) can guaranty the maximum pollutant abatement.

Piedmont has the highest number of plants (about 22%), followed by Lombardy (11%); both regions belong entirely to the River Po basin. The highest number of plants with secondary and tertiary treatment level is in the southern regions.

In spite of the incentives to install new treating plants in all the most crowded dwelling zones, several places are still without any facility for the domestic and urban wastewater, while several plants, already put into existence with public financial support, are not yet able to properly work due to unexpected technical and economic deficiencies. As consequence, several rivers and lakes, all over the national territory, still denote a perceptible level of pollution, which requests more and more the attention of the responsible authorities.

Beside the surface water bodies, several aquifers suffer now for a worrisome contamination due to the seepage of uncontrolled domestic wastewater, particularly in small dwelling places scattered in the countryside, where the construction of treating plants cannot be easily done.

9.3.2 Agricultural Pollution

Quite different is the pollution originating from agricultural activities, for which the contaminants reach the water body almost in a continuous way along the embankments or through numerous small streams not easy to identify and control. This gives rise to the agricultural pollution, which in Italy presents itself now as the most worrisome form of contamination (Benedini et al. 1998).

In recent years, the effect of agriculture has become appreciable on the quality of surface and underground water bodies that receive the residual water from crops. The farmers strew huge amounts of chemicals for increasing a productivity that is day by day threatened in the competition with other more remunerative activities.

The amount of fertilizers, herbicides and pesticides has become now very conspicuous, and their use is not always performed according to the rational criteria of an advanced agriculture that complies with the larger context of environmental protection. Consequently, the agricultural pollution is further aggravated by the fact that it occurs normally in form of "non-point source", thus in a way that cannot be easily identified. Moreover, the liberalization of the market for the agricultural products, already in operation within the European Union countries, imposes severe conditions on the agriculture that eventually stimulate the farmers towards a more intensive use of chemicals.

The consistence of this kind of contamination can be perceived taking into consideration the amount of the chemicals currently used in the agricultural activity, summarized for all the country as shown in Table 9.4. The values come from a survey relevant to year 2015 (ISTAT 2017).

Table 9.4 Chemicals used in Italian agriculture

| | Fertilizers | | Phytosanitaries |
| | Quantity | Per unit of cultiv. land | Quantity |
	(ton·10^3)	(ton/ha)	(ton·10^3)
Northwest	1246.60	0.60	20.57
Northeast	1558.09	0.60	45.71
Centre	647.39	0.30	16.40
South	556.07	0.20	32.98
Islands	220.81	0.10	14.33
Italy	4228.96	0.36	129.99

Nitrogen and phosphorus are the main fertilizer components, while the phytosanitary contains complex molecular compounds, very often persistent in water.

Agricultural activities include also the animal breeding, which, especially in northern and central zones, is done in concentrated farms with thousands of animals. The abundant wastes of these farms, rich of organic matter, request special plants with advanced treatment level before the discharge into a receiving body; very often they are used as fertilizers in nearby fields. These considerations hold also for the protection of groundwater, where a presence of pollutants due to the seepage of a contaminating discharge cannot be easily removed. Several cases occur, in various places of the country, in which some traces of pesticides persist for many years (Pisanello et al. 2015).

9.3.3 Industrial Pollution

Third source of pollution is the discharge from industrial activity, also present in many places. Small factories, often run at family level, are located inside the dwelling settlement, and the discharge of relevant wastewater occurs through the facilities already in operation for the urban and domestic use. Attention is then necessary for the productive lines the waste of which requests specific treatment, like those dealing with metals and advanced chemical compounds. Suitable pre-treatment process is recommended, which now is unfortunately adopted only in some restricted cases. The already mentioned survey of the Central Institute of Statistics has also evaluated the main pattern of the wastewater that reaches the treatment plants at the end of the urban sewerage (ISTAT 2017). In terms of equivalent population and referred to the 2015 situation for all the country, the total amount of organic biodegradable charge is of the order of 160,000,000 equivalent inhabitants, 38% of which of industrial origin. A comparison with the 2012 situation shows an appreciable reduction of such amount, partially justified with an increase of separate plants destined specifically to the industrial discharge. Many productive factories are now installing their own treating facilities that work under efficient control and possibly rely on separate line of disposal. Similar harnessing is recommended for the specific zones that host the grouping of several productive plants, which can benefit from proper facilities of water supply and wastewater disposal.

Some surface water bodies, especially in northern zones, receive the discharge of cooling plants operating both for industrial production and for energy generation. The reduction of the residual heat in the thermal processes occurs normally by means of exchangers supplied with large amounts of freshwater withdrawn from the natural bodies and returned to them with an increased temperature. Even though such increase is normally low (the actual law prescribes to keep it within 4 °C), the discharged stream can still alter the aquatic environment. The adoption of the cooling towers, which could avoid this impact, identified as "thermal pollution", is not frequent in Italy, because of many more or less acceptable constraints, among which the landscape preservation is included. In a quite general consideration, the pollution due to the sources described above can be ascertained through an accurate evaluation of the water quality in the receiving body, with a particular attention to the inlet of the affecting stream. The location of such inlet can provide the information necessary for identifying the possibility of intervention for pollution control and the place suitable to host a designed treatment plant.

The challenge of restoring completely the original status of the Italian rivers and lakes does not depend only on the efficiency of the wastewater treatment. The most advanced research in this field, at the end of last century (Viganò et al. 1997), has confirmed that, even in the most refined processes, a small concentration of pollutants can remain, able to alter eventually the original environmental aspects. To achieve better results, a more severe control on the usable water is necessary, in order to find working conditions able to reduce at the beginning the amount of polluting substances.

9.4   Actual Situation

The Italian monitoring structure is in a position to present a satisfactory picture of the actual water resources situation. With the exception of few restricted cases due to local organizing difficulties, the monitoring efficiency has noticeably grown during the last decades, also in comparison with similar structures existing in other European countries.

As concerns the Italian rivers, an evaluation of their chemical water quality, relevant to the 2009–2013 period, puts into evidence the consistence of the monitoring structure in Italy (MPAAF 2015), as shown in Table 9.5. The large number of measuring stations is grouped according to the simple way of classification described in above paragraphs. Putting together the number of stations classified as GOOD and NO GOOD, every region shows an acceptable level of monitoring potential. Similarly, the total length of the reaches allows estimating how much the measured value can signify the real river situation.

The available data give also a significant estimate of the overall status of the Italian rivers, as in Fig. 9.1, in which, for every region, the identifiable class is shown as percent of the relevant values. Clearly, an acceptable quality is predominant, with limited worrisome situations.

The northern regions, particularly Piedmont and Lombardy, show the highest number of measuring stations, which confirms the great stress impounding on their

Table 9.5 Chemical quality of Italian rivers

Region	Observation period	Number of measuring stations		Length of river reach (km)	
		GOOD	NO-OOD	GOOD	NO-OOD
Piedmont	2009–2011	210	25	3904	482
Aosta Valley	2010–2013	81	0	412	0
Lombardy	2010–2013	264	72	4477	968
Trentino-Alto Adige	2010–2013	114	4	757	37
Veneto	2010–2012	328	14	3912	172
Friuli-Venezia Giulia	2010–2012	24	0	224	0
Liguria	2009–2011	25	11	125	38
Emilia-Romagna	2010–2012	159	19	2099	289
Tuscany	2010–2012	103	47	2146	1752
Lazio	2011–2013	104	40	1372	619
Marche	2010–2012	80	7	1522	133
Umbria	2008–2012	34	0	697	0
Abruzzo	2010–2012	57	3	844	42
Molise	2011–2013	9	0	160	0
Campania	2012	88	4	1228	86
Apulia	2010–2013	31	6	1498	202
Sicily	2011–2013	32	1	462	9
Sardinia	2010–2012	62	30	1157	356
Italy		1805	283	26,998	5184

territory, due principally to the high population density but also to the concentration of industrial factories as well as to an advanced agricultural activity that requests a large amount of chemicals.

Similar considerations can be drawn for the lakes (Buraschi et al. 2005), as shown in Fig. 9.2.

Some regions can in fact rely on efficient treatment plants, able to control the polluted discharge of urban and industrial wastewater, while traditional agricultural practices enhance a rational use of chemicals.

Concerning groundwater (Fig. 9.3), the chemical status still denotes the existence of local contaminations, mainly due to the seepage of polluting discharge, taking also into account the difficulty of reclaiming the polluted aquifers.

Similar analyses have been conducted for the ecological quality indicators of rivers, as shown in Fig. 9.4, using some classification levels in view of different possibilities of water utilization.

The biological indicators emphasize the predominance of acceptable classes, even though at different level of pollutant concentration.

More detail is in Table 9.6, which confirms that also the ecological status prevails in the northern regions, with the existence of several cases of "high" classes. Similar considerations can be drawn for the high number of significant river stretches.

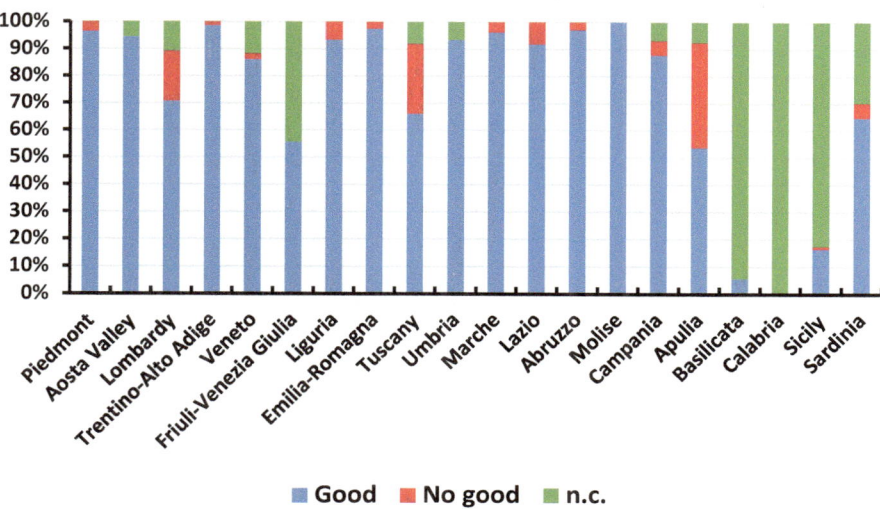

Fig. 9.1 Percent values of the chemical status of rivers in Italian regions. (*n.c.* not classified)

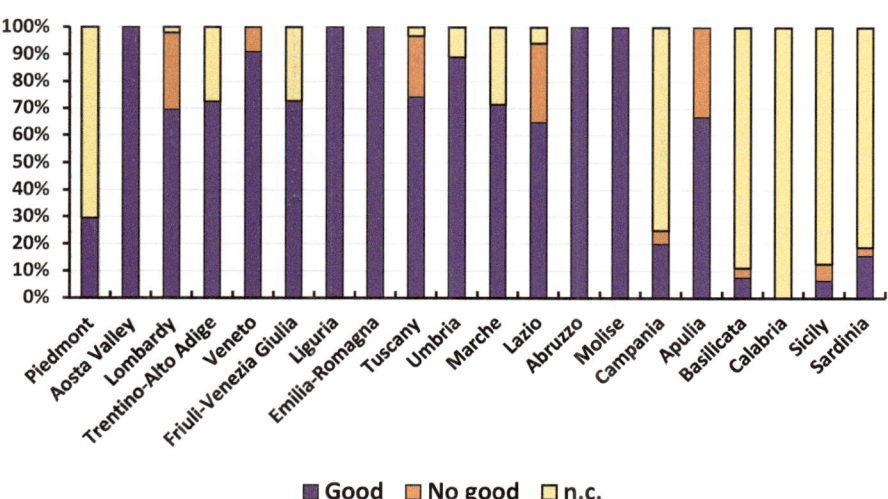

Fig. 9.2 Percent values of the chemical status of Italian lakes

Less significance characterizes the ecological status of the lakes (Fig. 9.5), due in particular to the high number of unclassified situations, but also in this case the acceptable classes are predominant.

Finally, also the ecological status of groundwater can be acceptable (Fig. 9.6), in spite of the small number of classes considered.

The above considerations can give an overall view on the water quality in the country, focusing the existing differences among the various regions.

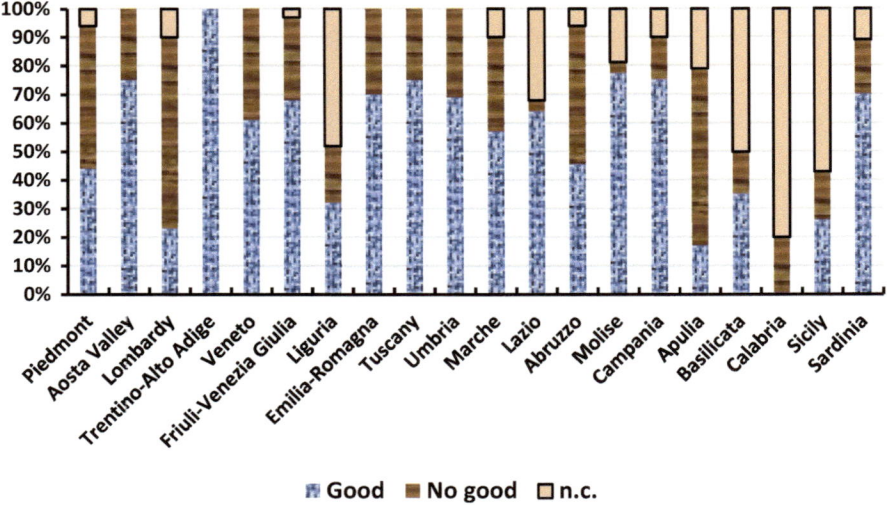

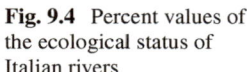

Fig. 9.3 Percent values of the chemical status of groundwater in Italian regions

Fig. 9.4 Percent values of
the ecological status of
Italian rivers

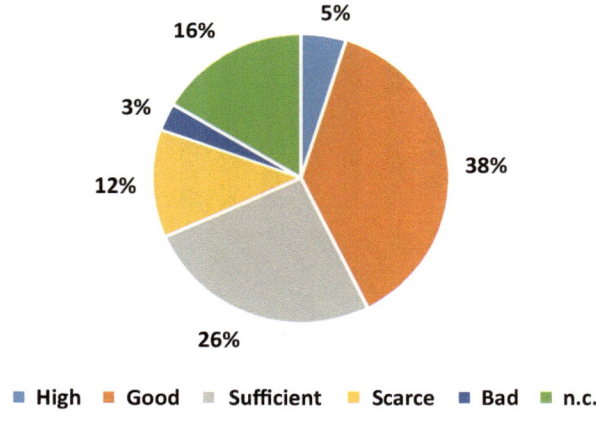

Table 9.6 Ecological quality of rivers

Region	Observ. period	Number of measuring stations					Length of river reach (km)				
		High	Good	Suff.	Scarce	Bad	High	Good	Suff.	Scarce	Bad
Piedmont	2009–2011	45	160	28	5	1	672	3033	598	96	17
Aosta Valley	2013–2013	32	160	9	23	1	178	211	23	1	0
Lombardy	2009–2011	7	76	119	82	19	64	1157	2031	1355	386
Trentino-Alto Adige	2010–2013	40	103	22	12	0	392	847	143	65	0
Veneto	2010–2012	88	55	107	38	9	527	773	1688	643	138
Friuli-Venezia Giulia	2010–2012	37	137	111	34	19	238	785	777	247	115
Liguria	2009–2011	1	31	21	7	0	4	155	79	21	0
Emilia-Romagna	2010–2012	0	40	69	55	14	0	417	984	746	241
Tuscany	2010–2012	15	68	63	52	18	269	1192	1427	1465	534
Lazio	2011–2013	8	33	39	36	23	53	464	631	574	268
Marche	2010–2012	0	38	28	19	2	0	728	564	355	8
Umbria	2008–2012	3	16	31	4	3	53	296	655	68	43
Abruzzo	2010–2012	1	33	41	26	10	6	475	616	460	135
Molise	2011–2013	0	4	5	0	0	0	67	94	0	0
Apulia	2010–2012	0	4	9	19	5	0	196	450	806	249
Sicily	2011–2013	0	7	5	5	1	0	85	58	87	7
Sardinia	2012–2012	5	20	29	13	2	123	577	834	298	20
Italy		282	985	736	430	127	2578	11,457	11,652	7286	2161

Fig. 9.5 Percent values of
the ecological status of
Italian lakes

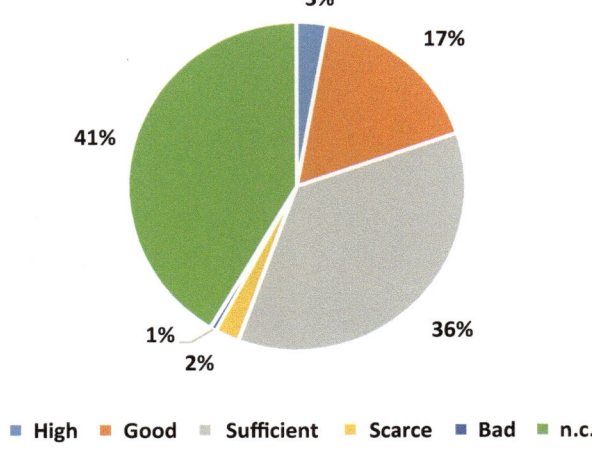

Fig. 9.6 Percent values of
the ecological status of
Italian groundwater

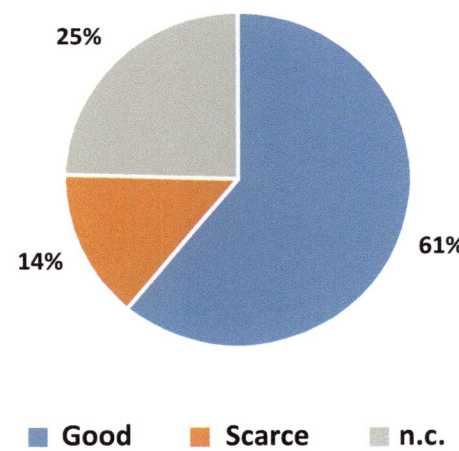

9.5 Remarkable Cases

9.5.1 *Monitoring the River Po*

The ecological preservation of River Po catchment has been always one of the main goals of the Italian environment policy. The importance of this catchment in the national economic development is known, not only because it covers about a quarter of the continental part of the country with a high number of population, but also for the high concentration of productive activities, in particular industry and agriculture.

At the end of the last century, the Po River Authority activated an intensive and extensive campaign of surface water monitoring, with the adoption of the most

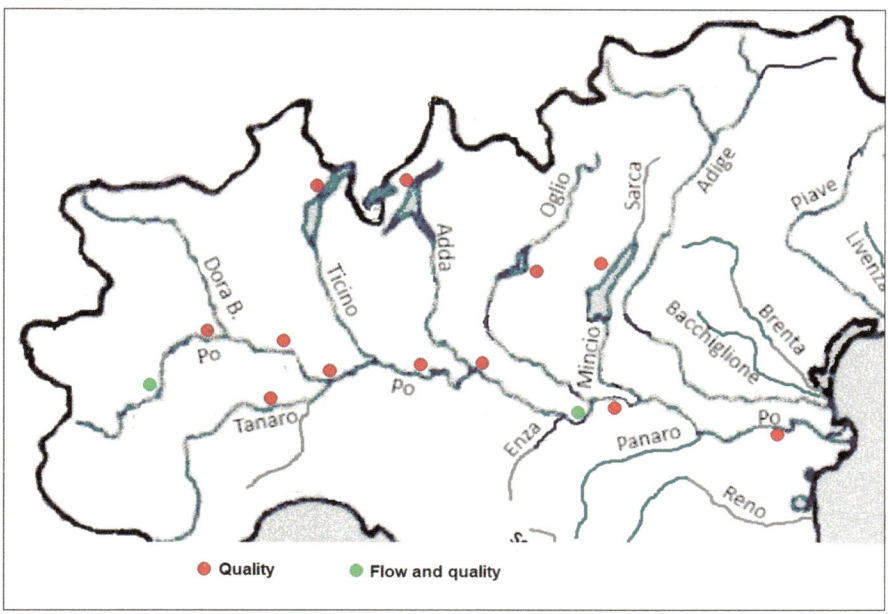

Fig. 9.7 Quality monitoring stations in the catchment of River Po

qualified methodologies. The campaign was one of the principal activities of the authority, in application of the Law 183/1989. The Water Research Institute played an important support with the participation of the Environment Agency of the United Kingdom and in cooperation with some Italian universities.

At that time, numerous gauging stations were in operation for the most significant ecological indicators, as shown in Fig. 9.7. These stations operated in accordance with the hydrological monitory system of the hydrographic office of River Po, which provided the necessary information relevant to the flow and level in the water bodies, particularly in relation to the morphological conditions related to the conspicuous sediment transport (Viganò et al. 2003).

In the River Po catchment, the interaction of the urban areas with the surface and underground water bodies is particularly perceptible where an extended conurbation disposes its waste. Typical is the case of Milan and the surrounding municipalities that discharge in the River Lambro, one of the main tributaries. The most precarious points for the concern of the River Authority have been the confluence of tributaries with the main river, in the downstream reaches of which several quality indicators undergo critical alteration.

The catchment of River Po contains the major lakes of the Alpine chain, which receive a considerable discharge of wastewater coming from urban settlements, agricultural and industrial activities, included the production of thermal power (Tartari et al. 2002; Pisanello et al. 2015). Thermal power plants are located also on the embankment of the main river, discharging the hot water of the cooling process.

River Po is therefore the main witness of the vital aspects relevant to this extended area, and the quality of its water reflects some precarious situations to be taken in due consideration, today and in the future.

The monitoring of the water quality has put into evidence the peculiarity of most indicators. Starting from conductivity, oxygen concentration and including nitrogen, phosphorous, organic carbon and trace metals, the analyses have always confirmed the presence of harmful impacts of man's activities. Moreover, several pesticides, in particular insecticides and herbicides, have been found downstream, till the closure of the river basin.

Several tests have put into evidence the high trophic potential of contaminated reaches, with a trend for toxicity. Even though sometimes low or not significant, the indicators characterized an increase of toxic levels, mostly reliable to the agricultural use of pesticides. Mutagenicity cases were found in fish exposed to the contaminants in the main river, confirming a genotoxic risk.

The computation of reference flow rate allowed defining the ecological stream, which became one of the main tools in the hands of the River Authority for water quality control. An overall picture of the situation in the basin was eventually drawn, able to emphasize the environmental aspect during the monitory campaign. In Fig. 9.8, the chemical and ecological status is defined adopting the classification schemes proposed by the Water Research Institute, applied to the measures conducted at the existing gauging stations.

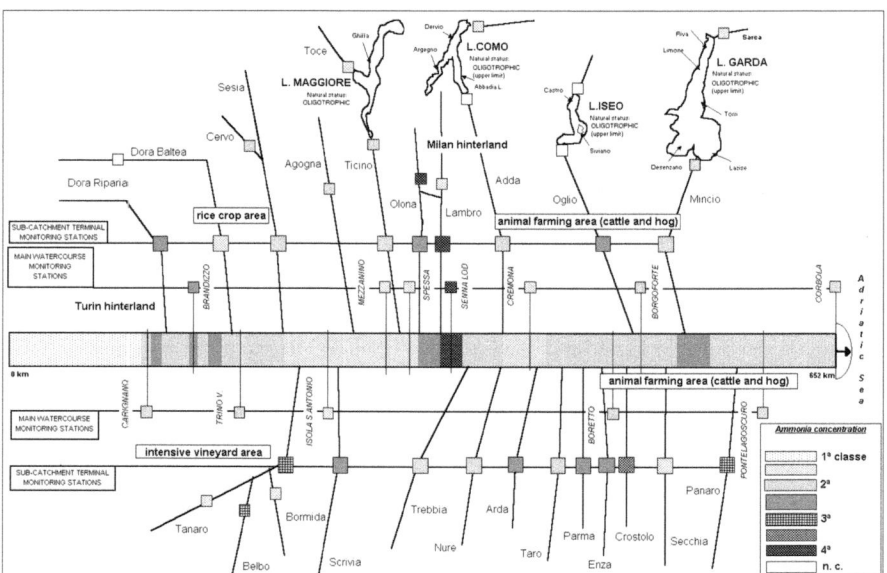

Fig. 9.8 Classification of the River Po and its main tributaries according to the concentration of ammonia. Location of monitoring stations is in Fig. 9.7

In the subsequent years, the attention of the Po River Authority was concentrated on particular situations in restricted places, concerning not only the quality aspect but also focusing the main source and the behaviour of the detected pollutants (Balestrini et al. 2016). The Po Hydrographic District, which has now inherited and enlarged the role of the River Authority, is now promoting an intense activity for the best evaluation of the water quality in the basin.

The monitoring campaign on the River Po catchment has been an opportunity to apply and test the methods and the tools suitable to application in several river catchments all over the country.

9.5.2   The River Tiber in the Urban Stretches of Rome

The water quality problems in the catchment of River Tiber are conditioned principally by the presence of Rome, an urban settlement with more than three million inhabitants, but also by the various utilizations of water, upstream and downstream, which affect the hydrological regime and sometimes amplify and worsen an already precarious situation. Mention must be made also to the sediment transport in the river, largely responsible of the pollutant behaviour. The exploitation of surface water and groundwater resources has had always a negative impact on water quality in the entire basin, where a massive use of fertilizers in agriculture, along with the municipal and industrial discharge, has increased the environmental degradation.

Since many decades, the water quality in the river has been the object of intensive control (Ubertini et al. 1996; Casadei et al. 2018). In particular, the reaches in the urban area of Rome have called the attention of the local authorities. With the support of the scientific community, a suitable set of gauging stations is now in operation, as shown in Fig. 9.9.

The monitoring activity has confirmed the precariousness of the situation, expressed in terms of the most significant indicators, which denote a clear trend of worsening. The content of ammonia, phosphorous and coliforms, already high due to uncontrolled discharges upstream of the urban stretches, remains very high and noticeably increases after the confluence of Aniene, the main tributary in the urban area, which upstream receives several untreated discharges. Worrisome value have also the principal biotic indexes. A significant concentration of heavy metals and organic compounds is present also today in the lower reaches and in some small tributaries. The status of the lakes is generally acceptable, as well as that of groundwater, with an exception of some zones of volcanic origin, where a natural concentration of arsenic has given rise to problems using that water for drinking purposes (Parrone et al. 2013; Preziosi et al. 2004, 2010).

This precarious situation threatens not only the environmental aspects of a large part of Central Italy but is also a serious and unacceptable damage for the city of Rome, worldwide known for its role in history, civilization and art. In order to improve the overall quality conditions, an adequate system of wastewater treatment plants is now in operation. Nevertheless, being of conventional biologic type, these

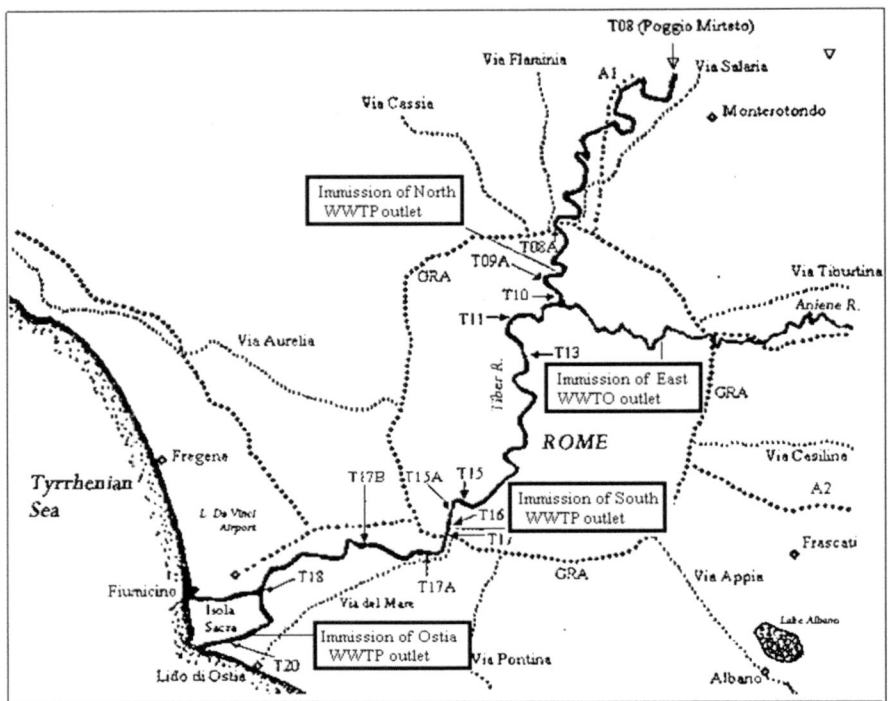

Fig. 9.9 Quality monitoring in the River Tiber around the urban area of Rome. (*WWTP* Wastewater treatment plant) (Source: ABT 2000)

plants have a beneficial effect only on the abatement of organic matter, as proved by the reduced amount of BOD_5 in the river, downstream from the inlet of the treated wastewater. Measures that are more efficient have been therefore necessary, not only relevant to the wastewater treatment but also aiming at the improvement of all the vital aspects in the interested communities (Patrolecco et al. 2015).

In application of Law 183/1989, at the end of twentieth century, the Tiber River Authority (ABT 2000) proposed a river basin masterplan, which contained also the outline for the activity of pollution abatement. The in-field evaluation of the water quality is mandatory of the Regional Environmental Protection Agencies, which operate in the territory that belongs to the river catchment.

After the merging of the Tiber catchment in the larger Central Apennines District (Cesari and Pelillo 2010), following the application of European Union Directive 42/2001, the water quality problems of the Tiber are dealt in a joint activity with those of the regions. One of the main actions in progress includes the planning for water resources protection.

9.5.3   Venice and Venetian Lagoon

Venice is one of the most renowned factors of the Italian landscape and historical patrimony, calling a worldwide attention for its preservation. The Venetian Lagoon is a basin closed by a narrow strip of land, in which some openings ("mouths") assure the connection with the open Adriatic Sea. The basin receives the water drained in a vast land (Fig. 9.10), through some natural and artificial channels, and the incoming water, combined with that entering from sea during high tide, is responsible of internal currents that affect the entire basin.

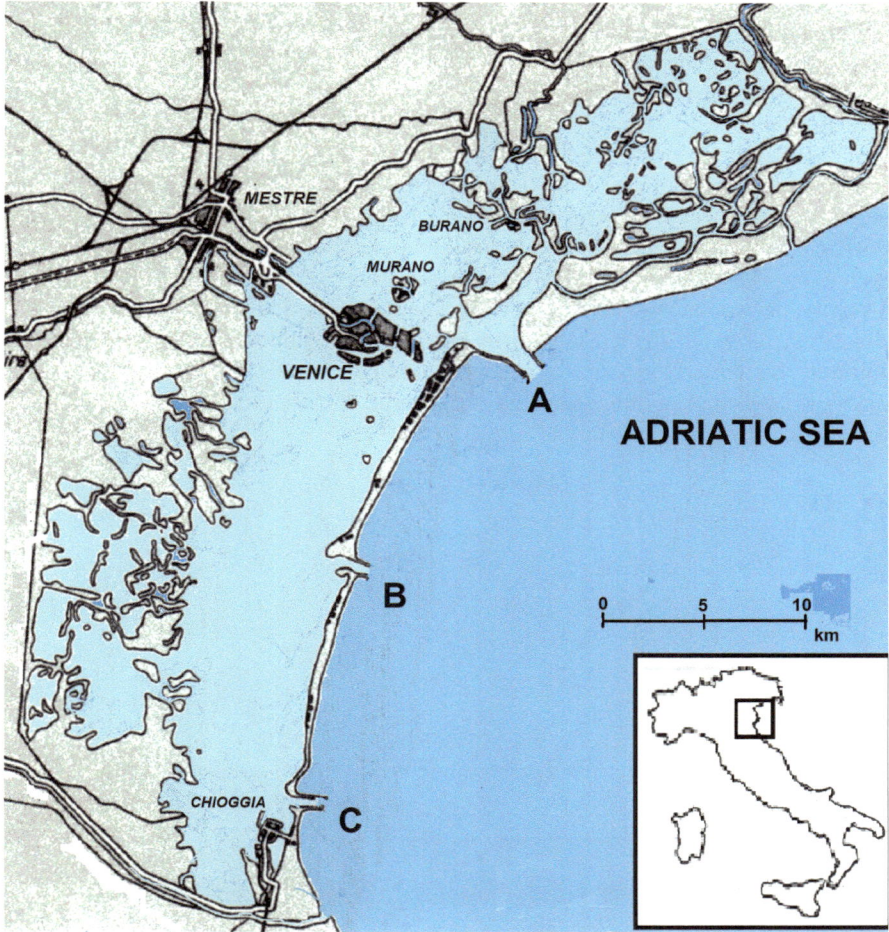

Fig. 9.10 The Venetian Lagoon and the "mouths" connecting to the open sea. (A = Lido; B = Malamocco; C = Chioggia), and part of the draining area (ISPRA 2015)

Until few years ago, the basin received entirely the wastewater coming from land and urban agglomeration of Venice. Industrial factories located at the border of the lagoon discharge predominantly chemical pollutants, while some thermal generation plants discharge hot cooling water. A large polluting load threatens the city and its lagoon, the presence of which may be attenuated if some polluted stream can reach the open sea through the "mouths" during normal conditions of tide.

Unfortunately, in autumn and early winter, the Venetian Lagoon undergoes the effect of the extreme high tide events that characterize the Northern Adriatic Sea, and the water in the basin rises to undesirable levels able to flood the city and its surrounding places. The "high water" occurrence has remarkable effects on the water quality in the basin. The need to protect the area and restoring the vital conditions have suggested to sever temporarily all the communications with the open sea, in order to avoid the high tide invasion (Ravera 2000). To achieve such situation, huge movable gates have been installed at the mouths, the "Mose system" (Fig. 9.11). However, such interception, even though for a short time, increases the pollutant concentration inside the basin. The entrance of polluted groundwater, although not easily measurable, may add a further contamination (Sharma et al. 2016).

An interpretation of the quality aspects in the lagoon has requested several terms, relevant to the specific pollution load and to the alteration of the pollutants in water. Besides the usual chemical and biological indicators, special attention has called the evaluation of algal content. Fish fauna is also an important component for the biodiversity, in relation with the temporal and spatial dynamics of the complex ecosystem (Franzoi et al. 2010; Mainardi et al. 2004). Reliable data are available for nutrients, particularly in terms of nitrogen and phosphorous, and for BOD. As concerns the land contribution, evaluated at the outlet of the various channels discharg-

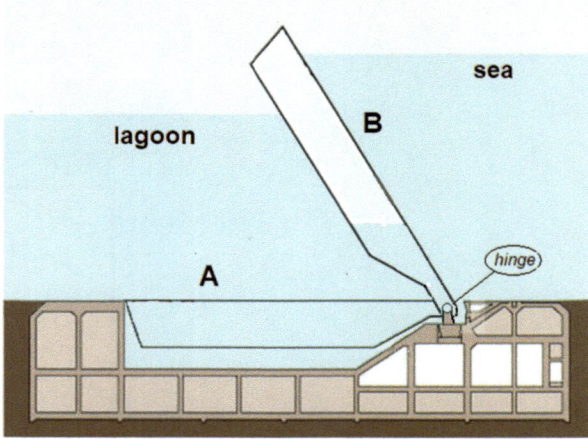

Fig. 9.11 The working principle of the "Mose system" gates in a "mouth": during the normal tide condition the gate, full of water, is sunk (**A**); in case of extreme tide events, inflated with air, the gate rises (**B**), preventing the entrance of high water in the lagoon

ing into the lagoon, nutrients are present mostly in autumn and winter, as consequence of intensive use of fertilizers. The highest BOD values are observed in May and June (Malavasi et al. 2004).

To improve the water quality status of the lagoon, an advanced system of treatment plants has been recently put into operation, able to treat a daily amount of wastewater estimated to be 155,000 m^3, with 400,000 equivalent inhabitants. About 50,000 m^3 of the treated water are used for cooling and productive process in the industrial factories, while the rest is conveyed into open sea through a 20-km-long pipeline.

The case of Venice and Venetian Lagoon can be considered a valid example of fruitful application of the most advanced methodologies for tackling the complex and multidisciplinary water problems. The pollution transport in the lagoon has been simulated by means of advection-dispersion models, which, coupling the hydrodynamic and the ecological aspects, can provide a very large set of information (Franco et al. 2009), suitable to understand the complex behaviour of the lagoon.

9.5.4   Groundwater Pollution in Apulia

Apulia (Southern Italy) is characterized by low availability of surface water, due principally to a scarce precipitation and to the karstic nature that enhances the immediate seepage of the rainwater fallen on the ground (Polemio and Casarano 2008). Very important is therefore the consistence of groundwater, while karstic and fractured aquifers are the most important water resource, located also in large coastal areas, where they are subjected to seawater intrusion (Cotecchia and Polemio 1998).

Among the numerous subjects concerning the behaviour of groundwater quality, the migration of pathogens has motivated some advanced investigations, taking in due account the degradation of the pollutants in the subsoil, along with their transport through the fissures of the carbonatic rocks. This is an important aspect, able to identify the distance from the injection point, at which the pollutant concentration can decrease to values acceptable for a safe use of the water (Masciopinto et al. 2008; Passarella and Caputo 2006).

To meet the demand for the various uses of water, especially agriculture, a massive withdrawal is in progress, by means of thousands of wells, with a remarkable lowering of the natural water table. Consequently, the amount of usable water is reduced and the seawater intrusion increases.

At the same time, various polluting sources give rise to a qualitative degradation of the aquifers that are greatly vulnerable (Polemio et al. 2006). This further decreases the availability of usable groundwater, especially for drinking purposes, which are now almost entirely satisfied with the large amount imported from nearby regions by the Apulian Aqueduct.

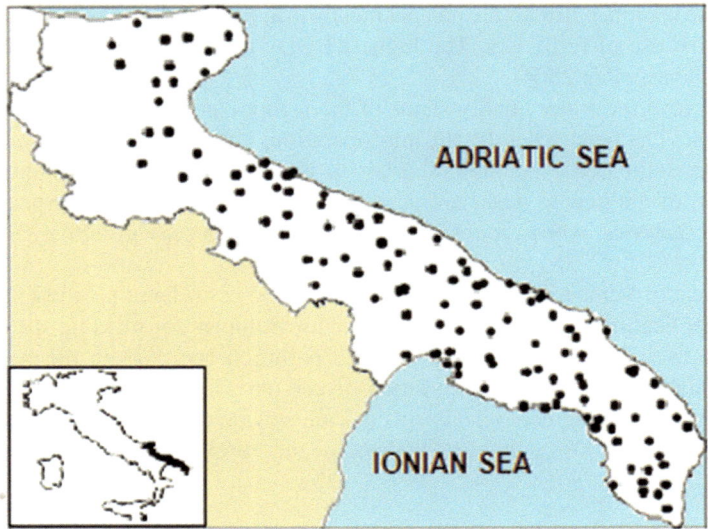

Fig. 9.12 Location of the main stations for monitoring the groundwater quality. (Source: Cotecchia and Polemio 1998)

Due to the extreme anisotropy and heterogeneity of the subsoil, understanding the behaviour of flow, solute transport and biological processes in these hydrogeology systems is a scientific challenge.

Numerous surveys, starting from 1995, both on coastal spring and on wells in the entire region (Fig. 9.12), have stressed the quality status of groundwater, in relation with the pollution sources and in view of possible ways of improving the actual and future situation.

The surveys have considered also the numerous submarine springs located around the coast, which can be also a way of identifying the peculiarities of the pollution source, located in the inner territories. Besides some untreated discharge from restricted urban areas, agriculture has been found the main polluter, due in particular to the use of fertilizers. Quite peculiar is the case of the industrial activity concerning the transformation of agricultural products, like olive mill, which often discharge the residual water directly into the natural chasms that characterize the carbonate territory. Detailed knowledge has been acquired about the behaviour and the pathways of pollutants, essential for defining the suitable measures for the protection and management of groundwater resources.

Among the several situations considered, the following case can be significant. During 2001, an area of 15 hectares, intensively utilized for corn productions, was subjected to illegal spreading of untreated sludge coming from local industrial factories. Consequently, the soil and an aquifer more than 400 metre deep were contaminated (Masciopinto and Caputo 2011). Toxic compounds like heavy metals (nickel, chrome, zinc and iron), arsenic, hydrocarbons, ammonium and nitrites were found at unacceptable concentration. The situation was examined by means of models able to simulate the wetting and non-wetting phases of pollution migration, according to the capillary pressure and global accessibility criteria in fractures with variable aperture. The model results confirmed the values recorded a few years later in monitoring wells positioned more than 10 km downstream from the injection point of the contaminants, providing also a clear identification of the pollutant pathways.

According to the enforced law, a large set of measured values has allowed the groundwater chemical status to be identified by means of the most significant pollutant. The ecological classification is in progress.

9.6   Conclusive Remarks

The preceding paragraphs put into evidence the peculiarity of the water quality aspects in Italy, in relation to the way the environmental problems are perceived and tackled. The development of economic and social conditions of the last century has increased the water demand and use and, consequently, the amount of wastewater has also increased. The principal commitment of the country is therefore for the actual status of the water bodies, in relation to their possibility of receiving a polluting discharge. This entails the enlarging of the water monitoring system, with the most reliable measurement tools and data processing. At the same time, an improved evaluation of the pollution sources is necessary, with their location in respect of the receiving natural body and in view of their foreseeable future development.

The existing structures allow already a satisfactory picture to be done, usable for an overall evaluation of the Italian situation, taking also in due count its role in the European and Mediterranean position. These considerations should be the starting point for any action of the responsible institutions.

A first consideration concerns the diversity existing among the Italian regions, in line with their diversity of social economic development. Northern regions, characterized by high social and economic development, denote high pollution level in the water bodies and, in turn, have fostered more efficient monitoring systems and measures for pollution abatement. Vice versa, some restricted areas still show a difficulty to overcome the precariousness of local financial resources that is partly responsible of unacceptable conditions.

The majority of rivers, lakes and aquifers, all over the country, host remarkable cases of acceptable classification, in terms of physical, chemical and ecological status. Cases of optimal classification are also present. This confirms the efficiency of the actual monitoring system and the existing treatment plants. Cases of scarce and bad classification can be found in restricted areas, downstream from the injection of uncontrolled polluting discharge, stressing the need of suitable intervention.

References

ABT (2000) Tiber River Authority La pianificazione del bacino del Fiume Tevere (The River Tiber Planning). Gangemi Ed., Rome, Dec. 2000. (In Italian)

Balestrini R, Sacchi E, Tidili D, Delconte CA, Buffagni A (2016) Factors affecting agricultural nitrogen removal in riparian strips: Example from groundwater-dependent ecosystems of the Po Valley (Northern Italy). Agric Ecosyst Environ 221:32–144

Barbiero G, Cicioni GB (2000) Report on pollutant loads generated in the pilot basin, Environmental Water quality Control Transnational Project, "EWAQC". 31 EWAQC-RAP-(2)-T009.0

Benedini M, Giulianelli M (2003) Urban water and climate change. In: Proceedings of International N.A.T.O. Research Workshop "Enhancing Urban Environment by Environmental Upgrading and Restoration". Rome, Italy, pp 311–320.

Benedini M, Passino R, Piacentini G (1998) The quality of receiving bodies in agriculture-dominated areas of Italy. In: EUEAQUA 5th Scientific and Technical Review, European Network of Freshwater Research Organisations, Oslo, Norway, pp 93–108

Bennett C, Owen R, Birk S, Buffagni A, Erba S, Mengin N, Murray-Bligh J, Ofenbock G, Pardo I, van de Bund W, Wagner F, Wasson J-G (2011) Bringing European river quality into line: an exercise to intercalibrate macro-invertebrate classification methods. Hydrobiologia 667:31–48. https://doi.org/10.1007/s10750-011-0635-2

Buffagni A, Erba S, Turse MT (2007) A simple procedure to harmonize class boundaries of assessment systems at the pan-European scale. Environ Sci Policy 10:709–724

Buraschi E, Salerno F, Monguzzi C, Barbiero G, Tartari G (2005) Characterization of the Italian lake-types and identification of their reference sites using anthropic pressure factors. J Limnol 64(1):78–84

Casadei S, Pierleoni A, Bellezza M (2018) Sustainability of Water withdrawal in the Tiber River Basin (Central Italy). Sustainability 2018:10485. https://doi.org/10.3390/su10020485

Cesari G, Pelillo R (2010) Central Apennines District: River Basin Management Plan of District (PGDAC) – problems and expectations. Paper presented at the European Water Association 6th Brussels conference, implementing the river basin management plans, Brussels/Rome, EWA/Tiber River Basin Authority

Cotecchia V, Polemio M (1998) The hydrogeological survey of Apulian groundwater (Southern Italy): salinization, pollution and over-abstraction. Hydrology in a changing environment, ISBN: 978-0-471-98660-7. pp 129–136. https://intranet.cnr.it/people/

Erba S, Buffagni A (2009) Ecological status definition: a focus on rivers and benthic invertebrates in relation to the EU Water Framework Directive (WFD, 2000/60/EC) requirement. CNR-IRSA, erba@irsa.cnr.it

Franco A, Torricelli P, Franzoi P (2009) A habitat-specific fish-based approach to assess the ecological status of Mediterranean coastal lagoons. Mar Pollut Bull 58:1704–1717

Franzoi P, Franco A, Torricelli P (2010) Fish assemblage diversity and dynamics in the Venice Lagoon. Rendiconti Fis Acc Lincei 21:269. https://doi.org/10.1007/s12210-010-0079-z

IRSA/CNR (2004) Metodi analitici per le acque. Manuali e linee guida 29/2003. (Water analytical methods) (in Italian). Agenzia per la protezione dell'ambiente e per i servizi tecnici. ISBN: 88-448-0083-7

ISPRA (2015) Annuario dei dati ambientali (Environmental data year-book) (in Italian). http://www.isprambiente.gov.it/files/pubblicazioni/statoambiente/annuario-2014-2015/9_Idrosfera.pdf

ISTAT (2017) Report-Censimento delle acque (Report-Water census) (in Italian). https://www.istat.it/files/2017/12

Mainardi D, Fiorin R, Franco A, Franzoi P, Granzotto A, Malavasi S, Pranovi F, Riccato F, Zucchetta M, Torricelli P (2004) Seasonal distribution of fish fauna in the Venice Lagoon shallow waters: preliminary results. In: Campostrini P (ed) Scientific research and safeguarding of Venice, Corila Research: Program 2002 results. Multigraf, Venezia, pp 437–447

Malavasi S, Fiorin R, Franco A, Franzoi P, Granzotto A, Riccato F, Mainardi D (2004) Fish assemblages of Venice lagoon shallow waters: an analysis based on species, families and functional guilds. J Mar Syst 51:19–31

Masciopinto C, Caputo MC (2011) Modeling unsaturated–saturated flow and nickel transport in fractured rocks. Vadose Zone J 10:1045–1057. https://doi.org/10.2136/vzj2010.0087

Masciopinto C, La Mantia R, Chrysikopoulos CV (2008) Fate and transport of pathogens in a fractured aquifer in the Salento area, Italy. Water Resour Res 44(1). https://doi.org/10.1029/2006WR005643

MPAAF (2015) Ministero Politiche Agricole Alimentari e Forestali. Rapporto Ambientale –novembre 2015 (Ministry for Agriculture, Food and Forestry. Environmental Report-November 2015) (in Italian). Ismeri Europa, p 316

Parrone D, Preziosi E, Ghergo S, Del Bon A (2013) Hydrogeochemical characterisation of volcanic-sedimentary aquifer in Central Italy. Rendiconti Online Soc Geol It 24:232–234

Passarella G, Caputo MC (2006) A methodology for space-time classification of groundwater quality. Environ Monit Assess 115:95–117. https://doi.org/10.1007/s10661-006-6547-3

Patrolecco L, Silvio C, Ademollo N (2015) Occurrence of selected pharmaceutical in the principal sewage treatment plants in Rome (Italy) and in the receiving surface waters. Environ Sci Pollut Res 22(8):5864–5876

Pisanello F, Marziali L, Rosignoli F, Poma G, Roscioli C, Pozzoni F, Guzzella L (2015) In situ bioavailability of DDT and Hg in sediments of the Toce River (Lake Maggiore basin, Northern Italy): accumulation in benthic invertebrates and passive samples. Environm Sci Pollut Res. https://doi.org/10.1007/s11356-015-5900-x

Polemio M, Casarano D (2008) Climate change, Drought and groundwater availability in southern Italy: Climate change and Groundwater. In: Dragoni W (ed) Geological Society, London, Special Publications, vol. 288, pp 39–51

Polemio M, Limoni PP, Mitolo D, Virga R (2006) Il degrado qualitativo delle acque sotterranee pugliesi (Quality degradation of Apulian groundwater) (in Italian). Giornale di Geologia Applicata 3:25–31. https://doi.org/10.1474/GGA.2006-03.0-03.0096

Preziosi E, Vivona R, Patera A, Barbiero G, Giuliano G, Petrangeli AB, De Luca A, Mari GM (2004) Integrating monitoring networks and groundwater vulnerability: a case study in Central Italy. In: International conference on groundwater vulnerability assessment and mapping, IAH – UNESCO – University of Silesia (PL), USTRON (PL), pp 111–113

Preziosi E, Giuliano G, Vivona R (2010) Natural background levels and threshold values derivation for naturally As, V and F rich groundwater bodies: a methodological case study in Central Italy. Environ Earth Sci 61:885–897

Ravera O (2000) The Lagoon of Venice: the result of both natural factors and human influence. J Limnol 59(1):19–30

Sharma S, Balhara A, Bedi M (2016) Experimental electromechanical module (MOSE) for flood control in Venice. A review. Int J Sci Eng Technol Res 5(1):31–33

Tartari G, Copetti D, Marchetto A (2002) Northern Italian lakes: regionalisation of limnological features and pressure factors relationships. Vehrhandlungen des Internationalen Verein Limnologie 28:223–227

Ubertini L, Manciola P, Casadei S, (1996) Evaluation of the minimum instream flow of the Tiber River Basin. Environ Monit Assess 41: 125–136 [CrossRef][PubMed].

Viganò L, Arillo A, Buffagni A, Camusso M, Ciannarella R, Crosa G, Falugi C, Galassi S, Guzzella L, Lopez A, Mingazzini M, Pagnotta R, Patrolecco L, Tartari G, Valsecchi S (2003) Quality assessment of bed sediments of the Po River (Italy). Water Res 37:501–518

Vigano' L, Benedini M, Badino G, Barbiero G, Buffagni A, Pagnotta R, Spaggiari R (1997) Quantity and quality aspects in the protection of the aquatic environment. In: Van de Kraats JA (ed) Let the fish speak: the quality of aquatic ecosystems as an indicator for sustainable water management, Fourth EURAQUA technical review, Koblentz, pp 149–171

Chapter 10
Ecological In-Stream Flows

Salvatore Alecci and Giuseppe Rossi

Abstract Since year 1989, an increasing environmental awareness introduced in Italian water legislation the concept of "minimum in-stream flow". The aim was to limit the abstraction of water from rivers for off-stream uses, in order to protect water quality and to preserve aquatic life in rivers. More recently, under the pressure of the European Water Framework Directive 2000/60/EC, more advanced rules have been issued to estimate the "ecological flow". After a review of the main methods developed to assess the minimum in-stream flow, the chapter analyses the evolution of the Italian legislation on this subject. Specifically, differences between rules issued to evaluate the minimum flow within a specific law objective, such as river basin planning, protection of water quality and regulation of diversion permit, are described. The methodological approaches provided by the guidelines issued by the Ministry for Environment, Land and Sea in the year 2017 are high-lighted. Further advances necessary to achieve a sustainable balance between needs of aquatic life and water abstraction for off-stream uses are also discussed.

10.1 Introduction

In the last decades, in the wake of increased environmental awareness, some principles have been identified in order to protect the river water quality and identify actions for rehabilitation and restoration of aquatic life.

These principles, which arose in the countries of Northern Europe and North America, aimed initially at protecting economic and social interests associated to the utilitarian human uses of river flow in water bodies (such as navigation, floatation, fishing, recreational activities), as well as at protecting fish species living in the river system. The principles of protection have been gradually extended to all the living

S. Alecci (✉)
Eastern Sicily Section, Italian Hydrotechnical Association, Catania, Italy

G. Rossi
Department of Civil Engineering and Architecture, University of Catania, Catania, Italy

© Springer Nature Switzerland AG 2020
G. Rossi, M. Benedini (eds.), *Water Resources of Italy*, World Water Resources 5,
https://doi.org/10.1007/978-3-030-36460-1_10

organisms, including minor and benthonic communities and plant species (ecosystems). In practice, these principles of protection have limited abstraction from watercourses, restricting the possibility to divert water only if abstraction does not damage the ecosystem. The minimum amount of water downstream of the section of any diversion that allows maintaining such a minimal safety condition has been traditionally called the *minimum in-stream flow* (MIF) or the *minimum acceptable flow* (MAF).

More recently it was understood that the protection of the community of living species (biocenoses) assumes greater value if it is considered in its complexity and variety (biodiversity). In fact, the natural balance between species is a useful condition for the conservation of each species. Finally, the recognition of the chemical-physical conditions of the water and the soil, for the survival of the biocenoses, has extended the protection of the water body to the riverbed, the banks and the areas in which water is occasionally present. To better represent this evolution, the term *ecological flow* (EF), in place of the *minimum in-stream flow*, has been established.

The protection has been extended to include ephemeral watercourses that present relevant flow variability during the year, with frequent very low values of the discharge, also with dry riverbed occurrences.

In this chapter, the analysis is limited to the criteria and methods used in Italy in order to estimate the minimum flow to be assured inside the watercourse by imposing constraints to water abstraction for off-stream uses. After a brief review of the criteria and procedures for the estimation of MIF (Sect. 10.2), the evolution of the concept of MIF in Italian legislation is illustrated (Sect. 10.3). In particular the conceptual bases and the rules established by law are discussed, referring to river basin planning and water quality protection, as well as to the revision of the diversion permits. Section 10.4 describes the introduction of the concept of ecological flow in the Italian legislation, in line with the European water rules. Finally, Section 10.5 provides conclusive remarks concerning specific aspects relevant to the applications of ecological flow in Italy.

10.2   Methods to Estimate Minimum In-Stream Flow (MIF)

Following the available scientific literature, the estimate of the minimum in-stream flow requirements in Italy has been conducted according to:

- *Empirical or regional methods*
- *Biological methods*, combining field biological results and hydraulic aspects linking aquatic habitats with discharge pattern
- *Hydrological methods*, based on processing of the flow records in the river, implicitly taking into account the biological conditions sustained by the instream flow regime.

10.2.1 Empirical or Regional Methods

These methods are based on the relationship between the in-stream flow required for the considered species and a number of geomorphological parameters (e.g. the area of the drainage surface) or hydrological characteristics (e.g. the average or median of monthly or yearly flow). The validity of the methods may be limited to the region where the field investigations have been carried out or to the specific living species considered.

One of the most common methods is that developed by Tennant (1976), for protection of Salmonidae. The oldest method of Baxter (1961) evaluates the minimum flow for salmonid preservation as a percent of the monthly average flow, while the method of Geer (1980) aims at the protection of trout. The method of Larsen (1981) estimates the in-stream flow for each month of the year based on the median value in a 25-year-long series of daily flow records. Renoldi et al. (1995) report a detailed review of empirical methods available from the literature.

In Italy this problem has been tackled during the 1970s at the Water Research Institute (Beccari et al. 1971; Marchetti et al. 1973) in view of the increasing pollution level in many rivers, particularly in River Po and its tributaries. Experimental investigations were carried out in laboratory and in the field, particularly concerning the trout as the most significant species. The High Council of Public Works and some regional administrations were involved for identifying suitable methods of practical intervention.

Actually the empirical method is still applied in some regions. For example, the percentages of the average annual flow rate (with implicit reference to the Tennant method) have been adopted by the Authority of Upper River Sele (2011) and, with the inclusion of corrective coefficients, by the Authority of Upper Adriatic for River Piave (1998), as well as by the Piedmont Region (2007). For the Authority of River Magra (1998), the estimates refer to the mean monthly flows of July, August and September, while for the Province of Trento, they refer to the minimum observed flow (1986).

10.2.2 Biological Methods

These methods require biological analyses (often long and expensive), for both the development and application phases, thus limiting the possibility of a generalized use of the results. The methods investigate the habitat conditions of one or more species living in the watercourse, chosen as the most representative, by measuring some significant parameters (velocity or depth of the stream, width of the cross section, wetted perimeter, temperature and water quality characteristics, riverbed substratum, sheltered areas for reproduction, etc.). The in-stream flow has been determined as the flow rate acceptable to maintain good habitat conditions.

Among the biological methods, the "In-stream Flow Incremental Methodology (IFIM)" (also known as "Weighted Usable Area (WUA)" or "method of habitats") has been widely applied, using the software package known as "Physical Habitat Simulation System" or "PHABSIM" (Bovee and Cochnauer 1977; Bovee 1982; Bovee et al. 1998).

The IFIM method was applied by the greatest Italian hydroelectric producer (ENEL spa) in many streams of Alps (Saccardo 1998; Maran et al. 2000) and by the Authority of River Arno Basin (Menduni et al. 2006); the results were compared with the duration curves. The application confirmed that the flow values corresponding to the weighted usable area are remarkably different since they depend on the fish species. The authority found that the value of discharge estimated for the species of interest was correlated with the discharge $Q(7d, 2yr)$ (the streamflow with duration 7 days and return time of 2 years). Different values result if they derive from in-depth knowledge of the watercourse or from experimental biological investigations.

10.2.3 Hydrological Methods

Hydrological methods are based on the idea to deduce the MIF from historical records of flow rate in the river. The in-stream flow requirements is estimated as a function of a flow rate series observed in the stream, considered representative of the conditions of low flow, according to one of the following statistics:

- Average flow rate
- Median flow rate
- Minimum value of mean daily flow rate
- Minimum flow rate in m consecutive days, with return time Tr: $Q(m, Tr)$
- Flow rate of duration d days (not necessarily consecutive) and return time Tr: $Q_d(Tr)$

For ungauged streams a "unitary regional contribution" has been adopted, usually expressed in $l/s\ km^2$, estimated on flow rate series, which present hydrologic homogeneity with the stream of interest.

Among the flows representative of the conditions of low flow, the $Q(m,Tr)$ can be considered the most significant, because it takes into account the number m of consecutive days, in which the flow persists in a low condition for a return time Tr. By accepting a small error, it has been approximated to the flow of duration $d = 365\text{-}m$ days and return time Tr: $Q_d(Tr) = Q_{365\text{-}m}(Tr)$. However, it should be noted that uncertainty exists on the choice of the two parameters of duration and frequency. Among the most frequently accepted values, literature reports those that consider the duration of 7 consecutive days and the return time of 10 years $Q(7d, 10yr)$ proposed in the method of Chiang and Johnson (1976). The choice of parameter values to estimate the MIF is justified by the assumption that the ecosystem could easily overcome the natural low flow of 7 days without severe impacts.

Several hydrological models have been developed by basin and district authorities, as well as by regions and autonomous provinces (Sect. 10.3.4).

10.3 Evolution of the Italian Legislation on the Minimum In-Stream Flow

10.3.1 Minimum In-Stream Flow in River Basin Planning (Law 183/1989 and DPCM 4/3/1996)

In Italy, the first law explicitly aiming at protecting the flow in watercourses involves the concept of MIF, named as "constant minimum vital flow". It has been introduced by the Law 183/1989, which regulates, in unitary way, land conservation, water quality protection and water resources management. In particular, Art. 3 of the Law provides that the activities of planning, programming and implementation of measures should achieve a rational use of water resources, thus assuring that the total amount of river abstractions do not reduce the in-stream flow below a threshold needed for river life. However, the application of this concept has been very slow. The Italian legislator and the ministerial offices that had the responsibility to draft the decrees for its application did not provide a timely implementation.

The technical addresses issued by the state for the implementation of Law 183/1989 (entrusted to the River Basin Authorities and regions) contain only a brief mention of the constant minimum in-stream flow, but gradually more and more significant. In particular, the act of address for forecasting and programming schemes (Decree of the President of the Council of Ministers, DPCM 23/3/1990) refers only to the need of protecting natural habitats in areas of particular value, by reducing withdrawals and maintaining natural in-stream flows. The technical rules on the preliminary investigations supporting the drafting of the river basin plan (Legislative Decree 7/1/1992) request only an overall assessment of withdrawals in the watercourse. Finally, the technical rules for drafting the basin plans (Presidential Decree 18/7/1995), although did not mention explicitly the *constant minimum in-stream flow,* require that:

1. The survey of current uses should include "naturalistic and environmental uses (parks, protected areas, fishing and bathing areas and areas of landscape and monumental interest)".
2. The identification of the imbalances should take into account the ecosystems alterations of flora and fauna.
3. The objective of "optimization of the various forms of water use" should include "both the withdrawal for off-stream uses (drinking, agricultural, industrial, hydropower)" and in-stream uses (inland navigation, maintenance of naturalistic, aesthetic and cultural sites).

Within the rules about water resources planning, the minimum in-stream flow is explicitly cited by the DPCM 4/3/1996 that implements Law 36/1994 on urban water services. It prescribes that the planned interventions must safeguard "the *minimum in-stream flow* and other downstream uses".

10.3.2   Minimum In-Stream Flow in the Laws on Water Quality Protection (DLgs 152/1999 e DLgs 152/2006)

The MIF has been considered by the Laws that deal with the quality of water bodies, since the Legislative Decree 152/1999, that states among its objectives the "maintaining ... the ability" of water bodies "to support large and well diversified animal and plant communities". It requires also that the "Plan for Water Protection", provided by the regions, should include the measures to ensure the water balance of the basin and the minimum in-stream flow.

It is also envisaged that the Minister of Public Works should define "the guidelines for the preparation of the water balance of the basin, including the criteria for ... the definition of the minimum in-stream flow". Later, the responsibility was transferred to the Ministry for Environment, Land and Sea Protection, which issued these guidelines after more than 5 years in 2004. These guidelines will be discussed in Section 10.3.5., while the new edition of the guidelines, issued in 2017, will be discussed in Section 10.4.2.

10.3.3   Minimum In-Stream in the Water Rights Rules for Diversion Permits (Updating of RD 1775/1933)

As already anticipated in Chapter 3, the regulations of the right for water surface diversion permits were issued with the Royal Decree 1775/1933, which coordinated several previous laws. The original text did not mention the minimum in-stream flow, but some successive acts have introduced specific rules for it.

Before changing the rules of rights for diversion permit in the whole country, a special law of 1990, referring to the reconstruction of Valtellina, in Central Alps, hit by a landslide in July 1987, required to modify the diversion permits for hydroelectric use in order to assure the minimum flow as introduced by Law 183/1989.

Starting from 1993, the Royal Decree 1775/1933 underwent some changes aimed at implementing the minimum in-stream flow. Legislative Decree 275/1993 prescribed that the new diversion permits to derive and use surface waters have to guarantee "the constant minimum in-stream flow in watercourses if previously identified in their basin". This rule was eventually valid throughout the national territory but concerned only the new derivations, while for the renewal of existing permits, it was limited to diversions for irrigation use.

Subsequently, other cases of application of the MIF to the diversion permits were introduced for abstraction in basins characterized by heavy withdrawals or transfers (both downstream and to another basin), and in protected areas, as well as to large diversions for hydroelectric use. The application of the last case was postponed after the issuance of general criteria by the Ministry for Environment. It was provided only in 2004.

Other important innovations have been introduced by Legislative Decree 152/1999. They include the consultation of the River Basin Authority on the compatibility of a new derivation with the Plan for Water Protection and with the water balance. They prescribe that the issue of the diversion permit is constrained by the achievement of the quality objectives of the watercourse and by the guarantee of the minimum in-stream flow. It is required also to check the possibility of reuse of treated wastewater or of collection of rainwater. Following these new rules, the maintenance of the minimum in-stream flow becomes a necessary condition for the maintenance of diversion permits, with the duty that quality and quantity of water to be returned into water body have to be cited in the contract of diversion permit.

10.3.4   Technical Procedures to Determine the Minimum In-Stream Flow and the Discharge Downstream from Intakes

The determination of the minimum in-stream flow was required in a first stage within the basin plans (under the jurisdiction of the River Basin Authorities) and within the Plans for Water Protection (issued by the regions and autonomous provinces). Until 2004, the lack of national technical rules has generated the proliferation of several criteria for the determination of the minimum flow, also very different from each other, even for neighbouring and similar watercourses.

The first local rule, preceding Law 183/1989, was issued in 1982 by the Province of Bolzano, in a very simplified way, requiring that downstream from a hydroelectric derivation the residual minimum flow, to be assessed on a case-by-case, must be not less than 2.0 l/s for any square kilometre of the reference catchment.

Among the numerous technical rules adopted by the local authorities, in a first stage, the preference was given to simple criteria referring to a unitary flow contribution, established for each territory or region considered, or to a characteristic streamflow value, e.g. the minimum observed flow rate, or the average annual flow rate, or a minimum flow of given duration. Several River Basin Authorities have proposed to correct the unitary flow contribution by means of coefficients depending from hydro-meteorological features (precipitation and/or runoff) and environmental needs.

Other Authorities have referred to flow duration curves, introducing also some return periods. More recently, a few rules require a time pattern, correlated to the time variability of the upstream flow rate of the diversion. In other cases, the minimum flow refers to the low flow in m day and Tr return period or to a flow volume corresponding to a fixed percentage of annual streamflow volume. Besides, some local rules do not define the MIF to be "guaranteed" in the watercourse but the "flow to be released downstream of the derivation". The release pattern can have different effects on MIF (Alecci 1998).

The Italian situation is summarized in Table 10.1, referring to the principal applications of the criteria mentioned above.

It is worthwhile to note that the rules issued by local Italian authorities can give very different values. This is confirmed by a comparative application of the technical rules to a Sicilian stream with great seasonal variability of flow, the Anapo River at San Nicola, with an 82 km² basin (Alecci and Rossi 2016).

Figure 10.1 shows the high variability of the obtained estimates. It depends on the large range of basic criteria adopted and on the significant differences in the climatic and hydrogeological contexts, as well as on the different objectives of the technical rules for assessment.

The maximum value is estimated by the rule of Veneto Region and minimum one by that of the Province of Trento. Almost all the methods give results smaller than that of annual value $Q(335)$ of the respective duration curve. Only Veneto Region, Calabria Region and the Province of Bolzano give greater values; Apulia gives exactly the $Q(355)$ value.

Table 10.1 Principal applications of the minimum in-stream flow

Adopted criterion	Responsible authority	Year
Unitary flow contribution	Bolzano Province	1982
	Veneto Region	2004
Minimum observed flow rate	Trento Province	1986
Average annual flow	Sele River Basin	2011
Minimum flow of given duration	Apulia Region	2009
Hydro-meteorological features (precipitation and/or runoff)	Po River Basin	1992
	Piedmont Region	1991
	Calabria Region	2007
	Abruzzo Region	2010
Flow duration curves	Alto Adriatico Basin (Tagliamento)	2002
Return period	Basilicata Region	2005
Variability of the flow rate	Piedmont Region	1995
	Magra River Basin	2000
	Serchio River Basin	2002
	Sarno River Basin	2002
	Right Sele River Basin	2006
	Calabria Region	2007
	Bolzano Province	2010
Low flow in m day and return period T_r	Tiber River Basin	1999
	Calabria Region	2009
Prefixed volume	Sardinia Region	2004
	Sicily Region	2007

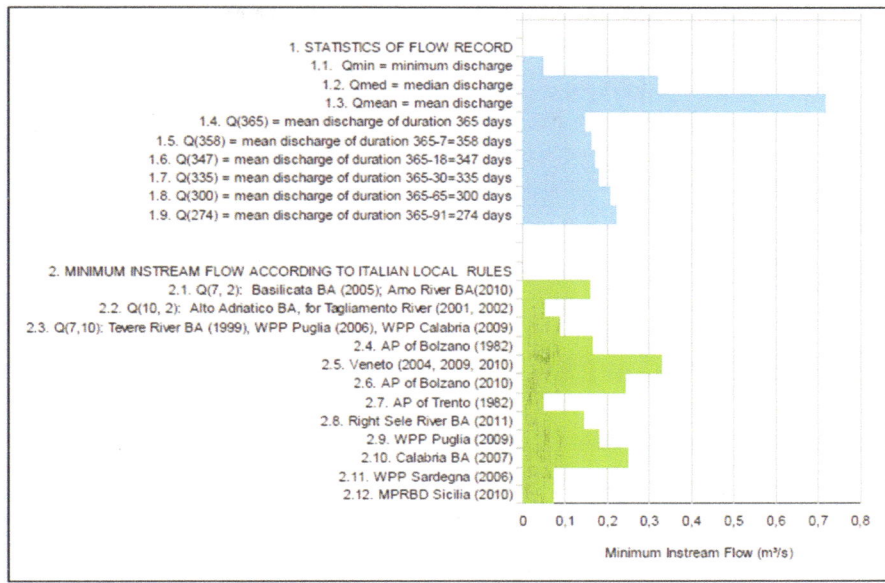

Fig. 10.1 Minimum in-stream flow values estimated by means of the various rules of Italian regions and River Basin Authorities applied to a stream of Sicily

10.3.5   Guidelines on MIF Definition Issued by the Ministry of Environment in 2004

The Guidelines of the Ministry for Environment, Land and Sea Protection (MATTM 2004) contain some limits on the heterogeneity of local regulations. The 2004 Guidelines do not formulate new criteria for the definition of the minimum flow and do not suggest preference among the local criteria. They state that the Regional Plans for Water Protection define the MIF for each section of watercourse with homogeneous conditions. These conditions refer to the hydrological, hydrogeological and geomorphological characteristics, including the naturalistic aspects describing the fluvial habitat and anthropic intervention, as well as the amount of abstractions and outflows. The guidelines highlight also that the aim of maintaining the typical biocenoses of local natural conditions is accompanied by the need of safeguarding the watercourse, the physical characteristics of the water body and the chemical-physical characteristics of water.

The prescriptions implicitly state that suitable water volumes should be released from reservoirs to ephemeral rivers when their discharge is lower than their minimum in-stream flow. This must be done taking in due account that releasing water in a watercourse in the season when it would naturally be dry may involve a significant alteration of the river ecosystem, with potential benefits for some species and detriment for others.

Finally, the 2004 Guidelines provide for the possibility of motivated derogations for limited periods, but only if one of the following conditions occurs:

- Supply needs for drinking use cannot be satisfied otherwise.
- Supply needs for irrigation use cannot be satisfied, but only in areas characterized by significant imbalances between resources and demands.
- Situations of water crisis, as declared by the civil protection authority.

Control actions downstream the abstraction sections must be done in order to check the impact on the ecosystem.

10.4 The Ecological Flow

10.4.1 The Water Framework Directive 2000/60/CE and the European Guidelines (CIS Guidance Document n. 31/2015)

The Water Framework Directive 2000/60/CE of the European Union has innovated the regulatory approach to water management, fully incorporating the environmental awareness and proposing environmental objectives significantly higher than those currently used in the member states. In fact, besides the consolidated objectives of reducing pollution and avoiding deterioration of water bodies, it emphasized the objective of improving the ecological status of water bodies.

Among the proposed guidelines published for a common implementation of the directive, the Guidance Document n.31/2015 (WFD, CIS 2015) deals with the *ecological flow*, defined as "a hydrological regime consistent with the achievement of the environmental objectives of the directive in natural surface water bodies".

10.4.2 The Guidelines on Ecological Flow Issued for Italy

In year 2017, the Ministry for Environment issued new guidelines (MATTM 2017) in order to adapt the Italian regulations to the European Guidance Document and to establish uniform methodologies for the ecological flow. The guidelines confirm that the District Basin Authorities are responsible for coordinating regional activities aimed at the implementation of the directive. It should be noted that this is only a matter of "proposed methodological approaches", since the definition of the operating rules is still entrusted to the regions. The decree also establishes at the Institute for Environmental Protection and Research (ISPRA) a National Catalogue of minimum in-stream flow calculation methods and provides a National Technical Table for the verification of technical-scientific consistency. This table has been established in May 2017.

ISPRA has already published some manuals and guidelines for the biological and environmental monitoring of watercourses and for the assessment of their ecological status (Agostini et al. 2017; Macchio et al. 2017; Fiorenza et al. 2018; Vezza et al. 2017; Potalivo et al. 2018).

In this manuals ISPRA makes use of results of a survey over 100 sites in watercourses of the Alps, Po Valley and Northern Apennines, for the calibration and validation of a "mesohabitat approach", useful for the evaluation of the ecological flow, also providing operational criteria and software for the application.

The methodology MesoHABSIM (Parasiewicz 2007; Vezza et al. 2017) is an improvement of the previously mentioned PHABSIM method. It refers to all the animal and plant species living in the habitat, instead of individual species. Besides, it adopts large spatial units, with homogeneous habitat conditions (pool, riffle, rapid), and classifies the suitability of the spatial unit to host the community by means of an integrity index of the river habitat, evaluated for various flows in the watercourse. Comparing the habitat trend in time before and after an anthropogenic intervention on the riverbed, indications are obtained for the definition of the "ecologic flow".

The new 2017 Guidelines provide updated definitions of minimum in-stream flow and ecological flow. The definitions focus, in particular, on the differences between the two terms, as one refers substantially to the hydrological regime, while the other deals with all the relevant components.

An important difference is also related to the objective to be achieved, as the ecological flow include both surface and groundwater and protected areas, while the minimum flow is based on achieving the objective of maintaining the environmental state of the watercourse. Moreover, the minimum flow concerns only the interventions for the abstraction from the water body and the regulation of volumes, whereas the ecological flow also involves interventions for river assessment and mountain stream stabilization.

The guidelines also provide methodological indications on the response of the biological communities to the hydrological and morphological alterations of the habitat, differentiated according to the community identified as most representative of the habitat (diatoms, macrophytes, fish fauna, macroinvertebrates, riparian species, species of commercial interest).

The 2017 Guidelines operate a classification of the methodologies for determining both the minimum and the ecological flow, according to three categories identified as:

(a) *Hydrological methodologies*, based on the analysis of the flow regime, in particular its variability over time and its flow duration curve
(b) *Hydraulic-habitat methodologies*, based on the relationships between the hydraulic parameters (mainly water depth, flow velocity and channel geometry), related to the flow rate and habitat characteristics available for biota
(c) *Holistic methodologies*, which, in addition to the components of the river ecosystem, take into account the socio-economic drivers related to the management of water resources. In these methods, experts from different disciplines are

involved (even far from the hydraulic and biological fields) and public participation processes are activated, in order to reach consensus on the flow regime
most suitable for maintaining or achieving environmental objectives.

The guidelines include a few appendices on hydrological and hydraulic-habitat
methods. Here the hydrological methods will be presented.

The first method is mainly hydrological, but it makes use also of biological evaluations. Its application requires the flow duration curve of the watercourse for several years, in terms of contribution of flow per unit surface of the basin $(m^3/s\ km^2)$.
Then, following direct biological observations, two groups of years can be identified, namely, that in which the ecological state of the watercourse is rated "good or
superior" and that in which the ecological state is rated "inferior to good". From
each of these groups, the flow probability distribution curves are extracted, and by
means of suitable statistical manipulations, two significant curves, as in Fig. 10.2,
are obtained, namely, that of good or superior (green line) and that of lower than
good ecological state (red line), for the considered water body.

Flow rates above the green line are compatible with conditions according to ecological flow. Flow rates below red line are not compatible. Values inside the two
lines need a thorough analysis. These two curves are used for verifying that the
changes in the duration curve induced by the water withdrawal may respect the
ecological flow.

The 2017 Guidelines propose also an empirical methodology for the evaluation
of the ecological flow in non-perennial watercourses (frequent particularly in southern regions and the main islands), where very low flow, sometimes null, occur for a
part of the year, contrasted by short episodes of high flow. This situation has been
deeply examined, with the definition of "aquatic states" (Gallart et al. 2012; De
Girolamo et al. 2015), which summarize the transient sets of aquatic *mesohabitats*
occurring on a given stream at a particular moment, depending on the hydrological
conditions.

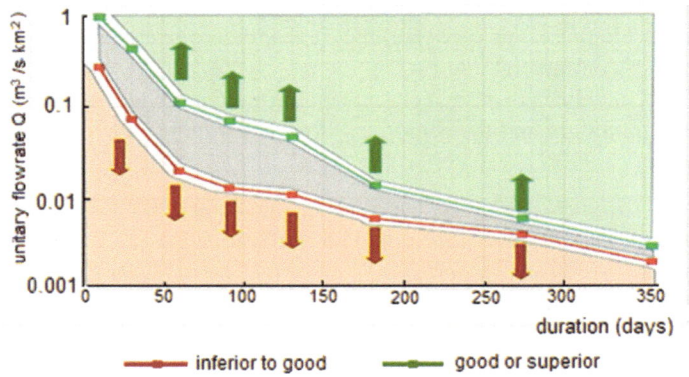

Fig. 10.2 Plotting of flow duration curves for identifying the ecological state. (Source: MATTM
2017)

Table 10.2 Classification of aquatic state in terms of hydrological and ecological features

Aquatic state	Hydrologic feature	Flow condition	Mesohabitats	Flow rate
Hyperrheic	Flood, overbank flow	Surface flow continuous	Drift of bed load and fauna	$Q \geq Q_1 = Q_{10}$
Eurheic	Normal flow	Surface flow continuous	All mesohabitat available and connected	$Q_{flex} = Q_2 \leq Q < Q_1 = Q_{10}$
Oligorheic	Low flow, pools connected by thin water threads	Flow only interstitial	Lentic fauna with most of lotic species present	$Q_3 \leq Q < Q_2 = Q_{flex}$
Arheic	Flow rate near zero, disconnected pools	Surface water in pools only	Only lentic fauna	$Q_{min} = Q_4 \leq Q < Q_3$
Hyporheic	No surface flow, flow in alluvium under riverbed	No surface water	Only hyporheic and terrestrial fauna active	$0 \leq Q < Q_4 = Q_{min}$
Edaphic	No surface flow, no flow in alluvium under riverbed	No surface water	Terrestrial fauna and edaphic fauna	$Q = 0$

Source: Gallart et al. (2012), modified

Six states are defined, starting from the occurrence of high flow or flood (*hyperrheic*) and ending at the occurrence of dry riverbed (*edaphic*). Any state is characterized by proper hydrological feature and by mesohabitat conditions, to which some flow and specific values of duration curves correspond, as shown in Table 10.2. The most significant "aquatic states" are shown in Fig. 10.3.

According to the 2017 Guidelines, the downgrading of the aquatic state of a watercourse, due to the abstraction, is considered acceptable if it is limited only to one stage. A conventional flow duration curve, which defines the boundary of the ecological state is obtained by means of an opportune sliding of the natural duration curve (Figure 10.4a). In case of a non-perennial watercourse, such curve presents zero values and therefore reaches the x axis for a duration $d < 365$ days (exceedance frequency $\leq 100\%$). Assuming, the following values:

$Q_1 = Q_{10} =$ flow rate with exceedance frequency 10%
$Q_2 = Q_{flex} =$ flow rate corresponding to inflexion point in duration curve
$Q_3 =$ flow rate extracted from observed conditions
$Q_4 = Q_{min} =$ minimum value > 0

According to the proposed procedure, the conventional duration curve (red in Fig. 10.4b) is obtained horizontally sliding Q_4 to the duration of Q_3, then Q_3 to the duration of Q_2, Q_2 to the duration of Q_1 and finally Q_1 to duration 0.

The obtained curve can be used to evaluate whether or not the changes induced by the withdrawal to the natural duration curve can be associated with conditions complying with the environmental flow.

Fig. 10.3 Typical appearance of aquatic states: (**a**) *edaphic*, (**b**) *arheic*, (**c**) *oligorheic*, (**d**) *eurheic*. (From De Girolamo et al. 2015)

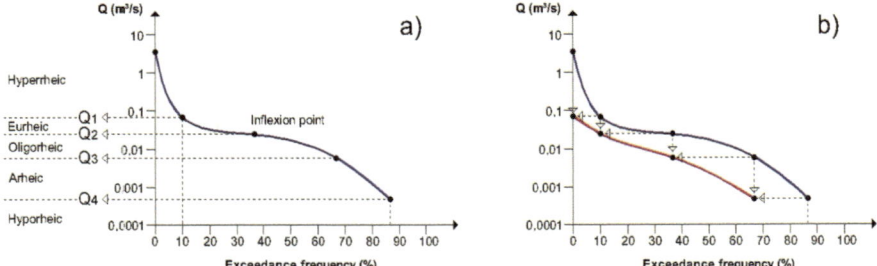

Fig. 10.4 Empirical method suggested by 2017 Guidelines for the evaluation of ecological flow in non-perennial watercourses
(a) Flow duration curve of non-perennial watercourse and thresholds of flow rate Q_1, Q_2, Q_3 and Q_4 defining the aquatic states
(b) Plotting of conventional flow duration curve representative of ecological flow (red line) according to 2017 Guidelines. If the flow duration curve resulting from derivation is above red line, the derivation is compatible with the ecologic flow
(Source: MATTM 2017, modified)

10.5   Concluding Remarks

The principle of "in-stream flow" in rivers has had normative acknowledgement in Italy only in the year 1989 with a slow implementation.

One reason of this delay is probably the lack, for most of the Italian watercourses, of those characteristics that, in other countries, have allowed the development of significant economic and social activities needing the presence of flows in rivers (e.g. navigation, floating, commercial fishing). Additionally, many streams, especially in the southern regions and the islands, show high variable hydrologic regimes that do not allow for the establishment of clear and persistent ecological requirements. The regime variability is a limitation for the stable development of water sport and recreational activities and affects the landscape value. Only after the increase in the environmental awareness, the value of in-stream flows and the need for river rehabilitation and restoration have been largely recognized in the Italian society.

The relatively slow progress of such concepts in the Italian legislation is probably due to several factors. One of them concerns the general difficulty in prescribing restrictions on the water use in the regulatory system, which is strongly marked by the private initiative. Another reason is the complexity of the institutional system, also due to the transfer of legislative and administrative functions from the central state to the regions since the 1970s.

However, other difficulties are not lacking. Among them, the limited availability of water resources for most common uses (municipal, agricultural and industrial) often produce severe shortage especially during the dry years, particularly in the southern regions. The high variability of the hydrological regime requires the development of specific strategies for water resources management, aggravated by the shortage of biological investigations and by the limited historical data of sufficient length to perform probabilistic processing.

However, despite these serious and objective difficulties, in less than 30 years, Italy has accomplished a significant cognitive and regulatory effort, and today the ecological flow is a consolidated principle oriented towards satisfactory applications.

References

Agostini A, Lucchini D, Genoni P, Martone C, Barbizzi S (2017) Qualità del dato nel monitoraggio biologico: macroinvertebrati delle acque superficiali interne [Data quality in biological monitoring: Water macroinvertebrate of inland surface water] (in Italian) – ISPRA – Manuali e Linee Guida 153/2017, Roma, p 69

Alecci S (1998) Effetti delle regole di rilascio del deflusso minimo vitale sul regime dei corsi d'acqua [Effects of the release rules on hydrological regime of watercourses]. In: Barbero G, Bertoli L (eds) L'influenza del deflusso minimo vitale sulla regolazione dei grandi laghi prealpini [The influence of minimum in-stream flow on the large lakes regulation] (in Italian)]. Guerini e Associati, Milano, pp 147–164

Alecci S, Rossi G (2016) Low flow and in-stream flow requirements. In: Eslamian S, Eslamian F (eds) Handbook of drought and water scarcity. CRC Press Taylor & Francis Group, New York, pp 353–373

Baxter G (1961) River utilization and preservation of migratory fish life. Proc Inst Civ Eng London 18:225–244

Beccari, G, Calamari D, De Fulvio S, La Noce T, Marchetti R, Mendia L, Merli C, Passino R, Provini A, Villa L, Volterra L (1971). Classificazione dei corpi idrici e requisiti di qualità degli scarichi. [Waterbody classification and quality requirements of outlets] (in Italian). Water Research Institute "Quaderni", n.5

Bovee KD (1982), A guide to stream habitat analysis using the in-stream flow incremental methodology. In-stream Flow information paper no. 12, U.S. Fish and Wildlife Service, Washington, D.C., 24p

Bovee KD, Cochnauer T (1977) Development and evaluation of weighted criteria, probability-of-use curves for in-stream flow assessments: fisheries. In-stream flow information paper no. 3. USDI Fish and Wildlife Service, Washington, D.C. FWS/OBS-77/63, p 39

Bovee KD, Lamb BL, Bartholow JM, Stalnaker CB, Taylor J, Henriksen J (1998) Stream habitat analysis using the in-stream flow incremental methodology. U.S. Geological Survey, Biological Resources Division Information and Technology Report USGS/BRD-1998-0004, p viii +131

Chiang SC, Johnson WF (1976) Low flow criteria for diversions and impoundments. J Water Res Plan Manag Div Am Soc Eng 102:227–238

De Girolamo AM, Gallart F, Pappagallo G, Santese G, Lo Porto A (2015) An eco-hydrological assessment method for temporary rivers. The Celone and Salsola rivers case study (SE Italy). Ann Limnol 5:1–19

Fiorenza A, Casotti V, Civano V, Mancaniello D, Marchesi V, Menichetti S, Merlo F, Piva F, Spezzani P, Tanduo I, Ungaro N, Venturelli S, Zorza R (2018) Linee guida per l'analisi delle pressioni ai sensi della Direttiva 2000/60/CE [Guidelines for analysing the pressure according to the 2000/60/CE Directive] (in Italian). ISPRA. Manuali e Linee Guida 177/2018, Roma, aprile 2018

Gallart F, Prat N, García-Roger EM, Latron J, Rieradevall M, Llorens P, Barberá GG, Brito D, De Girolamo AM, Lo Porto A, Buffagni A, Erba S, Neves R, Nikolaidis NP, Perrin JL, Querner EP, Quiñonero JM, Tournoud MG, Tzoraki O, Skoulikidis N, Gómez R, Sánchez-Montoya MM, Froebrich J (2012) A novel approach to analysing the regimes of temporary streams in relation to their controls on the composition and structure of aquatic biota. Hydrol Earth Syst Sci 16:3165–3182

Geer WH (1980) Evaluation of five in-stream flow needs methodologies and water quantity needs of three Utah Trout streams. Utah Division of Wildlife Resources. Publication 80-20, p 227

Larsen HN (1981) New England Flow Policy, Memorandum, interim regional policy for New England stream flow recommendations. US Fish and Wildlife Service, Region 5, Boston, 3 pp

Macchio S, Rossi GL, Rossi G, De Bonis S, Balzamo S, Martone C (2017) Nuovo indice dello stato ecologico delle comunità ittiche [A new index of the ecological state of aquatic communities] (in Italian). ISPRA. Manuali e Linee Guida 159/2017. Roma, luglio 2017

Maran S, Dell'Angelo F, Quaglia G, Gilli L (2000) Flussi minimi vitali a valle di derivazioni e ritenute [Minimum in-stream flows downstream from withdrawals and reservoirs (in Italian). Rapporto CESI Centro Elettrotecnico Sperimentale Italiano, Milano, p 137

Marchetti R, Gerletti M, Calamari D, Chiaudani G (1973) Elementi e criteri per la definizione del livello di accettabilità delle acque di scarico [Basic criteria for defining the acceptability of outlet] (in Italian). Water Research Institute "Quaderni2" n.24

MATTM (2004) Ministero dell'Ambiente e della Tutela del Territorio e del Mare. *Linee guida per la predisposizione del bilancio idrico di bacino* [Guidelines for water budget in a river basin] (in Italian) DM 28/7/2004

MATTM (2017) Ministero dell'Ambiente e della Tutela del Territorio e del Mare. Linee guida per l'aggiornamento dei metodi di determinazione del deflusso minimo vitale [Guidelines for revising of the methods for assessment of minimum in-stream flow] (in Italian) DDG 30/STA 13/2/2017

Menduni G, Brugioni M, Ceddia M, Nocita A (2006) Il calcolo del deflusso minimo vitale su base biologica mediante l'utilizzo di un modello idraulico monodimensionale [Computation of minimum in-stream flow by means of a one-dimension hydraulic model] (in Italian). XXX Convegno di Idraulica e Costruzioni Idrauliche, pp 1–14

Parasiewicz P (2007) The MesoHABSIM model revisited. River Res Appl 23(8):893–903

Potalivo M, Felluga A, Fonte A, Fioretti M, Moncalvo B, Usai MP, Abita A, Fiore M, Gerbaz D, Zanon F (2018) Il campionamento delle acque interne finalizzato alla determinazione dei parametri chimici e misura in campo dei parametri chimico-fisici di base per la Direttiva Quadro sulle Acque [Water sampling for inland water] (in Italian). ISPRA Manuali e Linee Guida 181/2018. Roma, luglio 2018

Renoldi M, Torretta V, Vismara R (1995) Gestione ecologica dei fiumi: Il problema della portata minima vitale. [Ecological river management: the minimum in-stream flow] (in Italian). Quaderni di Ingegneria Ambientale 22:138

Saccardo I (1998) Esperienze Enel nell'applicazione di metodi idraulico-biologici nella stima del deflusso minimo vitale [Experiment of the ENEL spa for estimating the minimum in-stream flow)] (in Italian). In: Deflusso minimo vitale. Metodi ed esperienze per un corretto utilizzo della risorsa idrica nel rispetto degli usi ambientali dei corsi d'acqua, Atti del Convegno 21/3/1997, Reggio Emilia, AGAC, pp 111–145

Tennant DL (1976) In-stream flow regimens for fish, wildlife, recreation and related environmental resources. Fisheries 1:6–10

Vezza P, Zanin A, Parasiewicz P (2017) Manuale tecnico-operativo per la modellazione e valutazione dell'integrità dell'habitat fluviale [Technical handbook for modelling and estimating the river habitat integrity] (in Italian). ISPRA Manuali e Linee Guida 154/2017. Roma, maggio 2017

WFD CIS (Water Framework Directive's Common Implementation Strategy) (2015) Guidance document no. 31. Ecological flows in the implementation of the Water Framework Directive, Luxembourg, Office for Official Publications of the European Communities, p 106

Part IV
Challenges

Chapter 11
Flood Risk Reduction

Giuseppe Rossi and Bartolomeo Rejtano

Abstract Italy is particularly vulnerable to water-related disasters (flooding, land-slides, drought) and to other related phenomena, such as soil and coast erosion, which affect land and human activities. Traditional strategies to face water-related hazards were based on structural measures, aiming at land reclamation by drainage networks, at soil conservation by agricultural and forestry practices and at flooding defence by hydraulic works such as structures against overflow and inundation (river banks, diversions) or for flood routing. The land transformation due to the growth of urbanized areas and infrastructures and the likely effect of climate change have increased the flood risk, while the structural measures began to be considered insufficient and costly and to be criticized for environmental reasons. Thus, non-structural measures such as constraints on land use, early warning systems and better information to increase public awareness and behavioural responses to floods have been emphasized. In this chapter, basic concepts of flooding risk and measures for its reduction are discussed. The most severe flooding disasters which occurred after the unification of the country (1861) and the development of policy about the flood risk reduction are analyzed. Then, a synthesis of flood hazard and risk assessments in Italian regions is presented. Finally, an attempt is made to identify priorities and trends to improve flood disaster resilience.

11.1   Introduction

Mitigation of flooding risk can be considered one of the most relevant challenges to be faced by a community in order to improve its resilience to water-related disasters. A severe flood is the result of several factors including (i) severity of precipitation, (ii) soil and vegetation coverage, (iii) geomorphological characteristics and land-use of watershed and (iv) extent and morphology of the flood expansion zone.

G. Rossi (✉) · B. Rejtano
Department of Civil Engineering and Architecture, University of Catania,
Catania, Italy
e-mail: grossi@dica.unict.it

© Springer Nature Switzerland AG 2020   251
G. Rossi, M. Benedini (eds.), *Water Resources of Italy*, World Water Resources 5,
https://doi.org/10.1007/978-3-030-36460-1_11

Nevertheless, the possibility that a flood becomes a disaster is strictly connected to various anthropogenic factors, such as the increased industrial and agricultural activities and the urban settling and development of infrastructures in the flood prone areas, the loss of a significant part of the floodplain as an expansion zone and the construction of hydraulic structures (dams and dikes), whose failure can increase the damage under exceptional circumstances. Besides, the impacts of climate change on rainfall intensity and on seasonal distribution of precipitation are other important factors. Furthermore community disaster preparation plays a very important role, particularly with respect to the risk of mortality, by means of early warning systems and by improving the population behaviour in response to threatening events.

Italy is particularly vulnerable to all water-related disasters and has been forced to adopt an adaptive approach to face the dramatic increase of frequency and damages of these disasters. In particular, the specific characteristic of a combined flooding and landslide, which seems a unique feature of the territory of many Italian regions in contrast to other European countries, explains the adoption of a more comprehensive view in Italian legislation in order to face both risks. Such a comprehensive approach has been the basis of Law 183/1989 and of the following regulations. While the European Flood Directive 2007/60/EC is focused only on flood risk, in Italy the activities aiming at understanding and managing the risks continue to concern both flood and landslide phenomena.

This is a positive feature of the Italian legislative framework, which in the last decades has been characterized by a closer attention to non-structural measures and to the role of a civil protection organization. However, the delays in the preparation and implementation of the planning tools, the limited financial resources available for works and the cumbersome bureaucracy have influenced negatively the flood risk mitigation policies based on structural measures. The application of non-structural measures has presented other difficulties, such as the constraints established by the flooding risk plans which often are not compliant with the land planning. Moreover, there is inadequate coordination in the decision-making process which aims at avoiding casualties and extensive damage, by meteorological/hydrological forecasting, early warning systems and real-time local decisions during severe events. Finally, there is a lack of public participation in the decision-making processes ranging from the strategy planning to the design of structural measures and to the operation of warning system

In the next sections of this chapter, the basic concepts and different types of flood risk are presented (Sect. 11.2), and an analysis of the severe flooding disasters that occurred in Italy since the unification in 1861 is carried out (Sect. 11.3). Section 11.4 outlines the major steps of the development of flood mitigation policy. The results of the recent assessment and mapping of flood hazard and flood risk are presented in Sect. 11.5. Then the expected trends in flood risk reduction are analysed in Sect. 11.6, and finally, in Sect. 11.7, the key approach to improve flood management is pointed out.

11.2 Basic Concepts and Types of Flood Risk

Flood risk management has been considered in Italian legislation within the more general frame of soil defence. A very comprehensive approach had already been adopted since the Royal Decree 215/1933, which established the rules for hydraulic reclamation by considering it as a part of an integrated land conservation strategy aiming at removal of the obstacles to an agricultural and socio-economic development of the land reclamation district by means of a general reclamation plan. Such a plan had to consider the needed measures for meeting the irrigation demand, improving rural roads and electric networks. It was parallel to other international experiences of integrated development of that period, such as the Tennessee Valley in the USA during the F.D. Roosevelt presidency.

The concept of soil defence has been proposed by the Inter-ministerial Commission for the study of hydraulic regulation and soil defence (Commissione Interministeriale 1970) established by the Ministry of Public Works and the Ministry of Agriculture and Forest soon after the dramatic floods occurred in November 1966 in Florence and in Veneto. "Soil defence" was defined as:

> the set of all activities for preserving and safeguarding the soil, its capability of production and the infrastructures from extraordinary assaults by intense rainwater, floods and sea water.

The Law 183/1989 approached the water and soil problems in a unitary way. It introduced the provision of a comprehensive river basin plan and gave a broader definition of "soil defence", as:

> the set of activities related to the knowledge, legal rules and land management aiming at conserving, defending and enhancing soil and appropriate use of water resources, including water quality protection.

Such a broad approach has been criticized for being too ambitious and it has been considered as one of the reasons for the delay that occurred in the implementation of planning provisions (as already discussed in Chap. 3). The Legislative Decree (DLgs) 152/2006 extended further the concept of soil defence, defining the "soil defence" and "hydrogeological instability" as follows:

> Soil defense is the set of actions and activities regarding the protection and safeguard of land, rivers, canals, lakes, lagoons, coastal areas, groundwater, with the goal of reducing hydraulic risk, eliminating the geologic instability, optimizing the use and management of water resources and enhancing the connected environment and landscape and the fight against desertification.

> Hydrogeological instability is the condition of areas, where natural or anthropogenic processes determine a risk condition to the land.

The more recent definitions, regarding "flood" and "risk of flooding", stated in the DLgs 49/2010, coincide with those of the Directive 2007/60/EC, which states:

> Flood means the temporary covering by water of land not normally covered by water. This shall include floods from rivers, mountain torrents, Mediterranean ephemeral water courses, and floods from the sea in coastal areas, and may exclude floods from sewerage systems.

Flood risk means the combination of the probability of a flood event and of the potential adverse consequences for human health, the environment, cultural heritage and economic activity associated with a flood event.

In order to provide homogeneity and reference about types of flooding, the E.C. *Guidance for reporting under the Flood Directive* (E.C. 2013) distinguishes different sources, mechanisms and characteristics of flooding as follows:

Source: Fluvial, pluvial, groundwater, seawater, artificial water-bearing infrastructures (or failure of such infrastructures)

Mechanism: Natural exceedance, i.e. water exceeding the capacity of the carrying channel, defence exceedance, i.e. flood waters overtopping flood defences, failure (breaching or collapse) of natural or artificial defence or infrastructure, blockage or restriction of a conveyance channel or bridges or sewage due to, e.g. landslide or ice jam

Characteristic: Flash flood, snow-melt flood, medium- or slow-onset flood, debris flow, high velocity or deep flood

In order to consider the prevailing types of floods occurring in Italy, the following simplified list of flooding categories is adopted in this chapter:

River flooding
Urban flooding due to river outbreak
Flooding originating from dam break and/or landslide
Flooding connected with debris flow or mud-slide.

A lot of flooding events occur each year in different regions of the country, but only a limited number causes deaths, missing persons, homeless and injured people and/or damage so serious as to be defined a flood disaster. Today, the awareness of an increased risk of flood disasters in many Italian regions is spread largely not only among the technicians and the politicians but also in the public opinion, due to the attention paid by TV and social media to natural disasters. Obviously, the increase in flooding risk is driven by different factors. First of all, the hydrological response of catchment to meteorological events (very heavy rainfall, worsened in some cases by snow melting) is becoming more severe. This derives mainly from the increase of impervious areas of the catchment, in turn due to the land consumption deriving from urban sprawl, construction of roads, occupation of floodplain by illegal buildings and upstream training works.

Also, the transformation of agricultural land, particularly the abandonment of agricultural practices, reduces soil infiltration and concentration time of the drainage basin, thus increasing peak flow and runoff volume.

Structural defence measures and river regulation infrastructures have become inadequate, e.g. since the floodplain areas have been modified, the dikes are poorly maintained, the sizing of the channel obtained by covering urban reaches of torrents is affected by mistakes in the design flood, etc.

Moreover, according to most of the scientific community, climate change must be considered as a contributing factor that increase the disaster risk, causing more frequent and higher intensity storms.

Besides the above physical causes, a more general but not less important factor should be mentioned, which has, perhaps, a greatest impact on the severity of disasters. This consists in the inadequacy of policy choices in the few last decades, which, in spite of advanced knowledge of the prevention and mitigation measures and of improvement of civil protection role for monitoring and acting during the events, has not improved significantly the resilience of several urban areas and of most of the basins to the flooding hazard. Finally the not adequate behaviour of people during the most severe events is another significant reason of many casualties.

11.3 Main Flood Disasters in Italy After the State Unification

Several historical documents keep alive the memory of water-related disasters that occurred in the past centuries in the Italian peninsula since the time of the Roman Empire to the various Italian states that existed before the unification. Examples of some of the most severe events include flooding of the Arno River at Pisa during 9 storms on September–November 1167, flooding of the Arno River at Florence on 4 November 1333 (which destroyed the Ponte Vecchio), flooding of Polesine (from the Po River) and Verona (from the Adige River) on Autumn 1348, flooding of Palermo on 27 September 1557 (about 7000 deaths), flooding of Pisa on 19 May 1680 due to the overflow of the Arno River, flooding along the Po River in 1705 (up to 15,000 killed people), disaster in the eastern Ionic coast of Sicily (Messina province) on February 1763.

Reports of these disasters generally do not provide enough information on the meteorological and/or hydrologic features of the events and on the number of fatalities (deaths and missing persons) and extent of the damage. Detailed research to gather technical details was started by the SGA (Storia Geofisica Ambiente) company of Bologna, funded by ENEA (Ente Nazionale Energia Atomica) in 1987 for the period 1000–1985, but unfortunately, it has not been completed.

Hence in the present section, the survey is limited to the events that occurred after the Italian kingdom was established (1861) as a unitary state. The most severe flood disasters after Italian unification, which are significant either for rainfall intensity (and for consequent flood discharge) or for serious human consequences and damages, are described in Table 11.6 of the Appendix to the present chapter. For each event, the prevailing category of flooding is indicated, according to the four types identified in the previous section and specified as follows.

River Flooding (RF)
This is the most severe type of flooding, deriving from long and intense storms striking most of the river catchments. It is characterized by the flooding of large part of floodplain and/or the urban area crossed by the river. It is due either to the insufficiency of the cross section of the river to contain exceptional flood runoff or to the rupture or overtopping of the dikes.

Urban Flooding due to River Outbreak such as Flash Flood (UFF)

This is generally due to the inadequacy of the cross section of urban reaches of a river to drain a flash flood (especially if the watercourse has been used as a road or has been covered to build a road, a square or a parking lot) or due to a discharge in the sewer network exceeding the design flow.

Flooding Originating from Dam or Landslide (FDL)

It includes both the disasters caused by a dam break (Gleno, Molara, Stava) and disasters due to the flooding consequent to landslide events (Tavernerio, Vajont, Valtellina). The first type of disasters, although destructive, had a positive impact for establishing more rigorous technical rules to improve design, construction and operation of dams.

Flooding Connected with Debris Flow or Mud-Slide (FDFMS)

This category includes the floods which caused most damages from the debris flows and/or mud-slides triggered by severe storm events.

The following information, if available, is given for each disaster: flood category, date, affected area, i.e. province and region hit strongly, main river basins, hydro-meteorological features, number of fatalities and number of evacuees and homeless, short description of the event and, whenever possible, a comment on the effects of the disaster on policy and/or technical provisions adopted to face the risk of new severe events.

According to the terms adopted by the Research Institute for Hydrogeological Protection (IRPI CNR), "fatalities" include the number of deaths and missing persons caused by a harmful flood event; "evacuees" indicates the number of people forced to abandon their homes temporarily; "homeless" indicates the number of people that lost their homes; "harmful" event indicates an event with human consequences, i.e. "casualties" (including fatalities and injured people), homeless people and evacuees.

The main sources of information are the reports prepared by the IRPI, beginning from the AVI (Vulnerated Italian Areas) Project, carried out within the activity of the group for Hydrogeological Disasters of the National Research Council (Guzzetti et al. 1994), till the Information System on Hydrogeological Disasters (CNR 1999–2019). Part of these data is now available in the website POLARIS (Salvati et al. 2016). Other sources of information, including books, scientific journals, technical reports and papers presented in congresses, have been examined in order to deepen the hydro-meteorological features, such as Piccoli (1972), Botta (1977), AII (2010), Accademia dei Lincei (2013) and Rosso (2017).

Since the original sources of information have different reliability, the data listed in the table are affected by high uncertainty, particularly with reference to the human consequences and to the damages, due also to the difficulty of distinguishing the effects of direct floods from the effects of landslides triggered by the same storms causing the floods.

According to the analysis carried out by IRPI (Salvati et al. 2015), the total amount of fatalities (deaths and missing persons) in the period between 1861 and 2013 is at least of 3268 people and the number of evacuees and homeless about 691,000 persons, as detailed in the Table 11.1. However, these estimates do not

Table 11.1 Human consequences of harmful flood events in Italy from 1861 to 2013

Period	1861–1910	1911–1960	1961–2013	1861–2013
Number of death	665	1782	735	3182
No. of missing person	–	18	68	86
No. of injured people	12	989	903	1904
No. of evacuees and homeless	123,918	275,743	291,012	690,677

Source: Salvati et al. (2015)

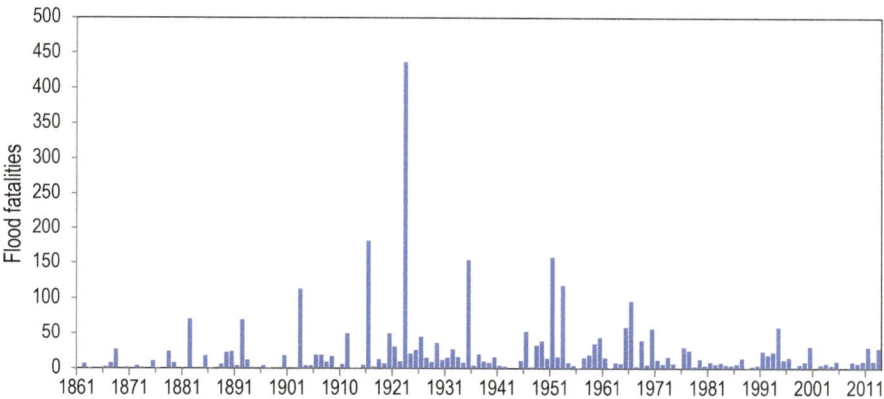

Fig. 11.1 Flood fatalities in the 1861–2013 period. (Source: http://polaris.irpi.cnr.it/)

consider the fatalities due to the floods that occurred in consequence of landslide (e.g. the Vajont disaster).

For the same period, Fig. 11.1 shows the number of flood fatalities per year. It is possible to ascertain that floods events with deaths and missing persons occurred almost every year of the period 1861–2013, though presenting a great variability. Also, excluding the Gleno dam collapse on December 1923, the worse years occurred between the 1950s and 1960s. The figure highlights the decreasing trend in the number of fatalities, which is evident in the recent decades.

In Italy, quantitative estimates of geohydrological risk to the population are estimated and made available by IRPI in the "Polaris" website (http://polaris.irpi.cnr.it/) (Salvati et al. 2016).

The risk posed by geohydrological hazard to the population is assessed commonly by means of mortality rates. Mortality rate is measured as the number of fatalities due to a specific hazard per 100,000 people in a period of 1 year. Based on data on landslide and flood fatalities, the annual flood and landslide mortality rates for all Italian regions are updated and published annually by IRPI. Table 11.2 shows, for each region, the number of fatalities and the average mortality rate in the period 1968–2017 due to landslides and floods (in separate columns) and the total number of damaging events, including both the disasters for the same period.

Table 11.2 Number of fatalities and of mortality rates, caused by landslides and floods in the Italian regions in the 50 year period 1968–2017 and number of damaging events in both types of disasters

Region	Landslides		Floods		Landslides and floods
	No. fatalities	Mortality rate[a]	No. fatalities	Mortality rate[a]	No. of damaging events[b]
Piedmont	130	0.060	138	0.064	140
Aosta Valley	24	0.406	6	0.102	35
Lombardy	116	0.026	31	0.007	231
Trentino-Alto Adige	324	0.732	11	0.025	159
Veneto	30	0.013	7	0.003	100
Friuli-Venezia Giulia	13	0.021	8	0.013	44
Liguria	38	0.044	96	0.101	119
Emilia-Romagna	49	0.025	19	0.009	97
Tuscany	56	0.032	50	0.028	157
Umbria	12	0.030	7	0.017	57
Marche	6	0.008	14	0.019	59
Lazio	18	0.007	16	0.006	123
Abruzzo	10	0.016	4	0.006	72
Molise	0	0.000	1	0.006	28
Campania	274	0.098	21	0.008	198
Apulia	3	0.001	37	0.019	54
Basilicata	14	0.046	11	0.036	63
Calabria	28	0.028	32	0.031	145
Sicily	66	0.027	92	0.037	160
Sardinia	7	0.009	40	0.050	90
Italy	1218	0.040	641	0.020	2131

Source: www.irpi.cnr.it

[a]Average no. of fatalities per year on 100,000 inhabitants; [b]Events with fatalities, evacuees and homeless

The damaging events listed in Table 11.2 can be considered as the ensemble of floods and/or landslides that occurred in a given geographical area (e.g. a catchment, a municipality, a province, a region) in a period, ranging from hours to weeks, triggered by the same meteorological conditions. It is very common that the same intense or prolonged rainfall generates widespread landslides and floods, human impacts and severe and widespread economic damage. In these cases, it is very difficult to assign the damage to a single landslide or flood phenomenon.

As an example, we can mention the Mediterranean cyclonic vortex originating from the Balearic Islands that on 1 October 2009 generated an intense storm cell dumping intense rainfall along the Ionian Coast of Sicily, southeast of the city of Messina, with a cumulated rainfall exceeding locally 220 mm in 7 h (Napolitano

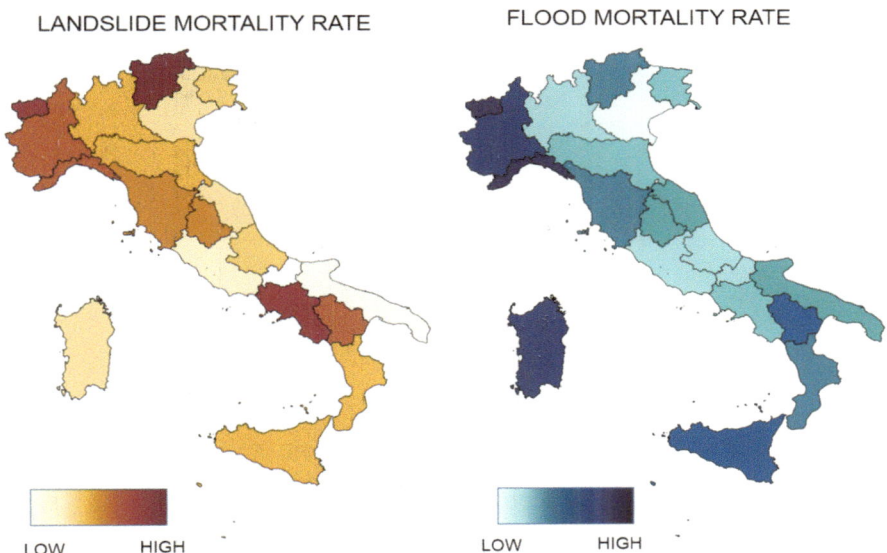

LANDSLIDE MORTALITY RATE   FLOOD MORTALITY RATE

LOW   HIGH   LOW   HIGH

Fig. 11.2 Average landslide and flood mortality rate in the period 1968–2017 in Italian regions. (Source: http://polaris.irpi.cnr.it/report/last-report/)

et al. 2018). The intense rainfall caused flash floods and widespread – mostly shallow – landslides and debris flows that affected public and private buildings and roads, in urban and rural areas. Landslides and floods caused 31 deaths and 7 missing persons. After the event, a total of 193.6 million euros were allocated by the national and the regional governments for the necessary recovery and risk mitigation actions.

Maps in Fig. 11.2 portray the average flood and landslide mortality rates for the period 1968–2017 for all Italian regions. The figure shows that the highest mortality was recorded in the northwestern Italian regions (Aosta valley, Piedmont and Liguria).

In this 50-year period, geohydrological events (floods and landslides) caused a cumulative number of 1858 fatalities in Italy (1796 deaths and 62 missing persons) and forced a more than 316,270 people to abandon their houses. Landslides were responsible for 1218 fatalities (65.5%) and floods for 641 (34.5%). Landslides killed more people than floods did. It could be a direct consequence of (i) the higher destructiveness of landslides compared to floods, (ii) the related larger vulnerability of the population to landslides than to floods and (iii) the generalized lack of effective landslide early warning systems (Salvati et al. 2017).

However, floods and landslides are not the only natural hazards that caused harm to people. Geophysical hazards (mainly earthquakes) are also frequent and mostly destructive in Italy. Table 11.3 lists the average mortalities calculated for the most impacting natural hazards in Italy for the period 1968–2017 and for the period 1861–2017.

Table 11.3 Average
mortality rates for different
natural hazards in Italy in the
period 1968–2017 and
1861–2017

	Average mortality rate*	
	1968–2017	1861–2017
Flood	0.022	0.048
Landslide	0.043	0.082
Earthquake	0.166	2.240
Volcanic activity	0.000	0.005

*No. of fatalities per year on 100,000 inhabitants (Source: www.polaris.irpi.cnr.it)

The available data shows that the amounts of the average mortality for all hazards computed for the last 50 years are significantly less than those of the whole 157 years period. In particular the difference is very high for mortality due to earthquake events, since the long period includes a few exceptional disasters with a very high number of fatalities, such as earthquake of Messina, 1908 (more than 80,000 deaths), of Avezzano, 1915 (more than 35,000), etc. In any case, the values referring to flood and landslide hazards in the recent period 1968–2017 are about one half of those ones referring to the long period. Besides, for both periods the mortality due to flood hazard is almost one half of that one due to landslide hazard.

11.4 A Summarized History of the Flood Mitigation Policy

Although the evolution of law on flood mitigation has been presented in detail already in Chap. 3 (Sect. 3.3.5) and a survey on flood disasters has been presented in previous Sect. 11.3, significant connections among disasters and development of law and consequent actions deserve to be recognized, in order to provide a better understanding of the recent history of the flood mitigation approach in Italy. Only a few significant steps will be recalled now, essentially those which happened within the memory of the present generation and were significant for their role, independently from magnitude, in order to give evidence to their generated impulses and shortcomings, so as to a few fore and back steps.

We will start recalling the flood of November 1951 in the Polesine area, generated from the Po River, mainly through several ruptures of levees (Fig. 11.3). The magnitude and impact of that immense catastrophe were such to impress and affect the whole nation. The Po River adjustments and the River Regulation Plan of 1952 established by Law 154/1952 have to be considered as an effect of that event. However its follow-up was limited, in time, in financing and in constructions, and floods continued to hurt the country.

The 1966 flooding from Arno in the city of Florence was another event which struck the public feeling, not only in Italy but worldwide, since, beside victims and economic losses, it affected an art patrimony of universal importance (Fig. 11.4).

This event gave impulse to the cultural and political process which at a later time led to a legislative recognition of the river basin size of the problem, in connection

Fig. 11.3 Views of the Po River flooding in November 1951

Fig. 11.4 Views of Florence during and after the flood of November 1966

with other types of water problems, and to legislation on river basin planning. This process went through the work of the De Marchi Commission and of the "Conferenza Nazionale delle Acque" (National Water Conference). It took many years of controversial political and engineering debates until Law 183/1989 was finally shaped and approved, thus introducing the concept of a comprehensive river basin planning approach into legislation.

That law, as it was discussed already in Chap. 3, established to plan at basin scale all activities concerning water use, defence from water and soil conservation and indicated the institutional and organizational scheme for the development of plans (with different sort of authorities for national, inter-regional and regional basins), including indication of preliminary studies to be carried out in order to support the planning phase. Although river basin planning as a practice had been already initiated in other countries, the Italian law must be recognized as an outstanding piece of legislation in its field; the legislative tradition which dates back to ancient Rome was producing still its fruits! However the gap between theoretical formulation and practical implementation was not easy to overcome, while in other countries, whose legislation and administration derived from Anglo-Saxons principles or empiric principles, things were running faster in practice.

The development of river basin plans was very inhomogeneous in the country. Mostly, National River Basin Authorities were more successful in their operation. In this sense, the Po River Basin Authority is an example. It took advantage of being born aside of the existing authority for regulation of the Po River (Magistrato del Po), beside of being located in the northern part of the country, where administrative organization was well rooted already. On the other side, mostly with respect to smaller and regional basins, and mostly in the southern regions of the country where collaborative and organizational activities face some political and social resistance, the establishment of river basin authorities and the development of river basin plans failed, while floods continued to hit heavily.

In order to overcome these difficulties, the national Law 493/1993, in a context aiming to accelerate public investments and to simplify many administrative procedures, established that river basin plans could be drafted and approved also for single subbasins and/or for specific sub-matters, such as the defence from floods, a sort of sub-plans. This provision was not intended to change the logic of river basin planning but only to allow an escape possibility to make things moving and to provide at least partial solution to urgencies. However, in a sense, it was a back step with respect to the comprehensive and basin-scale approach.

Nevertheless, this provision of Law 493/1993 did not produce the expected size of effects. And in the meanwhile, flooding events continued to occur and to require public financial commitment for repairs and refund of damages. At about the same time, a National Civil Protection Service was established by Law 225/1992, intended to face urgencies. Since a few years the government had already established the GNDCI, a national group for the defence from hydrogeological risks. It was entrusted to the National Research Council and to the academy in order to develop methodologies and strategies to face the water-related risks: floods, landslides, droughts and groundwater pollution.

In 1998, a combined flood and soil instability phenomenon hit the Sarno area. The event was not so severe as the Polesine and Florence events, but the times were mature to impose issuing a governmental decree to make the flood defence and the landslide defence plans as mandatory obligations for river basin authorities or for regions, according to the case (DL 180/1998). The decree established a deadline for the adoption of the plans for mitigation of hydrogeological risk and the obligation to establish safeguard measures, plus the provision of urgent action programs being entrusted to civil protection. That decree was soon approved by the parliament as Law 267/1998. It is indicated as the Sarno Law, from the name of the site where the disrupting event had occurred.

Also, times were mature to take advantage of the methodologies which had been envisaged by the GNDCI group about hydrological evaluations and flood risk mapping. So, soon after, in the same year, the government, by the decree issued on 29 September 1998, established detailed regulations and steps for the hydrogeological planning activity, with different specifications for the case of flooding and landslide risks. A step-by-step procedure was established with specific subsequent deadlines for each of the major steps: simplified "identification" of flood-prone areas, detailed mapping of hazard and risk in the selected areas, enacting of immediate safeguard measures, i.e. constraints, for the most risky areas and planning of long-term measures, with the option of revising the mapping and the safeguard measures once mitigation measures had been implemented.

Three orders of magnitude of flooding probability and consequent mapping were indicated: "high probability" areas (for return period Tr of about 20–50 years), "medium probability" areas (for Tr of about 100–200 years) and "low probability" areas (for Tr of about 300–500 years). Then, the mapping of four classes of risk was prescribed, ranging from "moderate" to "very high", in order to account also for activities, possible damages and human presence over the land. Stringent and immediate constraints on land use modifications were specified in general for areas and/or sites which would be mapped within "very-high-risk" and "high-risk" categories. Also, a mandatory scheduling of revisions and updating of the mapping and of the action plan was indicated, thus establishing a dynamic planning scheme.

We like to refer to the set of provisions of the Sarno Law and of the subsequent regulations of the 29 August 1998 decree as "the Italian methodology". It was a pioneer approach, since the European Directive on Flood Mitigation was still to come.

Another step in legislation was fostered by the harmful flood occurred in Soverato (Calabria) in September 2000. In fact, soon after, the governmental Decree 279/2000 (approved by the parliament as Law 365/2000) anticipated the deadline for the adoption of the sectorial plans, extended the validity of extraordinary mitigation plans in time, and in space, thus including all the riparian areas and the areas with flooding probability higher than once in 200 years in the safeguarding measures zone. Also it included provisions and additional resources for meteo- and hydrological monitoring and early warning systems and for the civil protection.

At a later time, in 2006, formally, the legislation about river basin planning was abrogated, but at the same time, it was reintroduced, substantially unmodified in the principles and methodology, into the Legislative Decree 152/2006, a broad comprehensive code which included the whole matters of water and environment. The only modification deserving to be pointed out is the adjustment of the administrative organization of the basin planning. In fact, according to European Directive 2000/60/EC, the Legislative Decree 152/2006 modified the administrative organization of the basin planning, thus substituting the several River Basin Authorities (national, interregional and regional) with only seven River Districts Authorities (plus the experimental district of Serchio basin) which had to group minor basins together.

In the following year, the European Directive 2007/60 EC on the management of flood risk was issued. It was more limited with respect to the Italian legislation, since it was dealing only with the flood risk, but its provisions and steps were reproduced from the Italian approach, although it established its own new deadlines. The Flood Risk Mitigation Plan, as introduced by the European Directive, was in the substance almost the same thing as the flooding part of the plan for hydrogeological asset of the Italian legislation.

However, Italy had to comply formally with the European Directive and had to transpose it into the national legislation, which Italy did with the DLgs 49/2010. It introduced the Flood Risk Management Plan to be developed for all districts. Since Italy had been working already on the same matter, Italian transposition law anticipated all deadlines with respect to the indication of the European Directive.

As the actual starting of the District Authorities was delayed, the responsibility of drawing up the planning tools established by the two European Directives has been ascribed to the National River Basin Authorities, by coordinating the regions within the districts. Since some duties of the Flood Directive had been carried out on the basis of previous laws (in particular the evaluation of flooding risk and landslide risk and the subsequent drafting of the Hydrogeological Asset Plan), Italy flew over the preliminary estimate of the flood risk and proceeded directly to map the flood hazard and risk and to develop the Flood Risk Management Plans.

While the National River Basin Authorities developed the planning tools in order to satisfy the European Directives, the national government initiated efforts to improve the actions concerning the flood risk mitigation, in particular to accelerate the use of financial resources for defence from floods (Law 116/2014), in order to ensure a better coordination between the Ministry for Environment, regional governments and local authorities and to improve the quality of the projects of the structural measures by specific guidelines.

The catalyst for these efforts has been the mission structure "Safe Italy" established by DPCM 27 May 2014, founded upon the idea that a body at national level could play a key role in pressing and coordinating the local responsibilities about flood defence. In this context other measures have been taken, e.g. the establishment of the plans for urban areas at high risk (DPCM 15/9/2015) and the issuing of an Act (Law 221/2015) which defines specific rules for the demolition of unauthorized buildings in high-risk areas and introduces rules at municipal level to reduce the vulnerability of buildings, including also a program for the management of sediments in river basins. Also, although in 2018 the new national government cancelled the mission structure, attention to the flooding defence has been confirmed with a recent program of investments of 11 billion of euros in 3 years (February 2019).

11.5 Mapping and Assessment of Flood Hazard and Flood Risk

The assessment of flood hazard and risk, carried out in the planning tools prepared by the River Basin Authorities and by regions in the context of the Hydrogeological Asset Plan and of the Flood Risk Management Plan, has been synthetized by ISPRA for the Ministry of Environment over all regions of the country (ISPRA 2018). In particular, according to the indications of DLgs 49/2010, the scenarios considered for flood hazard areas included floods with high probability P3 (with return period $T = 20$–50 years, where T is computed in terms of non-exceedance probability P, i.e. $T = 1/(1-P)$; floods with a medium probability P2 (return period 100–200 years); and floods with a low probability P1. Generally, the hazard maps indicate the extension of the areas affected by flooding, without details on water depths and flow velocity.

The areas with high flood hazard P3 in the whole of the Italian territory were estimated to be 12,405.3 km^2 (4.1% of the entire surface of the country), the areas with average hazard P2 were 25,397.6 km^2 (8.4%), and the areas with low hazard P1 were 32,960,9 km^2 (10.9%). The distribution of these areas among the different regions is indicated in Table 11.4.

Figure 11.5 shows the different probability percentages for the regions affected by flood hazard. The highest percentages regard Emilia-Romagna, Toscana, Lombardy, Piedmont and Veneto. The relevant extension of area with flood medium probability in Emilia Romagna derives also from the presence of a dense network of land reclamation channels. Figure 11.6 shows the areas affected by floods with medium probability P2 over the whole national territory.

Another interesting analysis, carried out by ISPRA by using also the landslide hazard data of the Hydrogeological Asset Plans, besides the data on the Flood Risk Management Plan, shows the areas of municipalities which are vulnerable to both flooding and landslide disasters (Fig. 11.7).

A number of 1602 municipalities out of the total of 7983 municipalities (20.1%) present a high or very-high landslide hazard, 1739 (21.8%) show a medium hydraulic hazard and a relevant number (3934, i.e. 49.3%) both hazards (see Fig. 11.8). The amount of surface area of municipalities under landslide hazard is 25,410 km^2 (8.4% of total area of Italy), the amount of areas with medium hydraulic hazard is 25,398 Km^2 (8.4%).

With respect to the flood risk maps required by the EU Flood Directive, the DLgs 49/2010 establishes that the potential adverse consequences associated with the flood scenarios (limited in a first stage to the medium probability range) be expressed in terms of the (i) approximate number of inhabitants potentially affected; (ii) stra-

Table 11.4 Areas affected by flood hazard in the Italian regions

Region	Total area (km²)	Area with high hazard P3		Area with medium hazard P2		Area with low hazard P1	
		km²	%	km²	%	km²	%
Piedmont	25,387	1148	4.5	2066	8.1	3272	12.9
Aosta Valley	3261	157	4.8	239	7.3	299	9.2
Lombardy	23,863	1860	8.0	2406	10.1	4599	19.3
Trentino-Alto Adige	13,605	52	0.4	79	0.6	114	0.8
Veneto	18,407	1231	6.7	1713	9.3	4635	25.2
Friuli-Venezia Giulia	7862	229	2.9	610	7.8	700	8.9
Liguria	5416	111	2.1	153	2.8	189	3.5
Emilia-Romagna	22,452	2485	11.1	10,252	45.7	7980	35.5
Toscana	22,987	1380	6.0	2791	12.1	4845	21.1
Umbria	8464	232	2.7	337	4.0	479	5.7
Marche	9401	12	0.1	241	2.6	35	0.4
Lazio	17,232	430	2.5	572	3.3	647	3.8
Abruzzo	10,832	97	0.9	150	1.4	179	1.7
Molise	4460	85	1.9	139	3.1	161	3.6
Campania	13,671	512	3.7	670	5.1	843	6.2
Apulia	19,541	651	3.3	885	4.5	1060	5.4
Basilicata	10,073	216	2.1	277	2.7	295	2.9
Calabria	15,222	563	3.7	577	3.8	601	3.9
Sicily	25,832	245	1.0	353	1.4	452	1.6
Sardinia	24,100	706	2.9	857	3.6	1602	6.6
Total	302,066	12,405	4.1	25,398	8.4	32,961	10.9

Source: ISPRA (2018)

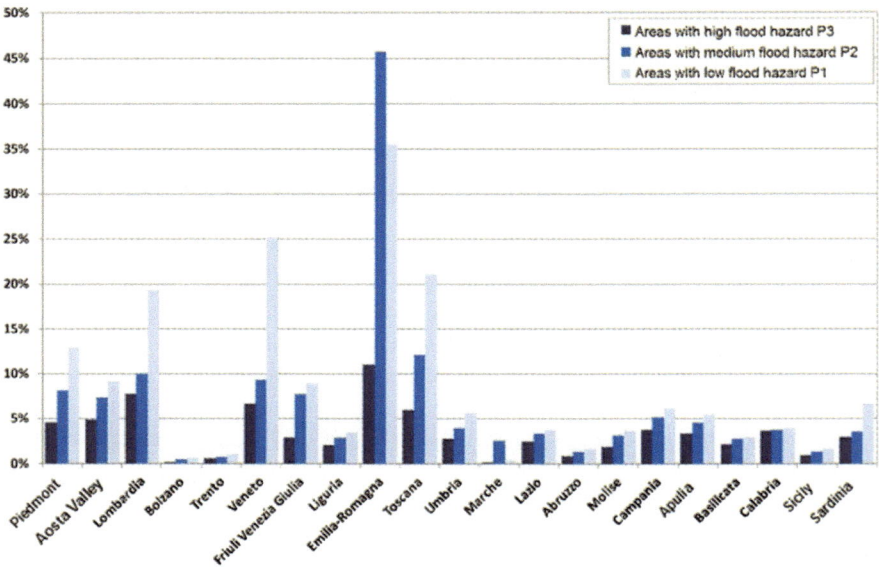

Fig. 11.5 Percentages of regional areas affected by the flood hazard of different probability. (Source: ISPRA 2018)

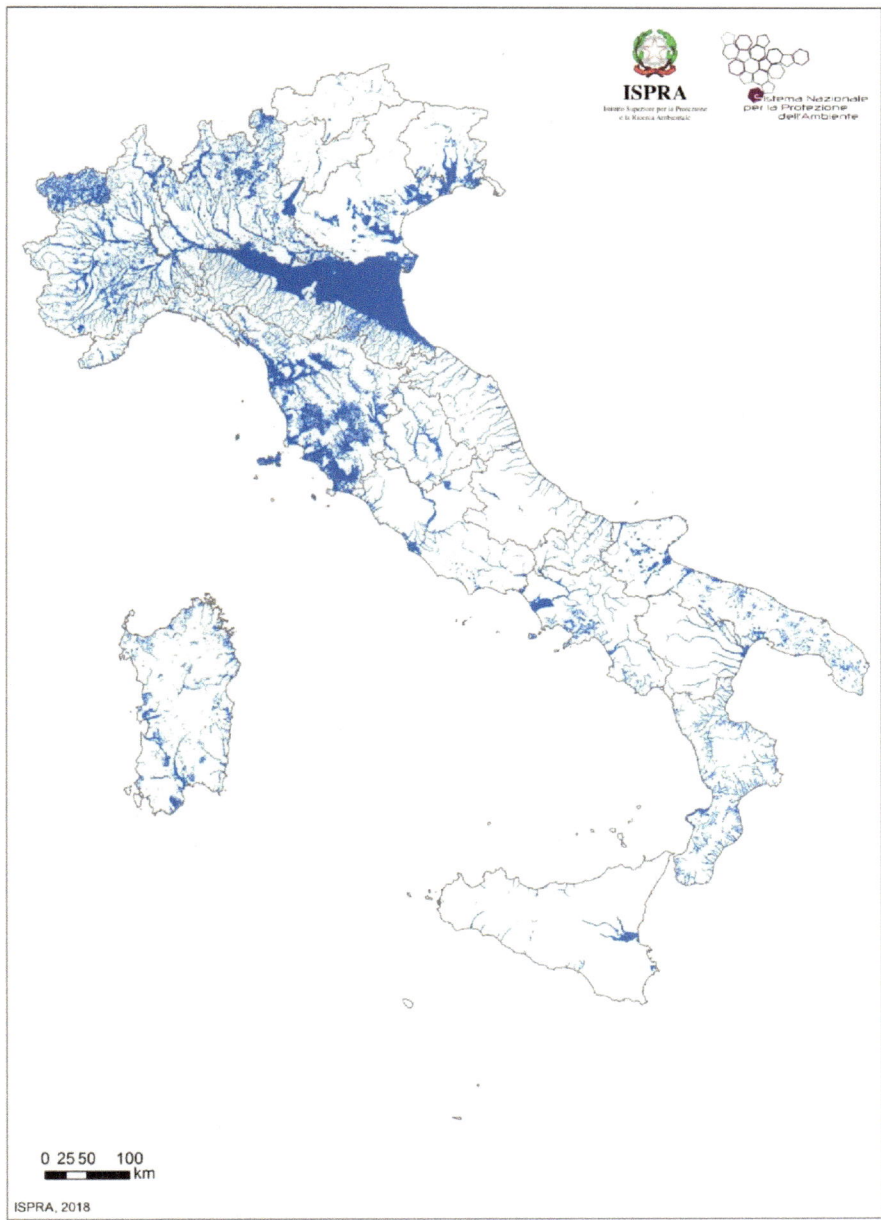

Fig. 11.6 Areas affected by medium flood hazard P2 (return periods 100–200 years). (Source: ISPRA 2018)

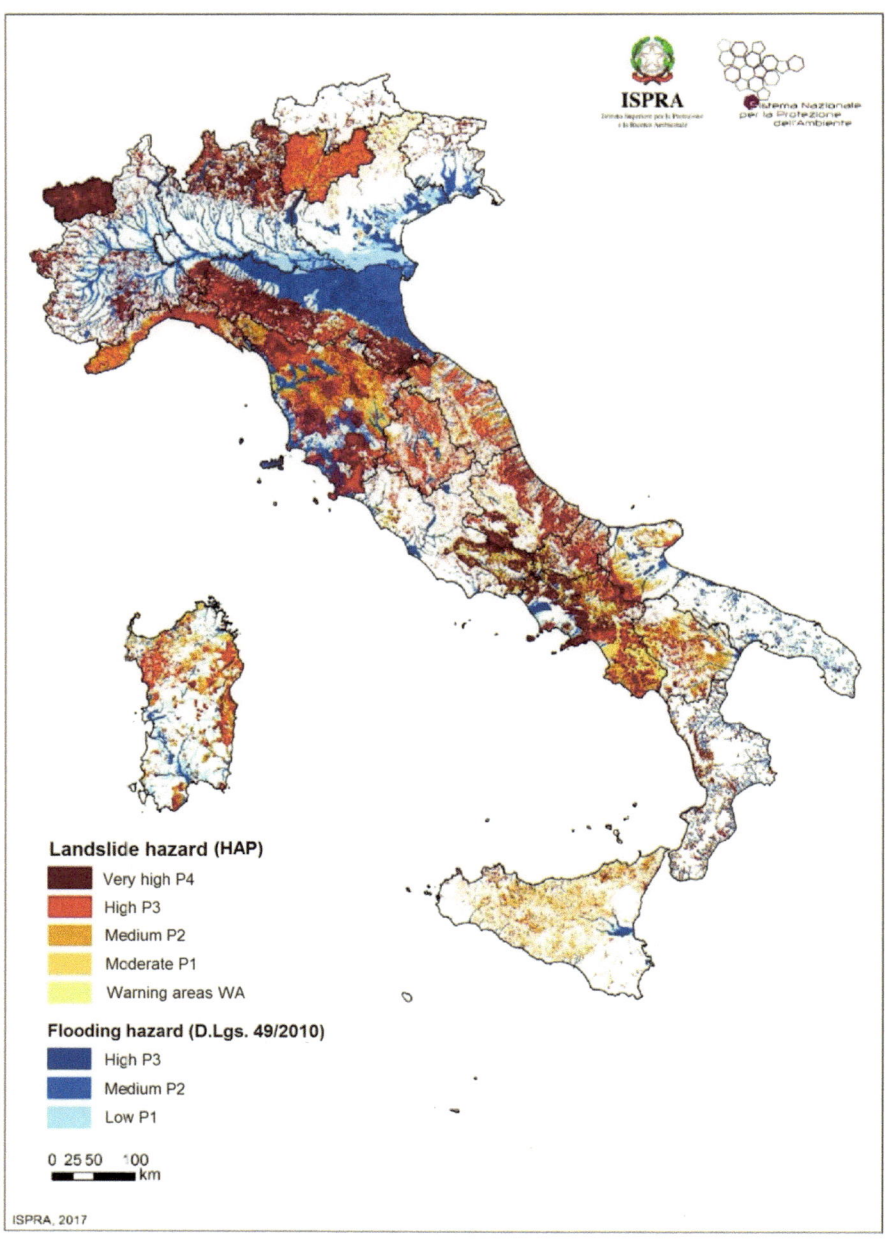

Fig. 11.7 Areas affected by landslide hazard (from Hydrogeological Asset Plan) and by flooding hazard (from Flood Risk Management Plan). (Source: ISPRA 2018)

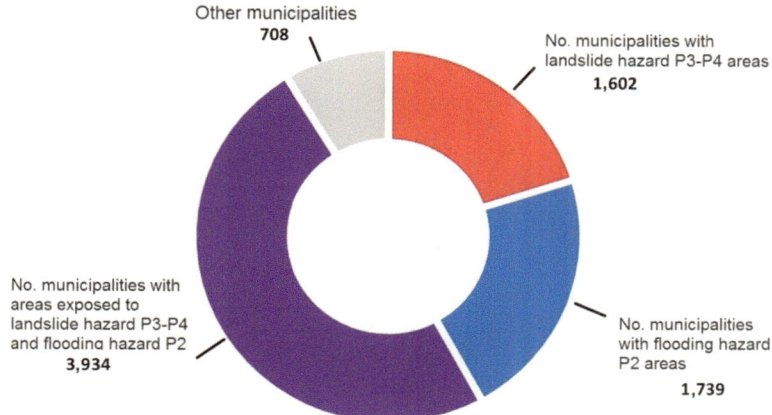

Fig. 11.8 Number of municipalities with areas affected by landside hazard and/or flooding hazard. (Source: ISPRA 2018)

tegic infrastructures and structures (motorway, rail-way, hospitals, schools, etc.); (iii) environmental, historic and cultural values of relevant interest; (iv) economic activities; and (v) installations concerning pollution prevention and control which might cause accidental pollution in case of flooding and protected areas (for withdrawal of water devoted to human consumption, for protection of aquatic species and habitat, etc.).

The mapping of the risk indicators, according to the synthesis by ISPRA (2018), provided many interesting results. The number of residential inhabitants affected by flood hazard has been estimated in 2,062,475 (3.5%) for high probability (T = 20–50 years), 6,183,364 (10.4%) for medium probability (T = 100–200 years) and 9,341,533 (15.7%) for low probability (T greater than 200 years). Figure 11.9 shows the number of residential inhabitants affected by the medium scenario in the Italian regions.

The number of buildings located in areas affected by flood hazard has been estimated in 487,895 (3.4%) for high probability P3, 1,351,578 (9.3%) for medium probability P2 and 2,051,126 (14.1%) for low probability. The number of industrial firms affected by flood hazard has been estimated at 197,266 (4.1%) for high probability, 596,254 (12.4%) for medium probability and 884,581 (18.4) for low probability. The related number of employed people has been evaluated at more of 2.2 million for the medium scenario.

The cultural heritage sites affected by flood hazard have been estimated in 13,865 (6.8 %) for high probability, in 31,137 (15.3%) for medium probability and in 39,426 (19.4 %) for low probability. Figure 11.10 shows the cultural heritage sites affected by the flood in the scenario of medium probability. The analysis carried out at local level has identified that the municipalities with the highest percentage for cultural heritage at risk (for medium flood scenario) are Venice, Ferrara, Florence, Genoa, Ravenna and Pisa. When considering the low probability scenario, Rome

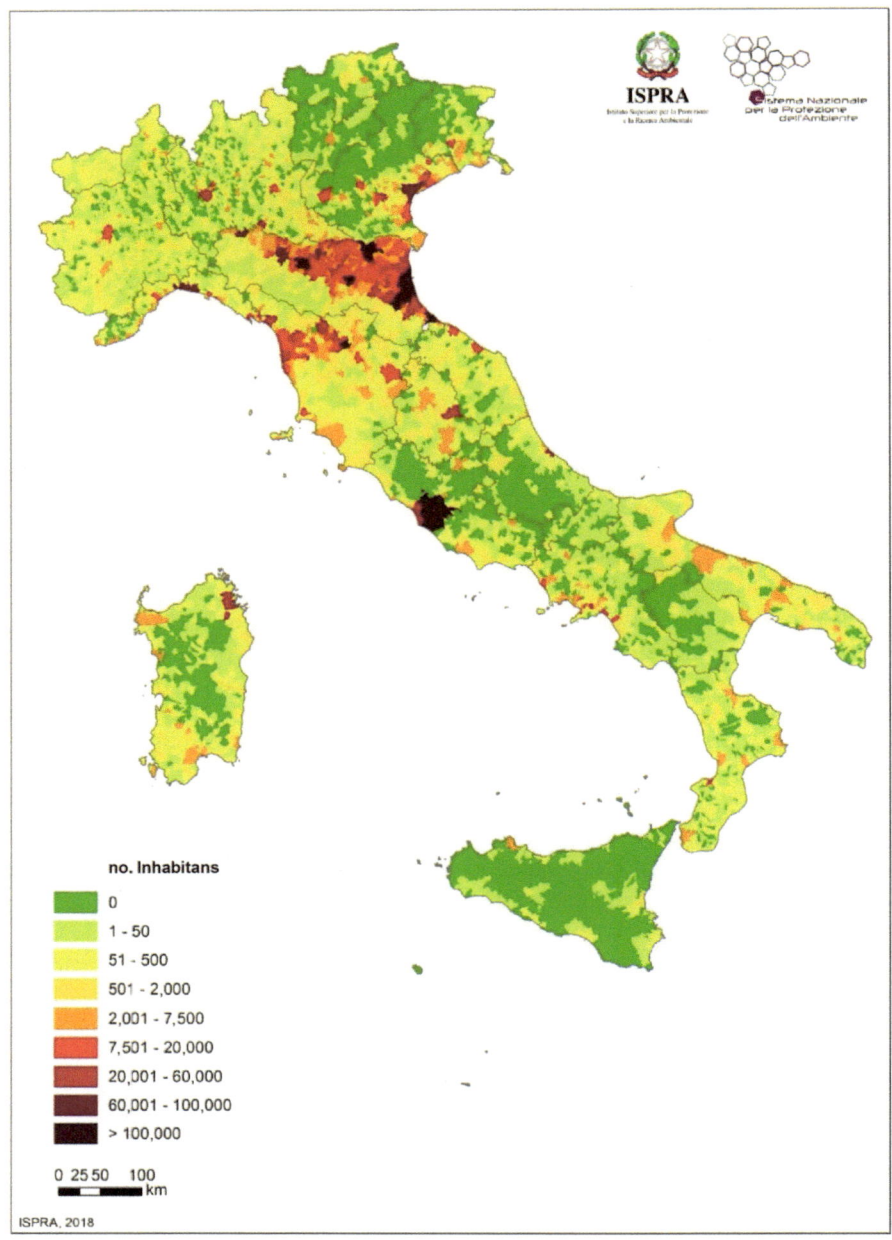

Fig. 11.9 Number of residential inhabitants potentially affected by medium flood hazard. (Source: ISPRA 2018)

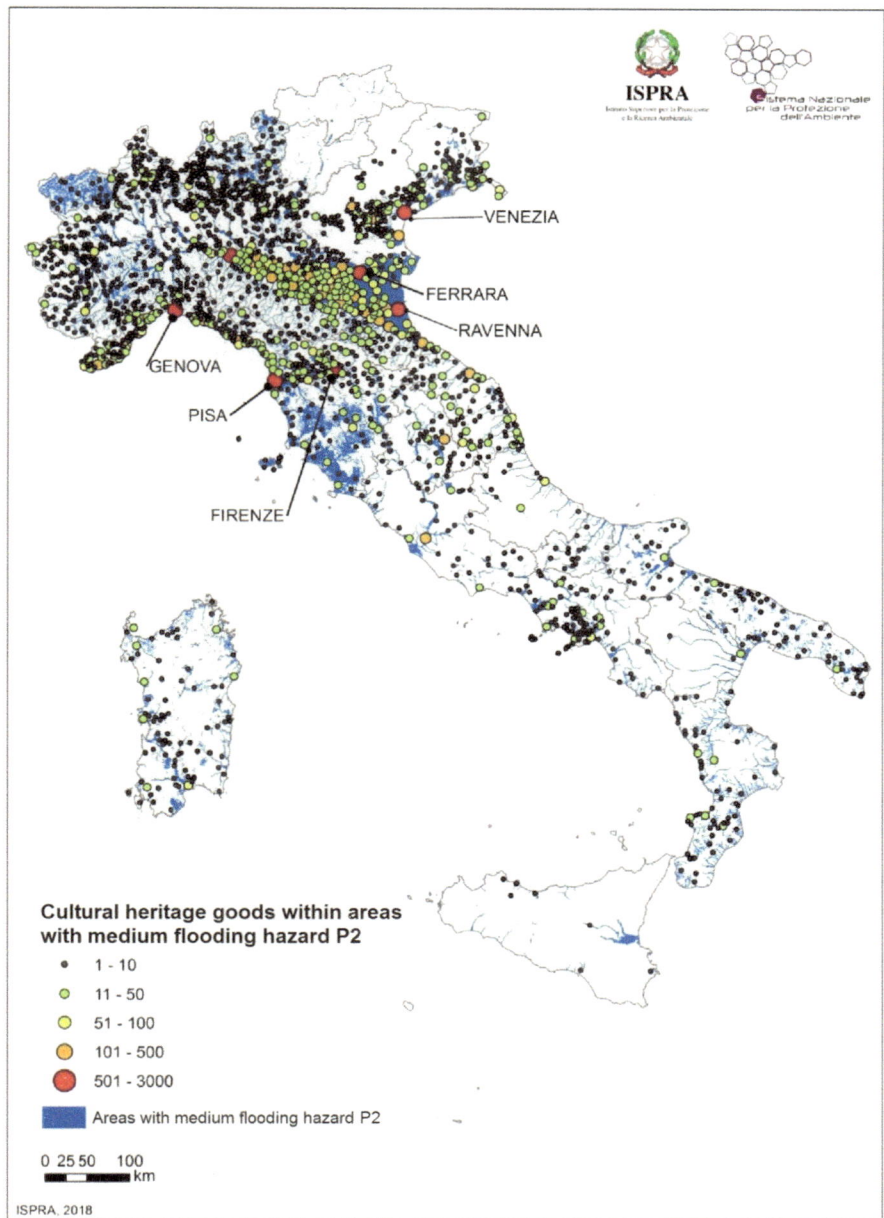

Fig. 11.10 Cultural heritage sites affected by medium flood hazard. (Source: ISPRA 2018)

must also be included. In particular the patrimony of architectonic, archaeological and monument sites at risk (medium scenario) in Florence has been estimated in 1259 (most of them have suffered the flooding event of 4 November 1966).

11.6 Expected Trends of Flood Risk Mitigation

Along many decades, defence from flood risk has been limited to the construction of hydraulic works. In particular, more ancient efforts, since the nineteenth century, were oriented to reduce damage in the valleys of the major rivers, completing or reinforcing the dikes along the watercourses (starting with the Po River) and recovering the marshes in coastal areas by means of land reclamation networks (e.g. valley of Reno, Pontine marshes, etc.). Only a few floodways were built (e.g. the Adige-Garda Lake). Also, the number of reservoirs devoted to flood routing is very limited (in total only seven) in the whole national territory.

Several hydraulic works were planned and built after the flooding disasters. The first example after the unification of Italy (1861) was the construction of the walls along the Tiber River in Rome after the flooding of 28 December 1870. The first planning tool referring to the whole country, namely, the River Regulation Plan (Law 184/1952), followed the Polesine disaster on Po River (14–18 November 1951).

Most of the financial resources to fund the works necessary to increase the safety of several cities were assigned after dramatic events which resulted in victims and severe damage (e.g. Trento and Florence, 4 November 1966). Later, the Law 183/1989 introduced the river basin plan as a comprehensive planning tool, thus giving a key role to the River Basin Authorities, similar in some aspects to the role of the *Agencies des basins* in France and *Water Authorities* in Great Britain.

If the Law 183/1989 emphasized the structural measures, the decrees established after the Sarno (1998) and Soverato (2000) disasters focused on the role of civil protection in order to reduce the timing for the implementation of actions aiming at safeguarding the more vulnerable areas. As a result, preparedness and recovery measures have increased, while inadequate funding and procedural delays in approving the projects of hydraulic structures and in entrusting contracts for public works have limited the construction of new structural measures.

While the positive results of civil protection actions are acknowledged, in particular for the development of warning systems based on the multi-functional centres and for emergency and recovery actions, the assessment of the flooding prevention and soil protection policies generally shows poor results in improving resilience to the flood risk in a major part of the territory. The reasons are the delays in the implementation of the District Authorities' role, the lack of coordination between the constraints imposed on the hazard and risk areas by the Hydrogeological Asset Plans and the choices of the urban and land planning, as well as the small and uncertain flow of funds for actions designed at prevention and protection.

What are the necessary modifications in the selection of measures aiming at achieving a more effective reduction of flood risk in the future?

Referring to the *Guidance for reporting under the Floods Directive* (E.C. 2013), it is convenient to adopt a classification of flood management measures into the following stages: prevention, protection by structural and non-structural measures, preparedness including the emergency response, recovery and review. This classification coincides substantially with that one suggested by the Chart of the Sendai Framework for Disaster Risk Reduction, established by the United Nations Office for Disaster Risk Reduction (UNISDR 2015). It includes many measures for each stage, as listed in Table 11.5.

Table 11.5 Measures to be adopted in the various stages of flood risk management (Source: E.C. 2013)

Stage/Measure	Description
Prevention	
Avoidance	Land use planning policies and regulation aimed at preventing the location of new additional "receptors" in flood-prone areas
Removal or relocation	Measures to remove or relocate "receptors" from flood-prone areas
Reduction	Measures to reduce the adverse consequences of flooding on building, public networks, etc.
Other prevention	Other measures, e.g. flood vulnerability assessment, maintenance programs, etc.
Protection	
Runoff and catchment management	Measures to reduce the flow into natural or artificial drainage systems (e.g. enhancement of infiltration, overland flow interceptors or storage, etc.) including in-channel, floodplain, etc.
Water flow regulation	Construction, modification or removal of water retaining structures (e.g. dams or online storage areas or flow regulation rules)
Channel, floodplain and coastal works	Construction, modification or removal of structures or alteration of channel, sediment dynamic management, dikes, etc.
Surface water management	Interventions to reduce surface water flooding, e.g. enhancing urban drainage systems
Other protection	Other measures, e.g. maintenance programs
Preparedness	
Flood forecasting and warning	Measures to establish or enhance forecasting or warning system
Emergency response planning	Measures to establish or enhance flood event institutional emergency response planning/contingency planning
Public awareness and preparedness	Measures to establish or enhance the public awareness for flood events or for reducing adverse consequences
Recovery and review	
Individual and societal recovery	Clean-up and restoration of building, infrastructures; health supporting actions (e.g. managing stress); disaster financial assistance; temporary or permanent relocation
Environmental recovery	Clean-up and restoration, e.g. mould protection, securing hazardous materials containers, etc.
Other recovery and review	Lessons learnt from flood events; insurance policies

The list proposed for the implementation of the European Flood Directive takes into account the recent directions of European policy concerning a high level of environmental protection in accordance with the principle of sustainable development and concerning the flexibility to be left to the local and regional authorities, according to the principles of proportionality and subsidiarity.

Priorities to increase the resilience of the Italian territory to the flooding risk include the following:

- It is necessary to achieve a more clear allocation of responsibilities among the district authorities (in terms of prevention measures), the regions, the civil protection and the local authorities (responsible for emergency measures) and to reinforce the key role of the District Authorities particularly for an authoritative coordination of the duties of regions within the district.
- The revision of the general river basin planning process should eliminate the current anomaly of two different plans aiming at coping with flood disasters: the Hydrogeological Asset Plan, established by the Law 267/1998 and the Flood Risk Management Plan. Today the presence of the two plans is maintained, since the second plan, according to the European Directive, does not include landslide risk;
- More homogeneous criteria should be adopted in the definition of the planning measures for flood risk mitigation among the districts operating in different parts of the country, and also more homogeneous criteria should be applied in the design of structural measures (according to the guidelines developed within the Safe Italy initiative) and in the choice of measures for emergency.
- Structural measures remain a key element of an effective flood defence policy, and the flood routing purpose should be achieved, if possible, in the reservoirs devoted to other purposes (e.g. agricultural supply, hydropower, ecosystem protection) by improving operation rules; a periodic maintenance is required on the old structures, and in some cases a re-assessment of the design flood should be carried out in order to take into account the new information provided by the available updated hydro-meteorological series as well as the consequences of expected climate change.
- An effort should be dedicated to consider the mapping of flood hazard not solely with regard to the probability of the hydrological event but also with respect to the reliability of existing structural measures. This is especially important, at the light of the insufficiency of monitoring and maintenance of existing structures such as levees, whose probability of failure may be higher than the probability of being overtopped by the river flow.
- More specific rules are required to implement the principle of hydraulic and hydrologic invariance, i.e. the provisions aiming to avoid the increase of the peak flow and flood volume notwithstanding the growth of impervious urban areas; also a larger use of flood-proofing measures, to be implemented by involvement

of private subjects, can contribute to reduce the vulnerability to the flooding hazard.

– In spite of the positive results obtained by the Multi-functional Civil Protection Centres in implementing the early warning system, a more effective management of emergency is required to better coordinate the meteorological/hydrological forecasting, the alert/alarm procedures and the operative actions under the municipal responsibility; in particular, a more accurate implementation of the emergency plan is required in the case of flash flood in urban areas, where the alert cannot be linked to the water depth in an above river cross section (as in streamflow forecasting of large rivers), but must be connected to the data of the short-duration high intense rainfall and weather radar information.

Besides the specific actions cited above, a few general changes of policy direction are necessary, such as the following:

– Particular attention should be paid to improve the functioning of the technical bodies of public administration by means of a careful revision of past reforms which led to the fragmentation of some national services (e.g. the Hydrographic Service) and transferred the Civil Engineering Offices (Genio Civile) to the regions with negative consequences on the homogeneity of the procedures.
– Sustainable land use should be achieved limiting the excessive transformation of agricultural land into urban areas and infrastructures; in some cases the reduction of the areas exposed to flood risk requires the demolition of unauthorized buildings in high-risk areas.
– It is necessary to establish once more a stronger link between the institutions which have responsibility for water governance and advise and the research institutions (universities, research centres) in order to improve the transfer of research results from the scientific communities to the bodies with operational duties and in order to direct some research efforts toward practical needs.
– Education and training activities in the water fields should be promoted by adopting a multidisciplinary approach but keeping most of contents (hydraulics, hydrology, geology, geo-mechanic, etc.), which have assured a high professional preparation for the development of water policies during the last centuries in Italy.
– A better awareness among the population about the correct behaviour to adopt during severe flooding events can contribute to reduce significantly mortality and economic damage. This objective requires that specific activities, devoted to informing the population on actions to take in case of floods, should be specifically financed and carried out, especially in schools, universities, etc., as well as through social media.
– It is not possible to postpone the initiatives aiming to guarantee the public participation in decision-making processes in the water field, through an effective transparency of the plans and programs and a real involvement of all stakeholders and citizens.

11.7   Key Approach to Improve Flood Management

Nowadays, an improved flood management is generally claimed as a guiding strategy to mitigate flood risk. This requires the contribution of a large range of disciplines to achieve a multidisciplinary approach and a coordinated effort at different levels of legislation and governance. Among the several shifts in the approach paradigms, invoked in Italy during last decades, a large consensus has been obtained on the following changes: (1) from the emergency management to the risk management, (2) from a simple structural approach to a combined structural and nonstructural actions approach and (3) from a top-down method to a shared responsibility between central government and subnational, regional and local authorities, according with the subsidiarity principle.

In spite of the differences between the Italian legislation framework on soil defence – aiming at facing both flooding and landslide risks – and the European Directive 2007/60/EC, which covers only the risk of flooding, both are inspired by the same principles: river basin is the territorial unit for planning; the risk concept includes the dimensions of hazard, exposure of persons and assets and vulnerability; and the choice of the solutions requires a careful understanding of the risk to be pursued by mapping of the different components of the risk and a well-calibrated combination of prevention, preparedness, recovery and rehabilitation measures.

Besides the specific measures referring to each stage of the flood risk reduction process, many general improvements are required in order to improve the institutional framework of water governance (political and technical branches of public administration) and the land use planning policies which should be able to reduce the land consumption and to remove the hazardous presence of buildings from flood-prone areas. The role of research and training, the increase of the awareness of people on correct behaviours to be adopted in advance and during the high-risk events and the public participation to the decision-making processes in water management are recognized at many levels as the key element to face the very complex goal of managing the flood disaster risk.

Appendix

Table 11.6 Main flooding disasters that occurred in Italy in the 1861–2017 period

Flood categories	Date	Affected area	Main river basins	Hydro-meteorological features	No. of fatalities/evacuees and homeless	Short description	Effects on defence policy
RF	26 December 1870	Rome city (Lazio region)	Tiber	Height of 17.22 m. and 3300 m³/s discharge in Tiber River at Ripetta	2/~6000	Overflow of Tiber flooded part of the centre of Rome	After the disaster, walls were built along the Tiber reach crossing Rome
UFF	26 September 1902	Modica, Ragusa province (Sicily region)	Scicli torrent	395 mm in 24 h (Giarratana); estimated peak flow 700 m³/s	112/	Flooding of three tributaries of Scicli torrent (two covered by urban roads) flooded the centre of Modica. Estimated damages 1–2 million of liras	
UFF	17 November 1908	Eastern Sicily region and Southern Calabria region	Riposto torrent	465 mm in 24 h, 150 mm in 25 min. (Riposto, Catania province)	10	Flooding of urban centres of Riposto and Giarre (Catania province)	
FDFMS	24 October 1910	Amalfi Coast and Ischia island (Campania region)			>250 (200 in Cetara, 20 in Maiori)	Severe storm and debris flows	

(continued)

Table 11.6 (continued)

Flood categories	Date	Affected area	Main river basins	Hydro-meteorological features	No. of fatalities/ evacuees and homeless	Short description	Effects on defence policy
FDL	1 December 1923	Valley of Scalve, Bergamo province (Lombardia region)			~500	The break of the Gleno dam in Pian del Gleno (BG), caused about 6 million cubic meters of water to flow out of the reservoir and covered the entire valley, until it ran out in the Iseo Lake about 45 min after the collapse	New rules for dam design
UFF	26 March 1924	Coast of Amalfi, Salerno province, (Campania region))			86/≥100		
UFF	21 February 1931	Palermo city (Sicily region)	Oreto, Papireto and Kemonia rivers	Max rain at Villa Pioppo (520 mm in 3 days, 411 mm in 24 h); peak flow of Oreto 248 m³/s	12/	Flooding of Palermo centre with max water height of 6 m.	Planning of regulation of river network near Palermo

FDL	13 August 1935	Valley of Orba, Alessandria province (Piedmont region)		400 mm in 8 h	>110/	A heavy rain affected the area on the border between the provinces of Alessandria and Genoa, causing the collapse of a portion of the Sella Zerbino dam (Molare, AL). The consequent flood wave from the reservoir produced fatalities and damages
RF	4 September and 12–14 September 1948	Asti, Alessandria, Cuneo, Torino and Vercelli provinces (Piedmont region)	Tanaro, Borbore and Belbo	About 250 mm in Borbore catchment; 350 mm in Belbo catchment.	23/≥790	Flooding of the Langhe (vineyards) with damages in the transport system. Estimated damage 20 billion of liras in Asti province
RF	1 October 1949	Benevento, Avellino, Caserta and Salerno provinces (Campania region)	Calore and Volturno	Max rain 246 mm in a day (S. Croce del Sannio). Peak flow 688 m³/s (h = 5.28 m) in Calore at Apice. Peak flow 3200 m³/s (h = 9.86 m) in Volturno at Ponte Annibale	32/1970	Flooding of Benevento and other municipalities. Overflow of Volturno river Estimated damage 7 billion of liras

(continued)

Table 11.6 (continued)

Flood categories	Date	Affected area	Main river basins	Hydro-meteorological features	No. of fatalities/evacuees and homeless	Short description	Effects on defence policy
RF	14–19 October 1951	Calabria, Sicily and Sardinia regions	Calabria: many torrents near Reggio Calabria. Sicily: Simeto, Sardinia: Flumendosa, Cetrino	Calabria: max rain 1770 mm in 4 days; 1595 mm in 72 h, 1068 in 48 h, 535 in 24 h (Petrace) Sicily: max rain 529 mm in 24 h, 166 mm in 3 h (Lentini Bonifica); Peak flow of Simeto 3300 m³/s (1547 km²) Sardinia 1014 mm in 2 days (1431 mm at Sicca d'Erba), 544 mm in 24 h	68/>8600 (Calabria); 10/>700 (Sicily); 6/≥280 (Sardinia)	Severe storms of very long duration caused main river flooding and landslides Aprico village (Reggio Calabria) was destroyed. Estimated damage in Reggio Calabria province 30 billion of liras	Embankments along Simeto River were designed for defence of the plain of Catania
FDL	8 November 1951	Tavernerio, Como province (Lombardia region)	Cosia torrent	120 mm in 24 h	18/	Heavy rains caused landslides and floods. At Tavernerio a landslide interrupted the flow of the Cosia torrent, leading to the formation of a temporary basin whose waters, at the breaking of the barrier, poured over the inhabited area	

	Date	Location	River/torrent	Rainfall/flow	Deaths/affected	Effects	Notes
RF	14–18 November 1951	Polesine, Rovigo province (Veneto region)	Po	About 300 mm over the Po river basin. Peak flow 12,800 m³/s at Piacenza	98/180,000	Break of banks of Po River in Rovigo province caused flooding of about 1170 km². 100 houses, 5000 buildings, 13,800 farms and 1130 km² of crops were damaged or destroyed. Estimated damages 400 billion of liras	A national plan for river regulation over the country was established (1954)
UFF	21 October 1953	Reggio Calabria, Cosenza, Vibo Valentia, Catanzaro provinces (Calabria region)	Valanidi	82.6 mm in 1 h. Overflow of Valanidi torrent (450 m³/s)	103/>3800		
UFF	25–26 October 1954	Coastal area of Salerno province (Campania Region)	Bonea and Cavaiola torrents	500 mm in 11 h, 95 mm in 1 h	325/12,000	Overflow of torrents and mud-slides from Cava dei Tirreni mountain hit urban areas of Salerno city and, Vietri and Maiori villages. Estimated damage 45 billion of liras	

(continued)

Table 11.6 (continued)

Flood categories	Date	Affected area	Main river basins	Hydro-meteorological features	No. of fatalities/ evacuees and homeless	Short description	Effects on defence policy
FDL	9 October 1963	Vajont valley, Pordenone province (Friuli-Venezia Giulia region) and Piave valley, Belluno province (Veneto region)	Piave	A flood wave of about 50 million of m³ overran the Vajont dam (261 m of height)	1917	A landslide (≥ 240 million m³) from Mount Toc fell into the Vajont reservoir and produced a wave that ran over the dam. The water destroyed some villages in the municipalities of Erto and Casso (Pordenone province), the town of Longarone and other villages in the province of Belluno. Estimated damage € 986 million	The study of landslide features have imposed an in-depth estimate of risk of side instability in the design of a reservoir
RF	4 November 1966	Florence and other towns of Tuscany region	Arno	Max rain 150 mm in 24 h. Peak flow of Arno in Florence 4000 m³/s. Rainfall during the previous October ≥200 mm	47 (38 in Florence and province)/>46,600	The heavy rains caused landslides and the overflow of the Arno, that flooded large part of the centre of Florence (4.85 m at the church of Santa Croce). Other villages in Tuscany were flooded. Young people (Mud Angels) saved artistic masterpiece and rare books. Estimated damage 300 billion of liras	After the disaster a national committee (presided by De Marchi) was established for flood defence over Italy. A plan to mitigate flood risk in the Arno basin through flood routing works was prepared

RF	2–3 November 1968	Biella, Asti, Vercelli, Novara, Verbano-Cusio-Ossola and Cuneo provinces (Piedmont region)	Tanaro, Bormida, Sesia, Ticino	Max rain 500 mm in 2 days. Peak flows: Sesia 3900 m³/s; Toce 2000 m³/s	86/>370	Main river flooding and landslides hit industrial areas of Biella and lands in Vercelli and Novara. Estimated damage 300 billion of liras.
RF	4 November 1966	Trentino-Alto Adige, Veneto, Friuli-Venezia Giulia and Lombardia regions	Adige, Piave, Livenza, TagliamentoBrenta-Bachiglione	Exceptional storms (>400 mm in 32 h, max 711 mm) and snow melting caused peak flows in Piave, Livenza, Tagliamento	87/>42,000	Overflow of Adige, Brenta-Bacchiglione, Piave, Livenza and tributaries caused many flooding events. High water at Venice (1.94 m) Estimated damage 15 billion of liras
UFF	7–8 October 1970	Genoa city and some municipalities of Genoa and Alessandria provinces (Liguria region)	Bisagno, Polcevera, Leiro	690 mm in 48 h, 480 mm in 24 h (Genova Molassana); peak flow of Bisagno 1000 m³/s; peak flow of Leira 510 m³/s	51/>1400	Max water height 2.0 m. Estimated damage 45 billion of liras

(continued)

Table 11.6 (continued)

Flood categories	Date	Affected area	Main river basins	Hydro-meteorological features	No. of fatalities/ evacuees and homeless	Short description	Effects on defence policy
RF	30 December 1972–3January 1973	Calabria region: Reggio Calabria and Catanzaro provinces Sicily region: Messina, Enna, Agrigento provinces	Torrents of Sila and Aspromonte mountains (Mesima, S.Agata, Bonamico and Corace) in Calabria. Torrents near Messina and South Imera and Platani rivers in Sicily	Calabria: max rain 433 mm in a day after previous heavy rains since 20 December 1972. Peak flow 430 m^3/s in Corace torrent Cumulated rain of almost 1500 mm over South Calabria (20 December 1972–3 January 1973) Sicily: Max rain 1000 mm in 4 days at Antillo; 640 mm in 5 days at Elicona. Peak flow 3036 m^3/s in Platani river (1247 km^2)	18/>5700	The torrents of Thyrrenian and Jonian sides of Calabria destroyed several bridges Landslides produced severe damages to building Also severe damages in many provinces of Sicily Estimated damages 900 billion of liras	Law 36/1973 provided 50 billion for people hit by the disasters and promoted the draft of plans for river basin assessment in Calabria and Sicily regions
FDL	19 July 1985	Valley of Stava, Trento province (Trentino-Alto Adige region		Mud flow of 160,000 m^3	268/~100	The break of two ponds for storing mine sediments caused a mud flow that destroyed Stava village. Estimated damage 8.5 billion Liras	A survey of small dams was made (Law 662/1985) and rules to estimate flooding areas below all dams was established

	Date	Location		Rainfall	Victims/Affected	Description	Notes
FDL	28 July 1987	Val Pola, Valtellina, Sondrio province (Lombardia region)		600 mm in 3 days (17–19 July 1987)	29 (due to the landslide)/~20,000	A landslide (30 million m³) produced a dam in the Adda river (lake capacity of about 20 million m³) with high risk down in the valley due to the break of dam (100 m. high) Estimated damage more of 1000 billion liras	An innovative method was tested for emptying the lake above the landslide dam: three pumping plants and pipes and a channel for controlled overflow (31 August 1987)
RF	5 November 1994	Alessandria, Asti, Biella, Cuneo, Torino, Vercelli and Verbano-Cusio-Ossola provinces (Piedmont region)	Tanaro	300 mm in 24 h	71/>5800	Main river flooding and landslides. The damages were estimated in 15,000–25,000 billion of Lire	
UFF	13 March 1995	Giarre, Mascali, Acireale municipalities, Catania province (Sicily region)	Urban roads	Max rain 376 mm in 12 h, 253 in 3 h, 144 mm in 1 h (Giarre-Jungo)	6/>60 (Other victims: 12 sailors of the sinked Greek ship *Pelhunter*	Flooding of centres of Giarre and Acireale; flooding of Presa (village of Mascali municipality)	
RF	19 June 1996	Lucca and Massa-Carrara provinces (Tuscany region)	Versilia	Max rain 478 mm in 11 h, 390 mm in 6 h, Rain peak 176 mm in 1 h	15/>700	Flooding of Versilia and debris-flows destroyed Cardoso village. Total damages estimated in 146 million Euro	

(continued)

Table 11.6 (continued)

Flood categories	Date	Affected area	Main river basins	Hydro-meteorological features	No. of fatalities/ evacuees and homeless	Short description	Effects on defence policy
FDFMS	5 May 1998	Sarno, Siano, Bracigliano, Salerno province and Quindici, Avellino province, (Campania region)		173 mm in 48 h after a rain of 614 mm in the previous 6 days	159/6000	Mud-slides of volcanic soils placed over saturated carbonate rocks struck the municipalities, located down the mountains. 178 buildings destroyed and 450 damaged	The Sarno Act (Law 267/1998) introduced the Hydrogeological Asset Plan (HAP)
FDFMS	16 December 1999	Cervinara and San Martino Valle Caudina, Avellino province, (Campania region)		325 mm in 48 h	6/1500	Mud-slides of volcanic soils located over saturated carbonate rocks with velocity of 14–15 m/s	
UFF	9 September 2000	Soverato and Jonian coast of Catanzaro and Reggio Calabria provinces (Calabria region)	T. Beltrame	630 mm in 3 days	13 victims/~200 (Soverato)	The flooding of Beltrame killed guests at a camping site	The Soverato Act (law 365/2000) introduced the procedures for Hydrological Asset Plans (HAP)
RF	13–16 October 2000	Aosta valley and Piedmont regions	Dora Baltea and Po	600 mm in 48 h	27/>15,500	Main river flooding and landslides	

FDFMS	1 October 2009	Giampilieri and other villages in Messina municipality, Scaletta Zanclea municipality and Itala municipality (Sicilia region)	225 mm in 9 h after severe rain in the preceding month (max 400 mm)	37/>2000	Several debris flows struck some villages of Messina and other municipalities of the eastern coast of Sicily. Estimated damages € 550 million including motorway, aqueduct etc.
UFF	25 October 2011	Cinque terre in La Spezia province (Liguria region), Lunigiana in La Spezia and Massa-Carrara province (Tuscania)	Max rain 542 mm in 30 h and 470 mm in 6 h (Brugnato)	14/>1500	Heavy precipitations have threatened the watersheds, causing floods and numerous debris flows and landslides
UFF	4 November 2011	Bisagno, Fereggiano, Sturla and Scrivia	Max rain 510 mm in 24 h (Rossiglione) and 469 in 24 h (Vicomorasso)	6/>150	Following heavy rains, widespread phenomena of geohydrological instability occurred, with extensive overflow, numerous flooding and landslides that caused extensive damages
UFF	18 November 2013	Olbia-Tempio, Nuoro, and Oristano provinces (Sardinia region)	Cedrino and Posada	Max rain 300 mm in 20 h. Peak flow of Posada 3000 m³/s	18/>1900	Flooding of Torpè village due to overflow from Posada

(continued)

Table 11.6 (continued)

Flood categories	Date	Affected area	Main river basins	Hydro-meteorological features	No. of fatalities/ evacuees and homeless	Short description	Effects on defence policy
UFF	8–9 October 2014 Genova (Liguria)	Genova city (Liguria region)	Bisagno, Fereggiano, Sturla	Max rain in Geirato (Genova) 754 mm in 5 days, 135 mm in 1 h. At Torriglia 513 mm in 5 days and 373 in 1 day	1/~200	Heavy rains caused the flooding of the watercourses of Genova resulting in serious damages, estimated at € 250 million of public goods and € 100 million of business goods	
RF	15 November 2014	Liguria, Piedmont, Lombardy regions	Polcevera, Bormida, Seveso, Niguarda, Lambro	Max 238 mm in a day at Alessandria	9/>2000	The heavy rains caused landslides in mountainous and hilly areas of Liguria (Genoa province), Piedmont (Biella province) and Lombardy (Varese province) and floods along the Bormida river at Alessandria city, Lambro and Seveso rivers at Milan city	
UFF	9–10 September 2017	Livorno city (Tuscany region)	Ardenza and Maggiore	Max rain 42 mm in 15 min (Quercianella). Total rainfall 260 mm.	8/20	Due to heavy rains the Ardenza and Maggiore torrents overflowed and flooded the south part of Livorno	The flood with 200 years return period adopted for a flood routing facility resulted inadequate

RF River flooding, *UFF* Urban flooding due to flash flood, *FDBL* Flooding due to dam break or to landslide, *FDFMS* Flooding connected with debris flow or mudslide

References

Accademia Nazionale dei Lincei (2013) Cosa non funziona nella difesa dal rischio idro-geologico nel nostro paese? Analisi e rimedi [What does not work in the hydrogeological risk reduction in Italy? Analyses and solutions](in Italian). Scienze e Lettere Editrice, Roma, p 299

AII Italian Hydrotechnical Association (2010) Le alluvioni in Italia [Floodings in Italy] (in Italian). Di Virgilio Editore, Roma, p 348

Botta G (1977) Difesa del suolo e volontà politica. Inondazioni fluviali e frane in Italia: 1946–1976 [Soil defence and political will. Flooding and landslides in Italy: 1946–1976] (in Italian). Franco Angeli, Milano, p 140

CNR (1999–2019) Sistema Informativo sulle catastrofi idrogeologiche [Information System on Hydrogeological disasters]. sici.irpi.cnr.it/storici.htm

Commissione Interministeriale per lo studio della sistemazione idraulica e della difesa del suolo (1970) Relazione conclusiva [Interministerial Commission for the study of hydraulic regulation and soil defense Final Report] (in Italian). Istituto Poligrafico dello Stato, Roma, p 900

Consiglio Nazionale delle Ricerche, Istituto di Ricerca per la Protezione Idrogeologica. www.irpi.cnr.it. Accessed 2Apr 2019

E.C (2013) Guidance for reporting under the Floods Directive (2007/60/EC), Guidance Document n. 29. European Communities, Luxembourg

Guzzetti F, Cardinali M, Reichenbach P (1994) The AVI Project: a bibliographical and archive inventory of landslides and floods in Italy. Environ Manag 18:623–633

ISPRA (Trigila A, Iadanza C, Bussettini M, Lastoria B) (2018) Dissesto idrogeologico in Italia: pericolosità e indicatori di rischio [Hydro-geological disasters in Italy: hazard and risk indicators] (in Italian). ISPRA, Rome

Napolitano E, Marchesini I, Salvati P, Donnini M, Bianchi C, Guzzetti F (2018) "LAND-deFeND" An innovative database structure for landslides and floods and their consequences. J Environ Manag 207:203–218. https://doi.org/10.1016/j.jenvman.2017.11.022

Piccoli A (1972) Esame delle piene verificatesi nel novembre 1966 e loro confronto con precedenti analoghi eventi [Analysis of floods in November 1966 and comparison with previous similar events]. In: Accademia dei Lincei, Piene: loro previsione e difesa del suolo [Floods: forecasting and soil defence] (in Italian). Bardi, Rome, pp 155–178

Polaris. http://polaris.irpi.cnr.it/report/last-report/. Accessed 2 Apr 2019

Rosso R (2017) Bombe d'acqua [Stormwaters] (in Italian). Marsilio, Venice, p 278

Salvati P, Bianchi C, Fiorucci F, Giostrella P, Marchesini I, Guzzetti F (2014) Perception of flood and landslide risk in Italy: a preliminary analysis. Nat Hazards Earth Syst Sci 14:2589–2603

Salvati P, Bianchi C, Fiorucci F, Marchesini I, Rossi M, Guzzetti F (2015) Chapter 22: Flood risk to the population and its temporal variation in Italy. In: Moramarco T, Barbetta S, Brocca L (eds) Advances in watershed hydrology. Water Resources Publications, LLC, Chelsea, pp 433–456

Salvati P, Pernice U, Bianchi C, Fiorucci F, Marchesini I, Guzzetti F (2016) Communication strategies to address geohydrological risks: the POLARIS web initiative in Italy. Nat Hazards Earth Syst Sci 16:1487–1497. http://www.nat-hazards-earth-syst-sci.net/16/1487/2016/. https://doi.org/10.5194/nhess-16-1487-2016

Salvati P, Petrucci O, Bianchi C, Rossi M, Guzzetti F (2017) Gender, age and circumstances analysis of flood and landslide fatalities in Italy. Sci Total Environ 2018:867–879. https://doi.org/10.1016/j.scitotenv.2017.08.064

UNISDR (United Nations Office for Disaster Risk Reduction) (2015) Chart of the Sendai framework for disaster risk reduction. www.unisdr.org

Chapter 12
Coping with Droughts

Giuseppe Rossi

Abstract Drought represents a serious threat to agriculture, water supply and environment. Nevertheless, only recently Italy has started to adopt an effective response to drought risk, shifting from a crisis management (reactive) approach towards a risk management (proactive) strategy. In this chapter, the drought features within the hydrological cycle and the principles suggested at international level for disaster reduction and the guidelines issued by the European Union on drought and water scarcity are described. Then, a summary of the main strategic and operational drought mitigation measures is presented together with the description of the role of drought monitoring. Afterwards, a recall of the Italian legislative framework to cope with droughts and to prevent or mitigate water shortage in supply systems is presented, accompanied by a short description of the most severe drought events occurred in Italy in the last century, which allows to draw a few lessons for the future. Some indications on the methods to assess drought-related water shortage risk in water supply systems in Italy are finally provided.

12.1 Introduction

Several drought events occurred in Italy have caused relevant damages to many of its socio-economic sectors and environment. Nevertheless, it is only in recent years that Italy has started to pay particular attention to drought issues. Crisis management has been the prevailing approach to respond to drought in the past, and in fact, since the 1990s, the National Civil Protection Agency has been the main actor in charge for drought response. As a consequence, there has been a lack of interest towards the implementation of measures aimed at preventing drought impacts, especially on water supply systems. In fact, only recently a risk management approach has been adopted, fostered by European recommendations, such as the

G. Rossi (✉)
Department of Civil Engineering and Architecture, University of Catania, Catania, Italy
e-mail: grossi@dica.unict.it

© Springer Nature Switzerland AG 2020
G. Rossi, M. Benedini (eds.), *Water Resources of Italy*, World Water Resources 5,
https://doi.org/10.1007/978-3-030-36460-1_12

Drought Management Plan Report (EC 2007) and the *Report on Water Scarcity and Droughts in the European Union* (EC 2011).

In the first part of this chapter, the evolution of the approaches to drought management is addressed (Sect. 12.2), and the criteria for risk management suggested by the International Strategy for Disaster Reduction, as well as the specific recommendations for drought risk management issued by the European Union (Sect. 12.3), are presented. An analysis of the main typologies of measures that can be adopted for drought prevention and mitigation is carried out (Sect. 12.4) and the role of drought monitoring is discussed (Sect. 12.5). In the second part, the Italian legislative framework regarding drought prevention and mitigation is recalled (Sect. 12.6), the most severe drought events that occurred in Italy in the last century are described (Sect. 12.7), and a few lessons learnt from recent droughts in Italy are discussed (Sect. 12.8). Objectives and methodologies to assess the risk of water shortage due to drought in supply systems are presented, and Italian experiences are described (12.9). Finally, a few concluding remarks are drawn.

12.2 Principles of Drought Management

Drought is defined as a temporary condition of a severe reduction of water availability compared to normal values, lasting a significant amount of time and affecting a large region (Rossi et al. 1992). Although drought is a natural phenomenon, as it stems from the variability of meteorological conditions – in particular, precipitation – it can be considered a disaster. Thus, similar to other disasters, the severity of its impacts on society depends on the vulnerability of water supply systems and of economic and social sectors, as well as on the preparedness to implement appropriate mitigation measures (Mishra and Singh 2010). Drought phenomena exhibit different features in the various components of the natural hydrological cycle, which in turn cause different impacts on the water resource systems (Fig. 12.1). In particular an initial reduction of precipitation with respect to normal conditions (*meteorological drought*) affects different processes within the hydrological cycle, thus determining soil moisture deficit (*agricultural drought*), as well as streamflow and groundwater deficits (*hydrological drought*).

According to the strategies for natural disasters mitigation, *drought risk* refers to the expected loss (in economic and/or social terms) caused by a drought event. It is evaluated as function of drought hazard, exposure and vulnerability. *Drought hazard* is taken into account through the probability of drought occurrence, whereas the *vulnerability* describes the degree of loss for a given element exposed to drought risk.

The risk of water shortage in water supply systems differs from natural drought risk because *water shortage* results from an imbalance between water supply and demand, caused by a meteorological/hydrological phenomenon. Moreover, anthropogenic factors – such as demand pattern development, supply infrastructures and management strategies and especially types of measures adopted for coping with drought – could exacerbate or reduce water shortage.

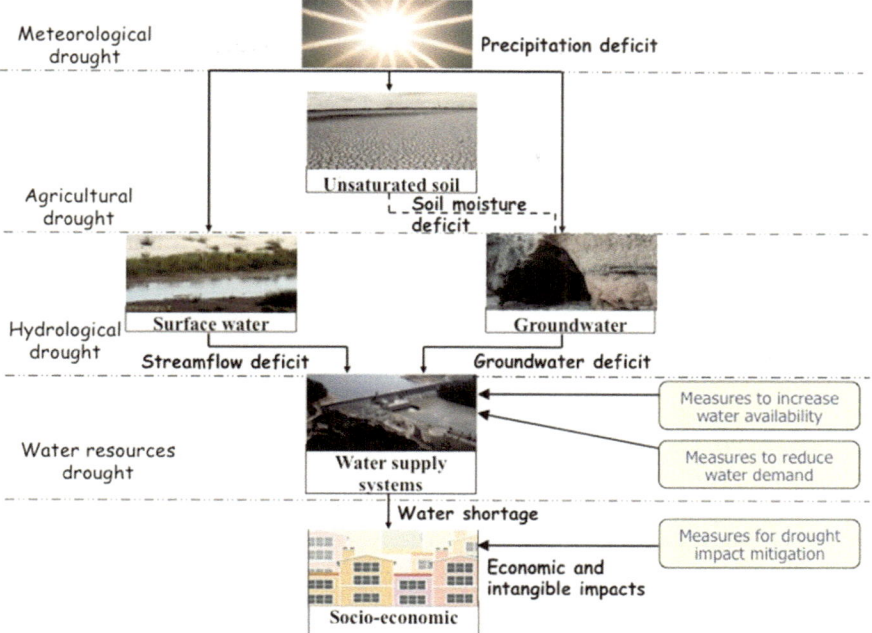

Fig. 12.1 Evolution of drought within the hydrological cycle and water shortage in water supply systems. (Source: Rossi 2017)

As a consequence, drought management presents some differences with respect to managing other natural disasters: (1) prevention actions may be effectively planned since drought effects evolve slowly along a large time span; (2) strategic measures for improving drought preparedness are generally more complex, since the spectrum of potential long-term actions is very large; and (3) operational measures, to be implemented at drought inception, require an adaptive response, due to the dynamic feature of phenomenon. In particular the operational measures should take into account the uncertainty in drought evolution, which can yield a duration and a severity different than those considered in the planning stage (Rossi 2017).

As mentioned in the introduction, drought risk has been traditionally managed by a *crisis management* approach. Although this approach still represents the most common response to drought at local, national and international levels, there is an increasing awareness about its weaknesses. In fact, since it is based on last-minute decisions, it generally leads to expensive actions, with unbearable environmental and social impacts. Thus, a shift towards a *risk management* approach, based on measures planned in advance, has been progressively advocated (Yevjevich et al. 1983; Wilhite 1987; Rossi et al. 1992; Wilhite 2000). Today, such a shift is emphasized in policy instruments adopted in drought-prone countries such as Australia (Botterill and Wilhite 2005), South Africa and USA (Wilhite et al. 2005). It is also suggested by international or European recommendations (UNISDR 2014; EC

2007, 2011), as well as advocated by research projects on drought (see, e.g. MEDROPLAN 2007). Nonetheless, for several countries, including Italy, it is not yet adequately transferred into the legislative and institutional framework on water resources management.

12.3   Drought Risk Management: International and European Recommendations

Risk management is the approach suggested and adopted by the United Nations International Strategy for Disaster Reduction (UNISDR). In particular, the general procedure, established at the World Conference on Disaster Reduction within the "Hyogo Framework for Action 2005–2015", has identified the following priorities to build resilience of nations and communities to drought (UNISDR 2014): (1) policy and governance; (2) drought risk identification and early warning; (3) awareness and education; (4) reducing underlying factors of drought risk; and (5) mitigation and preparedness, as well as cross-cutting issues. The most recent UNISDR recommendations have been included in the Sendai Framework for Disaster Risk 2015–2030 (UNISDR 2015), highlighting four priorities for action: (1) understanding disaster risk, (2) strengthening disaster risk governance, (3) investing in disaster risk reduction for resilience and (4) enhancing disaster preparedness for effective response and to "Build Back Better" in recovery, rehabilitation and reconstruction. The guiding principles emphasize the primary responsibility of states to prevent and reduce disaster risk, the empowerment of local authorities and communities through resources, incentives and decision-making responsibilities, the coherence between disaster risk reduction actions and sustainable development principles, and the consideration of local specific features of disaster risk.

Several countries, affected by frequent severe droughts, have developed national drought policies in recent decades. In many cases, the planning legislation to cope with drought refers to principles and criteria developed either by the UN Convention for Combating Desertification (UN Secretariat General 1994) or from the UN International Strategy for Disaster Reduction (UNISDR 2009). Also some laws refer to the Integrated Water Resources Management (IWRM) paradigm, introduced in the UN Conference on Water in the Mar del Plata (1977), and emphasized during several international water events, such as World Water Forums. Nonetheless differences in drought policies implementations are strongly affected by the national legal framework and by the structure of the institutions which share water resources governance. A comparison of drought policies implemented in Australia, South Africa and USA by Wilhite et al. (2005) points out that, despite the differences, common strategies are used to address the goal of reducing societal vulnerability to drought. These common strategies include monitoring and early warning of droughts, assessment of drought risk and adoption of a mixture of preparedness measures.

Until recent years, European water policies paid little attention to drought issues in terms of technical and financial instruments and legislative acts. The Water Framework Directive 2000/60 (WFD) promoted a complex water resources planning process at basin level, aimed at preserving or improving water quality for eco-system and human use protection. However, the WFD treated drought only marginally, in spite of mentioning drought as one of its objectives. Indeed, drought events are considered only as one of the exceptional cases that allow a derogation of good ecological status requirements of the affected water bodies. In order to overcome these weaknesses in EU water policy, the EU Water Scarcity and Drought Expert Network developed a guidance document on drought preparedness and miti-gation (EC 2007) with the proposal of drafting a "Drought Management Plan" (DMP), as an Annex to the "River Basin Management Plan" (RBMP). Such a DMP should be prepared by the same body responsible for basin planning, i.e. the River Basin (or District) Authority. In spite of the fact that the DMP is not mandatory for member states, it aims at extending goals and criteria of WFD (in a similar way to the EU Flood Directive 2007/60) to improve drought management and in particular to reduce the vulnerability of the water supply systems, as well as to mitigate drought impacts. Its specific objectives are (1) to ensure sufficient water availability to cover essential human needs to safeguard population's health and life, (2) to avoid or minimize negative drought impacts on water bodies and (3) to minimize negative effects on economic activities (see Rossi 2009).

On the basis of the successive work of the EU Water Scarcity and Drought Expert Network, the Report from the Commission to the European Parliament and the Council on these issues (EC 2011) has suggested to revise the related European policy including some specific topics listed in Table 12.1.

At EU level, discussions are ongoing on how member states should incorporate climate change issues into the implementation of EU water policy, in order to mini-

Table 12.1 Recommendations to revise European policy on water scarcity and drought

Improving water efficiency
Introducing water-saving devices and practices in buildings and improving water-efficient construction.
Reducing water leakages in supply distribution systems.
Improving efficiency in agricultural use of water .
Achieving better planning and preparedness to deal with droughts
Integrating actions against water scarcity and drought into other sectorial policies (agriculture, households, industry).
Assessing the adequacy of the River Basin Management Plans on water scarcity and drought issues.
Further developing the prototype of an observatory and early warning system on drought.
Defining a more comprehensive list of indicators on water scarcity and drought and of vulnerability of water resources.
Developing adequate implementation instruments of financing, water pricing, water allocation research and education
Encouraging EU funding of natural risks through Cohesion Policy, Regional Policy, Solidarity Fund and Economic Recovery Plan and reforming Common Agricultural Policy.
Making the national rules more restrictive to authorize water abstractions.
Developing new research projects on vulnerability and increased drought risk.
Introducing new educational programmes and awareness-raising campaigns .

Source: Rossi and Cancelliere (2013)

mize vulnerability to future climate and to fight possible emergencies by means of specific response actions. According to the White Paper on *Adapting to climate change* (CEC 2009), one of the strategies proposed to increase resilience to climate change consisted in the improvement of the management of water resources through the enhancement of water efficiency in agriculture, households and buildings. Also, the expected revision of the WFD and of the water scarcity and droughts strategy includes options to increase drought resilience. Furthermore, revisions of the River Basin Management Plans (due in 2021) could take advantage from the incorporation of climate change effects within basin planning, through analysis of the pressures on water bodies, definition of the phenomena to be monitored and verification of the resilience of the action program to climate change. Among European countries, Spain has probably developed one of the most advanced legislation systems for drought management, capitalizing on the existing central role of River Basin Authorities in water management. In particular, according to the National Hydrological Plan Act (2001), Drought Management Plans have been adopted by the River Basin Authorities for all districts, and a national drought indicator system has been established (Estrela and Vargas 2012). More recently, the newest Drought Management Plan, approved in 2018, introduced two national indicator systems: (1) the drought indicator system and (2) the water scarcity indicator system. The first system is aimed at the detection of prolonged drought in rivers and the definition of the actions to be taken for limiting temporary water quality deterioration. The second system aims at the timely detection of reduced availability of water resources, in cases where there is a risk of water shortage. The final aim of this system is then to define the measures to delay or avoid the most severe phases of water shortage, by distinguishing for a specific sub-basin the scenarios (i.e. normal, pre-alert, alert and emergency) using a set of hydrological variables (streamflow and groundwater level) and reservoir storage indicators.

12.4   Drought Mitigation Measures

In literature, several classifications of drought mitigation measures are available. A consolidated classification of drought mitigation measures distinguishes three main categories of measures, oriented at (*i*) increasing water supply, (*ii*) reducing water demands and (*iii*) minimizing drought impacts (Yevjevich et al. 1983). Other traditional classifications are based on the type of approach to drought management, either reactive or proactive, or relate on the timing of their implementation (Rossi 2000). In particular, *long-term measures* include the measures aimed at improving drought preparedness through a set of structural and non-structural adjustments to an existing water supply system. *Short-term measures*, defined within a contingency plan, are designed to mitigate ongoing drought events, through actions oriented to improve water supply by using additional water resources or to reduce water demand (Dziegielewski 2000).

In Table 12.2, a list of long-term and short-term mitigation measures is proposed, distinguishing the main categories of actions above-mentioned and the different sectors of application (urban, agricultural, industrial and recreational).

Table 12.2 Long-term and short-term drought mitigation measures

Category	Long-term actions	Affected sectors			
Demand reduction	Economic incentives for water saving	U	A	I	R
	Agronomic techniques for reducing water consumption		A		
	Dry crops instead of irrigated crops		A		
	Dual distribution network for urban use	U			
	Water recycling in industries			I	
Water supply increase	Conveyance networks for bi-directional exchanges	U	A	I	
	Reuse of treated wastewater		A	I	R
	Inter-basin and within-basin water transfers	U	A	I	R
	Construction of new reservoirs or increase of storage volume of existing reservoirs	U	A	I	
	Construction of farm ponds		A		
	Desalination of brackish or saline waters	U	A		R
	Control of seepage and evaporation losses	U	A	I	
Impacts minimization	Education activities for improving drought preparedness and/or permanent water saving	U	A	I	
	Reallocation of water resources based on water quality requirements	U	A	I	R
	Development of early warning systems	U	A	I	R
	Implementation of a Drought Contingency Plan	U	A	I	R
	Insurance programs		A	I	

Category	Short-term actions	Affected sectors			
Demand reduction	Public information campaign for water saving	U	A	I	R
	Restriction in some urban water uses (i.e. car washing, gardening, etc.)	U			
	Restriction of irrigation of annual crops		A		
	Pricing	U	A	I	R
	Mandatory rationing	U	A	I	R
Water supply increase	Improvement of existing water systems efficiency (leak detection programs, new operating rules, etc.)	U	A	I	
	Use of additional sources of low quality or high exploitation cost	U	A	I	R
	Overexploitation of aquifers or use of groundwater reserves	U	A	I	
	Increased diversion by relaxing ecological or recreational use constraints	U	A	I	R
Impacts minimization	Temporary reallocation of water resources	U	A	I	R
	Public aids to compensate income losses	U	A	I	
	Tax reduction or delay of payment deadline	U	A	I	
	Public aids for crops insurance		A		

Source: Rossi (2017)
U urban, *A* agricultural, *I* industrial, *R* recreational

Regardless of the adopted classification, after a set of potential mitigation measures are identified, selection of the best combination should be based on a comparison and ranking of the performance of each measure in mitigating negative impacts of drought as well as of the main economic, environmental and social consequences of the adopted measures. To this end, multi-criteria approaches have been proposed since the pioneering work by Duckstein (1983). The NAIADE model has been applied to different complex water systems, in order to rank the alternative combinations of long-term and short-term measures for reducing shortage risk in a water supply system (see Munda et al. 1998; Rossi et al. 2005).

12.5 Role of Drought Monitoring

Several methods have been developed for drought identification and the estimation of its severity, as well as for the subsequent assessment of drought vulnerability referred to a specific area and/or a water supply system. An objective procedure to identify the onset and the end of a drought and to evaluate its characteristics (duration and severity) was proposed by the pioneering work of Yevjevich (1967), based on the *run method* to be applied at the time series of the variable of interest. The run method has been largely used for probabilistic characterization of drought by means of univariate, bivariate and spatial-temporal analyses (Bonaccorso et al. 2003; Cancelliere and Salas 2010; Mishra and Singh 2011).

Assessment of ongoing drought conditions and water shortages in water supply systems is a crucial step for drought management. To this end, a set of indices measuring the anomaly from a "normal" condition in terms of one or more meteorological or hydrological variables can be employed. In literature, several indices for drought monitoring have been proposed, which differ for the selected variable, the time scale of analysis, the definition of "normal" conditions, the way of computing the anomaly (e.g. difference, ratio, etc.) and the standardization method.

Several reviews and classifications of drought indices have been made in the last decades. Examples can be found in MEDROPLAN (2007) and on websites of drought monitoring and information services such as National Drought Information Center (NDIC 2011). Generally, drought indices are categorized in meteorological, hydrological, agricultural, remote-sensing-based drought indices. Some indices also attempt to combine different data related to different variables (e.g. precipitation, soil, water content, etc.) and/or to merge the information from several indices into an indicator that takes into account also the status of water reserves.

A drought monitoring system has a key role to assess drought risk and to define preparedness and mitigation measures. The indices to be used in a drought monitoring system must satisfy several requisites in order to provide an effective early perception of the severity of the phenomenon and its impacts. A few requisites have been identified in Rossi (2017). In particular, the indices should be appropriate for:

1. Representing the complex interrelation between meteorological and hydrological components of a significant reduction of water availability.
2. Making use of real-time easy available hydrometeorological data.
3. Being able to describe drought conditions even in a drought's early stage.
4. Providing comparability of drought events both in time and space.
5. Describing in some way drought impact.
6. Assessing the severity of the current drought so to induce decision-makers to effectively activate drought mitigation actions.

Regardless of the selected indices, effective drought early warning must rely on a network of meteorological and hydrological stations, including remote sensing devices, an advanced DSS running on adequate computer facilities (adequate computational power and ease of use).

12.6 Legislation Framework to Cope with Drought in Italy

In spite of several drought events which hit several Italian regions with dramatic impacts, drought management has been marginally covered by Italian legislation until recent years. The Law 183/1989, regulating soil conservation and water supply, listed the long-term measures to improve drought preparedness among the measures to be included in the river basin plan by the authority of basin. However, this provision did not find application in most of the river basin plans. Actually, the actions to mitigate the drought effects have been carried out by the civil protection system according to Law 225/1992. Most of the measures were emergency actions, generally implemented by local commissioners that operated according to a deliberation drafted after an occurring drought was recognized as a natural disaster to be faced through emergency measures.

An attempt to introduce drought mitigation concepts with reference to municipal water supply was established by the Prime Minister Decree 47/1996, which, with reference to the Law 36/1994, required to introduce the assessment of water deficiency risk in municipal supply systems and the proposal for water crisis prevention within the Plan of Optimal Territorial Unit for Integrated Water Service. Unfortunately, most of the plans did not consider drought risk, likely due to a lack of specific indications about the methods for risk assessment in the guidelines for the drafting Optimal Territorial Unit plans.

Another attempt to implement a preventive approach to drought risk was done by the Legislative Decree 152/1999 and the Deliberation 21/12/1999 of the Interministerial Committee for the Economic Planning (CIPE), which required the identification of areas vulnerable to drought and desertification, as well as the program of action in the framework of the international convention against drought and desertification. However, no guidelines on technical standards or operational steps to identify areas vulnerable to drought areas were issued.

As a consequence, for several years, only a reactive response to drought has been adopted in Italy. In particular, two main action lines have been implemented: (1) emergency actions by the Department of Civil Protection and (2) subsidies to farmers for covering the agricultural damages caused by drought, under the provisions of national acts on natural disasters. The implementation process of emergency measures in Italy includes the following phases:

1. Emergency declaration by national government due to drought, by request of local authorities through the regional government.
2. Appointment of a Commissioner for Water Emergency by the Prime Minister.
3. Establishment of the Office of the Commissioner for Water Emergency at regional level.
4. Approval of a list of water emergency measures, including the funding of new hydraulic works, the authorization of water exchanges between users and the simplification of administrative procedures for the design and the realization of the planned works.

Furthermore, local authorities can be authorized to implement specific actions such as rationing of supply or transfer of private sources to public use.

With reference to the subsidies issued to farmers for covering drought damages in agriculture, they are the result of a joint action by national and regional governments. Usually, Regional Department of Agriculture and Forest request a drought declaration to the national Ministry of Agriculture and Forestry Policies based on evidenced impacts on different crops in the territory of some provinces or municipalities. After the Decree of the Ministry, the regional department publishes the rules to be followed for subsidies, which can consist of financial contributions for revenue losses and of loans with reduced interest rate. Funding is provided through a specific national fund for natural disasters.

A significant shift from emergency management approach to preventive approach occurred in Italy as a consequence of the Guidance Document on drought preparedness and mitigation, drafted by the EU Water Scarcity and Drought Expert Network (EC 2007). The Agency for Environmental Protection and Technical Services (APAT) and the Institute for Environmental Protection and Research (ISPRA) since its establishment in 2008 (see Chap. 4) contributed to the preparation of a Guidance Document and have developed several activities to improve drought management.

The most recent initiative aimed at drought and water scarcity monitoring and the regulation of water resources management during drought events is the establishment of the Observatories on water resource uses within Hydrographic Districts (ISPRA 2018). The Observatories, in operation since February 2016 on a voluntary basis, include, besides the District Authorities, representatives from various ministries (Environment and Protection of Land and Sea, MATTM; Agricultural Food and Forestry Policies, MPAAF; Infrastructures and Transport, MIT), as well as from regions within the district, the Department of Civil Protection (DPC), ISPRA, the Institute of Statistics (ISTAT), the Council for Research in Agriculture and Economy (CREA), the National Research Council (CNR), the Association of Land Reclamation Consortia (ANBI), the Lake Regulation Consortia and companies for water and electric services. A Technical Committee for coordinating the Observatories, established in October 2016 at MATTM, has prepared guidelines for selecting proper indicators to monitor drought and water scarcity.

12.7   Severe Drought Events in Italy in the Last Century

Italy has experienced several drought events, both in its northern regions, characterized by humid climate and abundant water resources, and in semiarid southern regions where the higher variability of the hydrometeorological conditions and the reduced amount of water resources combined with an increase of water demand lay the basis of more frequent water shortage conditions. In what follows, some of the most severe documented droughts that occurred in the last 100 years are briefly described.

12.7.1   Drought of 1921

A very severe drought affecting most of the Italian regions, with significant agricultural impacts, occurred in the last trimester (October–December) of 1921, following similar dry periods occurred in summer over other European countries (Great Britain, France, Switzerland) (Eredia 1922). The Po River Basin and Liguria Region were affected by the maximum deficit: total precipitation in the trimester at Piacenza was 7 mm (precipitation deficit of −98% of the long-term average), at Milan 8 mm (−97%) and at Genoa 95 mm (−81%). Minor deficits were reported for the islands: precipitation at Cagliari in Sardinia 241 mm (−41%) and in Catania in Sicily 289 mm (−40%).

12.7.2   Drought of 1938

Another severe drought occurred in the first 4 months (from January to April) of the 1938 in North and Central Italy, Sardinia included. The available analysis (Marchetti 1938) includes both precipitation deficit and low-flow deficit in several rivers. The most severe precipitation deficits were recorded at Trento in Trentino-Alto Adige (−93%), Genoa in Liguria (−93%), Belluno in Veneto (−91%) and Turin in Piedmont (−91%), while minor deficits occurred in Central Italy (Rome, −39%) and Sardinia (Cagliari, –43%). The amount of precipitation from January to April 1938 was the minimum since 1866 at Turin (21.5 mm) and Trento (17.6 mm). In other stations precipitation was very close to the minima recorded in very long series such as Padua (203 years from 1713 to 1915) and Milan (158 years from 1764 to 1922). Also severe streamflow deficits in the first 4 months of 1938 occurred in several rivers, particularly in the Northern and Central Apennines (Trebbia −72%, Magra −70%, Reno −60%), as well as in Sardinia (Flumendosa −65%). Minor deficits were registered in the Veneto rivers (Isonzo −62%, Po −44%), while streamflow deficits in Southern Italy river were even lower (Ofanto in Apulia −36%, Simeto in Sicily −31%).

12.7.3 Drought of Years 1988–1990

A severe 3-year drought occurred in the 1988–1990 period, affecting almost the entire national territory, and with the most severe deficits observed during the wet season, which led to severe impacts on water supply. A survey of such a drought was promoted by the Department of Civil Protection, which organized two round tables at Rome at the National Research Council (on February 1989 and February 1990) and established a commission, with the aim to monitor and to analyse the drought process, so to identify the best drought mitigation measures. The commission included representatives of the involved ministries, hydrometeorological services and water supply management companies. The commission produced a detailed report on this drought event (Rossi and Margaritora 1994). In Table 12.3 the long-term average (annual and seasonal) precipitation values and the estimated related deficits per year and geographic area are shown. Annual precipitation deficits over the whole Italian territory were −21% in 1988, −24% in 1989 and −16% in 1990. Deficits in the period from September to March were more significant: −44% in the 1988–1989 and −43% in the 1989–1990 period.

Precipitation deficit generally determined great water deficit in rivers, in Alpine lakes, and caused groundwater deficits in aquifers. Table 12.4 shows annual deficits and mean deficits on 1988–1990 period related to streamflow observed in selected rivers. The mean 1988–1990 deficits range from −12% (Adige in Veneto) to −74%

Table 12.3 Precipitation deficit during the 1988–1990 drought event in Italy

	Area	Annual values				Seasonal values (Sept–March)			
		Avg. prec	Deficit (%)			Avg. prec	Deficit (%)		
	(10³ km²)	(mm)	1988	1989	1990	(mm)	1988–1989	1989–1990	1988–1990
Northern Italy	106.8	1116	13	18	12	567	38	42	40
Central Italy	79.5	977	27	16	18	716	53	39	46
Southern Italy	65.2	1106	20	30	20	733	40	50	45
Insular Italy	49.8	723	25	30	12	612	44	39	41
Italy	301.3	1012	21	24	16	645	44	43	43

Source: De Vito and Rusconi (1994)

Table 12.4 Runoff deficit (shortfall) in some significant Italian rivers

River	Catchment area (km²)	Avg. runoff (mm)	Annual deficit (%)			Mean 1988–1990 deficit	
			1988	1989	1990	(mm)	(%)
Brenta	1567	1143.1	12	2	42	199	19
Adige	11,954	536.3	4	2	29	708	12
Sieve	831	576.8	38	64	41	814	48
Ombrone	2657	316.9	37	67	66	540	57
Fiora	818	269.8	25	53	51	375	43
Pescara	3125	573.2	33	34	39	597	35
Tevere	16,545	443.7	23	39	47	480	36
Biferno	1290	281.1	60	32	52	388	48
Cervaro	657	134.4	43	89	89	297	74
Ofanto	2716	160.2	2	67	59	232	43
Oreto	76	375.6	33	67	11	407	37
Tirso	587	214.9	48	70	75	406	64

Source: De Vito and Rusconi (1994)

(Cervaro in Apulia). Lake Como and Lake Maggiore reached almost their minimum level on record in both 1989 and 1990, and many aquifers displayed an unusual drop in water table levels. Also the increase in average temperature, recorded in several stations, contributed to reduce snowfall and snow coverage in the Alps and in part of the Apennines.

Because of the severe water deficits, extreme impacts on water supply and economic activities occurred. Southern regions and major islands, particularly hit by unexpected water shortage, had to rely on tank trucks to supply drinking water, as a last option. Irrigation agriculture was also badly struck by the drought. For example, in several districts of Sicily and Sardinia in 1989, a very severe decrease of irrigated surface was observed (see Table 12.5). Also, hydroelectric production exhibited a significant reduction, and persistent low flows caused heavy damages on aquatic ecosystems.

The 1988–1990 drought found the communities unprepared to cope with the impacts on water supply, as no mitigation measures were planned in advance. Only emergency actions were adopted with funds provided by the national government, especially through the Department of Civil Protection. The total cost was estimated to be of 1133 billion of Italian liras for both short-term and long-term measures (Cittadino and Landrini 1994).

12.7.4 Drought of 2017 in Various Italian Rivers

According to the data published by the ISTAT for World Water Day 2018 (ISTAT 2018), and to the analyses carried out for different parts of the country (AA.VV 2017), the drought that hit the main Italian rivers (Po, Adige, Arno and Tiber) in

Table 12.5 Reduction of irrigated surface in some regions of Southern Italy during the 1988–1989 drought

Region	Irrigation districts	Irrigable surface (ha)	Irrigated surface (ha) 1988	1989	Comments
Abruzzo	Right and Left Pescara	13,300	11,050	8300	Reduced supply
Campania	Sannio, Right and Left Sele	36,280	33,945	29,675	Reduced supply
Apulia	Tavoliere, Tara, Arneo	115,540	55,143	41,860	Reduced or no delivery
Basilicata	Bradano, Val d'Agri	59,300	31,400	18,680	
Calabria	Catanzaro, Lao Abatamarco, Sibari and Crati	53,180	14,570	23,150	Irregular delivery
Sicily	Catania plain, Caltagirone, Lentini, Scicli, Acate, Gela Salso, Belice Carboj, Delia Nivolelli, Birgi	100,500	28,850	28,375	Reduced supply to selected crops or no delivery
Sardinia	Southern Sardinia, Oristano, Central Sardinia, Liscia, Nurra	96,580	31,670	10,670	

Source: Leone (1994)

2017 has been severe, due to the scarce precipitation of the autumn 2016 and of the entire 2017, characterized also by high temperatures.

The analysis carried out by ISTAT, by using the SPI (Standardized Precipitation Index) at 12 months over the river basins, shows much dry or extreme dry conditions from May to December for all rivers (where "much dry" refers to $-1.50 \leq SPI \leq -1.99$ and "extreme dry" to $SPI \leq -2.00$). A moderate dry condition ($-1.00 \leq SPI \leq -1.50$) was registered in the Adige basin since February and in the Tiber basin from April.

The annual streamflow deficit, computed with reference to the average value of the 1981–2010 period, presents significant values: -41% for Po river at Pontelagoscuro (with monthly peak of -61% in October); -33% for Adige river at Boara Pisani; -27% for Arno at S.Giovanni alla Vena (with monthly peak of -88% in October, while positive deviation occurred in the months of February, March and September); and $-39.\%$ for Tiber at Ripetta (with maximum deficit of $-55.\%$ in November). The drought which hit the Lazio region in 2017 had severe effects on the water supply of Rome, especially after the emergency withdrawals from Bracciano Lake were stopped for environmental reasons.

12.7.5 Droughts in the Po River Basin in the Last 20 Years

Several drought events occurred in the Po river basin in the last 20 years, generally due to anticyclone conditions in the Mediterranean leading to low precipitation and high temperatures over the Italian peninsula. The most severe events occurred in 2003–2008, 2011–2012 and 2016–2018, which caused heavy reduction of water withdrawals from Po river and its tributaries for the different uses, as well as severe impacts on the water quality in the lower reach of the river.

The severity of the droughts in the Po river is estimated by using the available series of monthly flow observed at the Pontelagoscuro gauge (basin surface 70,091 km^2) (Pecora 2019), whose mean flows computed on the 1923–2018 periods and minimum monthly flows on the same period are shown in Fig. 12.2. Figure 12.3 shows the monthly flow deficit or surplus, computed with reference to long-term means (1923–2018) for each month, as well as the 12 months Standardized Flow Index (SFI) (12) referring to the same means. Each graph covers a 3-year period of the most severe droughts of last 20 years comparing their characteristics with those of the two previous worst drought events (1943–1945 and 1988–1990).

During 2003, low flows occurred from February to November with mean monthly deficits greater than 50% from May to October and with a maximum deficit of 70% in June. The mean monthly flow in June, July and August was lower than 600 m^3/s, which is considered a warning threshold for the upstream flow of brackish water from Adriatic Sea in the Po delta mouth, although the extreme critical low flow below which water quality is not acceptable for withdrawals is 450 m^3/s. Despite in December 2003 and during the first months of 2004 the monthly flows exceeded the mean flows, the value of SFI (12) continue to indicate a "moderate dry condition" ($-1.00 \leq$ SFI ≤ -1.50) or a "much dry" condition ($-1.50 \leq$ SFI ≤ -1.99).

The 2005–2008 drought was very long and severe. According to the recorded monthly deficits, it lasted for 41 months (flows below average flows from January 2005 to May 2008, with the exception of September 2006). Based on the time-distribution of SFR (12), duration is estimated 39 months, with peaks of "extreme dry" (SFI ≤ -2.00). The monthly flows in June and July 2006 were lower than all previous historical minima. The 2011–2012 drought was not too severe, while the 2016–2018 drought covered 23 months (on the basis of monthly deficits) or 26 months (on the basis of SFI(12) amounts).

However the comparison with the previous past droughts shows that the droughts that occurred in the Po river during the last 20 years have been more severe than the

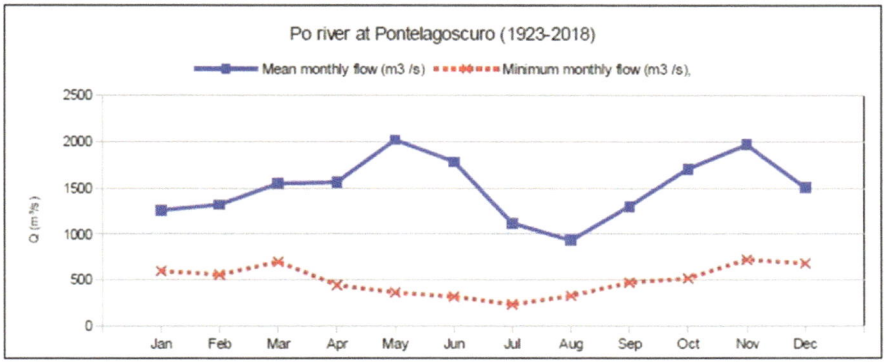

Fig. 12.2 Mean and minimum monthly flow in the 1923–2018 period in Po river at Pontelagoscuro

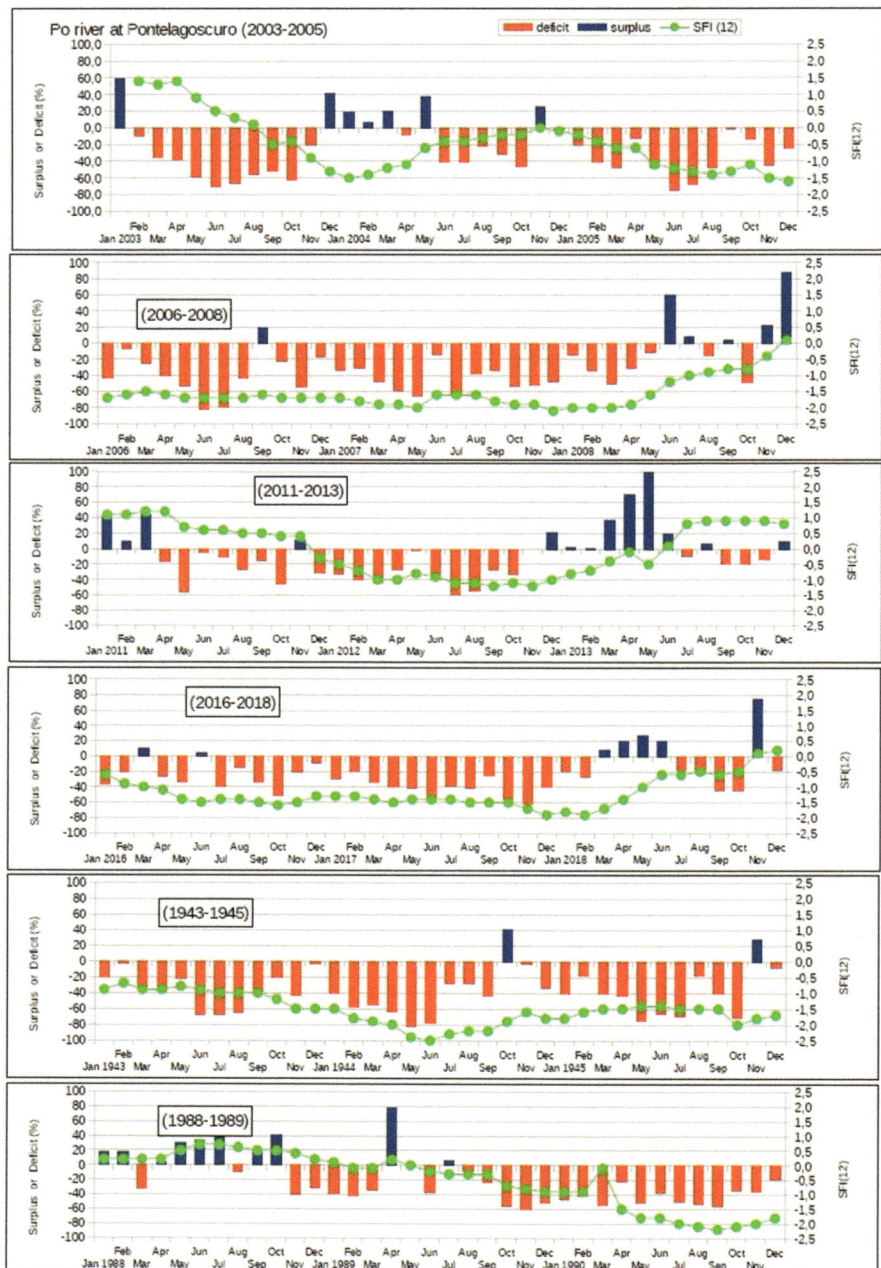

Fig. 12.3 The most severe droughts in Po river at Pontelagoscuro

1989–1990 drought, but not as much as the 1943–1945 drought. In fact, two moderate droughts occurred in the 1988–1990 period: length of 5 months (from November 1988 to March 1989) and length of 17 months (August 1989–December 1990). Instead, during the 1943–1945 period, a drought of 34 months occurred (from January 1943 to October 1945 with the exception of October 1944).

Obviously, the effects of the more recent droughts were more severe due to the increase of water demands for different uses in the basin. Also the consequences of extreme low flows (lower than 450 m^3/s) in the valley reach of Po caused more severe impacts due to the presence of withdrawals for municipal use (Aqueduct of Ferrara and Aqueduct of the Po delta), which cannot derive water with high salt content. Thus the occurrence of events with flows lower than 450 m^3/s during the years 2003 and 2005 (five events per year with duration of 6 and 8 days) caused more damaging impacts than those ones of the years 1944 and 1945 (six and five events per year with duration of 8 and 12 days).

12.7.6 Italian Droughts Versus European Droughts

By using investigations on the most severe droughts occurred in Europe, a comparison of Italian droughts with those observed in other European countries can be helpful. Since the beginning of the twenty-first century, a large part of Europe has experienced a series of exceptionally severe drought events, affecting a wide range of socio-economic sectors. Most of these events have been the results of heat waves in combination with a lack of precipitation during the summer months (2003, 2010, 2013, 2015 and 2018).

Several studies have investigated drought events over Europe (Bonaccorso et al. 2013; Gudmundsson and Seneviratne 2015; Spinoni et al. 2015). In general, a different drought regime is observed in the Mediterranean than the rest of Europe, with the only exception of drought in 1949 which is evident both in Southern and Central Europe. Bonaccorso et al. (2013) conclude that besides Euro-Mediterranean regions, North Western and Central Eastern regions appear more drought-prone than the rest of Europe, in terms of low values of return periods. Other studies on drought hazard frequency and intensity seem to find increasing trends in Southern and Western Europe (Gudmundsson and Seneviratne 2015; Spinoni et al. 2015).

Spinoni et al. (2015) identified the most severe drought events in Europe between 1950 and 2012, by means of three indicators: Standardized Precipitation Index (SPI) (McKee et al. 1993), Standardized Precipitation Evapotranspiration Index (SPEI) (Vicente-Serrano et al. 2010) and Reconnaissance Drought Index (RDI) (Tsakiris et al. 2007). These indicators, computed using the E-OBS gridded data ($0.25° \times 0.25°$), have been averaged into a combined indicator X at 3 months and 12 months scale for 13 European regions. The values of the X12 series from 1950 to 2012 for Italy are shown in Fig. 12.2, where drought events are marked in red. A drought event starts when the indicator falls below a threshold (computed as the mean minus the standard deviation) for at least three consecutive months and ends when it turns above the mean of the series (Fig. 12.4).

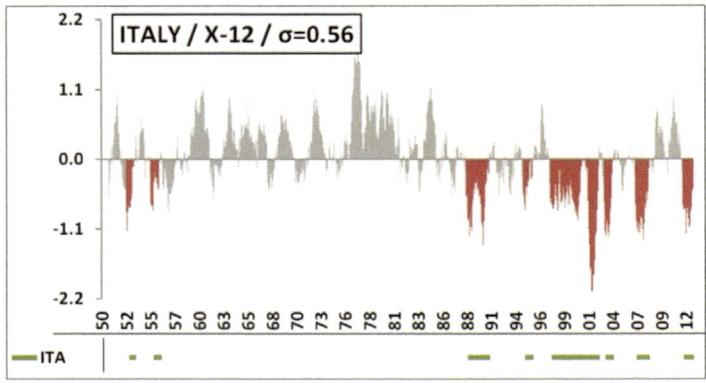

Fig. 12.4 Drought events identified through the indicator X12 (computed as average of SPI, SPEI and RDI) in 1950–2012 period in Italy. (Source: Spinoni et al. 2015)

Table 12.6 Characteristics of droughts in some European/Mediterranean countries in the period 1950–2012

Region	Countries	No drought events in:			Average duration (months)			Average severity		
		1951–1970	1971–1990	1991–2000	1951–1970	1971–1990	1991–2010	1951–1970	1971–1990	1991–2010
Italy	Italy	2	1	4	11.0	30.0	26.3	1.3	5.7	6.9
Aegean	Cyprus, Greece and Turkey	3	4	3	10.7	18.3	27.7	1.7	1.8	5.3
Balkans	Yugoslavia, Albania	5	3	4	9.0	8.3	16.3	2.7	3.6	4.2
Central Europe	Austria, Germany and Switzerland	3	2	3	16.7	25.0	13.7	6.0	8.0	2.6
France Benelux	France, Belgium the Netherlands and Luxembourg	1	2	3	13.0	26.0	33.0	2.6	7.8	7.0
Iberian	Spain, Portugal	2	3	5	14.5	9.7	18.6	2.5	1.3	5.5

Source: Spinoni et al. (2015)

Table 12.6 shows the main characteristics of the identified droughts in Italy and in the Mediterranean European regions for the periods 1950–1970, 1971–1990 and 1991–2010: number of events, duration and average severity (computed as the sum of dimensionless differences between indicator values and threshold).

This study confirms that the most severe events occurred in between 1991 and 2010. The features of Italian droughts are similar to the events that affected other parts of Europe. Droughts heavier for duration and severity occurred in France and Benelux in the 1991–2000 period.

12.8   Lessons Learnt from Recent Droughts in Italy

Following the severe droughts, several monitoring systems have been set up in Italian regions. The monitoring systems aim at providing the necessary information to implement adequate mitigation measures useful for the agencies responsible for water government and water supply system operation. For example, drought monitoring bulletins have been developed in Sicily (Rossi and Cancelliere 2002; www.osservatorioacque.it), in Emilia-Romagna (www.arpae.it), ISPRA (www.isprambiente.gov.it/pre_meteo/siccitas/), and Sardinia (www.sar.sardegna.it/servizi/agro/monit_siccita.asp).

In the recent guidelines issued by the Technical Committee for coordinating the Districts Observatories (ISPRA 2018), already mentioned in Sect. 12.6, the set of suggested drought indicators include Standardized Precipitation Index (SPI), Standardized Runoff Index (SRI), Standardized Snow Pack Index (SSPI), Standardized Precipitation Evapotranspiration Index (SPEI), Spring Anomaly Index (SAI) and Fraction of Absorbed Photosynthetically Active Solar Radiation (FAPAR). In addition, to monitor water stress it is suggested to adopt the Water Exploitation Index Plus (WEI+), defined as the ratio between water consumption (WC) (computed as withdrawal minus outflow in water body) and renewable water resource:

$$WEI+ = \frac{WC}{RWR} = \frac{\text{Withdrawal} - \text{Outflow}}{RWR} \tag{12.1}$$

The main activities carried out in each Italian district for drought monitoring and for defining alert triggering levels for the activation of the Water Emergency Plan are summarized in Table 12.7.

The activities of the Authority of the Po River Basin are particularly significant among the notable examples of the ongoing shift towards a proactive approach for coping with drought in Italy.

The severity of the water crisis which hit the Po plain during spring and summer of 2003 pushed the National Civil Protection Department and the Po River Basin Authority to establish a Technical Committee and an Agreement among the bodies responsible of the water resources management. The aims were the following: to mitigate the effects of drought, to guarantee withdrawal and to avoid impacts on cooling needs for thermoelectric production as well as on water quality due to the upstream flow of brackish water. The Agreement included, besides the National Civil Protection Department and the Po Basin Authority, the Dam Directorate of the Minister of Infrastructures and Transports, the regions within the Po basin, the manager of the national electric energy network (today TERNA), the bodies which provide the regulation of Pre-alpine lakes (Garda, Como, Maggiore, Iseo, Idro), the

Table 12.7 Features of drought monitoring and thresholds for emergency measures in the Italian districts

Hydrographic district	Drought monitoring	Drought-related documents	Decision-making process for emergency measures
Po river	Early warning system (from 2010) based on SPI, SFI, SWSI and "run method"	Plan of the water balance	Four phases (normal status, and ordinary, moderate, high criticality with increasing detail and more severe emergency measures)
Eastern Alps	Current hydrometeorological network operated by Province of Trento, by ARPA in Veneto and regional bodies and civil protection in Friuli-Venezia Giulia (where flow reduction curve is observed by the Regional Hydrographic unity) The District Observatory is planning to set up two monitoring systems (surveillance and operative)	Water balance in Trento, monthly water resource report in Veneto (with warning system, snow water equivalent, storage volume and aquifers levels in Piave river basin)	
North Apennines	Drought indicators: SPI (1–24 months), WEI+ (June–September), DMVf and DMVf (June–September); reservoir storage; triggering levels in aquifers	Water Management Plan	
Central Apennines		Water Management Plan	
Southern Apennines	No specific drought indicators. DMV and WEI+ estimated in the Water Management Plan	Water Management Plan merging specific plans of river basins or regions	
Sardinia	Drought bulletin including SPI (3–24 months) and storage volumes in reservoirs		Contacts between region and management bodies for defining long-term and short-term measures
Sicily	Drought bulletin including SPI (3–24 months) and storage volume in reservoirs. The previous bulletin since 2003 included also rainfall deficit and Palmer index.		

Association of Land Reclamation Consortia (ANBI) and the companies of electric energy production. The Technical Committee facilitated the exchange of information to improve water resources monitoring, which included storage levels in the reservoirs managed by hydroelectric companies, lake volumes and releases, withdrawals from rivers for irrigation, industrial and municipal uses and the minimum flow for acceptable water quality in the Po river.

During the 2005 drought, a new Cooperation Agreement was established with the objectives of improving analysis and control of water balance and preventing exceptional low flows and water shortage for all water users of the Po river. Among the activities developed at the Drought Steering Committee, one of the most important consisted in real-time monitoring and assessment of drought evolution scenarios. This activity contributed to reduce the conflicts between different water uses. For instance, opposing interest exists between Land Reclamation Consortia or regional authorities and Lake Regulation Consortia. This happens because the former are interested to ensure the satisfaction of irrigation requirements through water release from lakes and hydroelectric reservoirs, while the latter are interested in preserving high water levels for environmental and recreational purposes in the lakes.

The publication in July 2015 of a Water Balance Plan contributed to reinforce the cooperation among the several institutions responsible for water resources management and the main stakeholders, as well as to minimize the impacts of water shortages. The key element of this Plan is the assessment of available water reserves during spring and summer, in order to optimize the allocation across different sources and water uses, as well as to reduce water shortage risk during months with higher water demands. For example, Table 12.8 lists, for the most severe recent droughts, the available storage in Pre-alpine lakes, in hydroelectric reservoirs and in terms of snow water equivalent (SWE) at the beginning of August of each year.

In addition, an early warning system was developed with the objective of optimizing the use of water resources and avoiding water quality deterioration. It comprises four possible criticality phases:

Table 12.8 Estimated water reserves for supplying the Po plain demands during the recent severe droughts (at the beginning of August) (Pecora 2019)

| Year | Stored volumes (hm³) | | | |
	In lakes	In reservoirs	Snow water equivalent	Total
2003	195	722	127	1044
2005	147	632	131	910
2006	262	637	16	915
2007	352	694	49	1095
2011	892	790	164	1846
2015	445	657	63	1165
2017	497	740	85	1322
2018	573	668	116	1357

- *Normal*: monitoring of meteo-climatic conditions and of the state of the available resources according to the provisions of the Water Balance Plan.
- *Pre-alert* (vigilance): reinforcement of monitoring and definition of measures aimed at water saving with particular reference to irrigation uses.
- *Alert* (danger): improved monitoring, daily updating of the state of water reserves, assessment of impacts of drought on sensitive uses (municipal needs, touristic requirements and water quality in protected areas of the Po delta); start-up of the measures foreseen by the Drought Directive, established by the Authority of the Po River Basin.
- *Emergency*: several measures take place, under the responsibility of the civil protection system, in order to satisfy priority emergency requests, to reduce threats to population and to minimize environmental impacts.

12.9   Water Shortage Risk Assessment

Assessment of water shortage risk due to droughts can be carried out with reference to two distinct objectives, namely: (1) increasing the robustness of the system within the strategic planning stage through appropriate prevention strategies and (2) improving the performance of water supply during droughts within the operation stage. When dealing with strategic planning, risk assessment should be *unconditional*, i.e. not referred to a particular state or condition of the system, and aimed at selecting the best long-term measures (Bonaccorso et al. 2007). In the case of the operation of water supply systems under drought conditions, the risk assessment has to be *conditional*, i.e. taking into account the current state/condition of the system. In this case the main problem is defining the status of the system with respect to predefined operational levels (e.g. normal, alert, alarm) in order to decide *when* and *how* to activate predefined sets of mitigation measures (e.g. rationing policies and/ or use of additional water resources) able to prevent future severe shortages (MEDROPLAN 2007).

Since risk represents a probabilistic measure of economical or intangible damages consequent to a disaster, when dealing with water supply systems, water shortages are generally assumed as proxy of damages, and drought risk is generally assessed in terms of the probabilistic features of water shortages due to droughts. To this end, Monte Carlo simulation of the system, where the probabilistic features of droughts are implicitly taken into account through the stochastic generation of hydrological inputs and where the vulnerability of the system is considered in the simulation stage, may constitute a more efficient tool to assess water shortage risk, as compared to other approaches based on the concept of design droughts (Cancelliere et al. 1998). As shown in Fig. 12.5, such procedure can still be framed within the more traditional risk assessment based on combination of hazard and vulnerability.

Monte Carlo simulation can also find application for the selection of different alternative mitigation measures in combination with multi-criteria techniques (Rossi et al. 2008). For example, such procedure has been applied in Italy with reference to

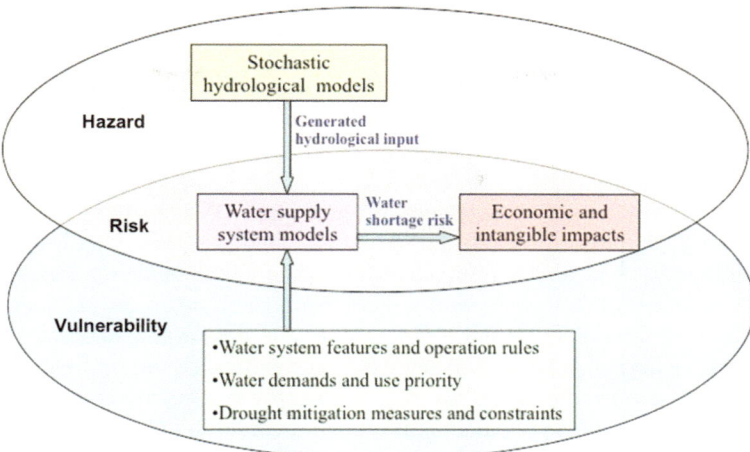

Fig. 12.5 Assessment of water shortage risk due to drought in a water supply system through Monte Carlo simulation. (Source: Rossi 2017)

Palermo water supply in Sicily (Munda et al. 1998), Simeto water system in Sicily (Rossi et al. 2005) and Flumendosa-Campidano water system in Sardinia (Rossi et al. 2006).

Many papers have been devoted to evaluate the risk of water shortage with the aim of defining operating policy or storage allocation in a single reservoir and/or complex water supply system in response to drought. Some studies in Italy addressed the use of early warning system information within drought management of water supply systems, for example, with reference to the Simeto water system in Sicily (Cancelliere et al. 2009), Acate system in Sicily (Nicolosi et al. 2009) and Basilicata-Apulia water system (Nicolosi et al. 2008). In some cases, the probability of shortages within a short-term time horizon has been assessed to support the selection of different mitigation measures, for instance, with reference to the Acate water system in Sicily (Rossi et al. 2011).

Early warning systems could also benefit from the use of forecasting methods based on past observations of the drought indicators or hydrometeorological variables, also including exogenous forcing mechanisms in the forecasting scheme. Indeed, the links between precipitation and large-scale circulation patterns such as El Niño-Southern Oscillation (ENSO), Sea Surface Temperature (SST) and Geopotential Height (GpH) observed in some regions in the last decades have sparked interest about the possibility to use such indices to improve mid-long-term precipitation forecasting, which in turn can be used to make seasonal drought predictions. The relation of large-scale circulation patterns with rainfall remains very much region-dependent. Several studies have established links between NAO and climate in Europe and the Mediterranean basin. Some studies have investigated whether encompassing the influence of NAO into models structure improves drought forecasting (Cutore et al. 2009; Chen et al. 2013; Santos et al. 2014; Bonaccorso et al. 2015).

12.10   Concluding Remarks

Based on the Italian legislative and institutional framework as well as on the recent experiences in coping with drought events, some concluding remarks on the possible issues related to the implementation of an effective drought proactive approach in Italy can be drawn.

First, effective drought management in water supply systems represents a complex challenge due to several specific weaknesses and gaps affecting drought preparedness and mitigation planning, organization and implementation. Besides, still the differences between *water scarcity* (i.e. a permanent unbalance between available water resources and demands) and *drought* (i.e. a natural temporary negative deviation from normal precipitation amount for a significant duration and extension, which leads to a temporary water shortage in a supply system) are not fully clear to many stakeholders and the population in general. In order to overcome these issues, specific drought preparedness and mitigation planning activities should be adopted within the general framework of water resources and water quality protection planning.

In addition, the need of an appropriate institutional structure for coping with drought risk is not fully recognized in Italy. Indeed, there is a lack of coordination between River Basin Authorities and the civil protection. The former plans to water shortage prevention, while the latter is in charge for drought risk mitigation and recovery. Therefore, it appears necessary to entrust the bodies responsible for each water supply system (similar to what has been done in Spain) to define in advance the measures needed to face water shortage risk, with a specific focus on primary needs such municipal water use. Furthermore, general criteria for declaring drought as a natural disaster should be developed at the national level, while emergency management organizations should intervene at local level within a predefined emergency plan. Also advanced drought early warning systems, which are being developed in different Italian districts, can improve drought management, at the condition that standard criteria, drought indicators and triggering thresholds for a timely implementation of contrasting measures are available.

Beyond the interesting outcomes of several research projects on drought, which recently have fostered a renewed interest towards this hazard also in Italy, future research to improve resilience to drought should be oriented to address the following aspects (Rossi 2017):

• Better modeling of drought occurrence and characteristics, both in terms of their stochastic nature and their links with global atmospheric circulation patterns, especially when the analysis aims at taking into account climatic changes.
• Thorough analysis of past experiences in drought monitoring and mitigation as well as a greater knowledge exchange on the measures adopted in different contexts, in order to implement best practices in drought management.
• Advanced assessment of economic, environmental and societal impacts of droughts and mitigation measures, preferably based on multi-criterion tools.

- Development of appropriate models and integrated packages of techniques, which can be easily understood and applied by decision-makers. This may foster the inclusion of the most advanced scientific tools in drought preparedness planning (e.g. stochastic hydrology for drought characterization; economics and environmental sciences for a comprehensive impact evaluation; social sciences for selecting the way to reduce users conflicts).
- Development of advanced tools for an "adaptive" drought management in water supply systems, based on drought early warning systems, drought monitoring and forecasting systems and advanced DSS that use modern technologies.

In conclusion, Italy is still struggling to adopt a comprehensive drought risk management approach. However, fostering good practices at the European level can certainly contribute at improving the situation, as shown by the recent promising efforts at the national and the river basin district scales.

References

AA.VV (2017) Siccità 2016–2017 [Drought 2016–2017] (in Italian). Ecoscienza 4:56–64

Bonaccorso B, Cancelliere A, Rossi G (2003) An analytical formulation of return period of drought severity. Stoch Env Res Risk A 17:157–174

Bonaccorso B, Cancelliere A, Nicolosi V, Rossi G, Cristaudo G (2007) Methods for risk assessment in water supply systems. Options Méditérr 58:115–127

Bonaccorso B, Peres DJ, Cancelliere A, Rossi G (2013) Large scale probabilistic drought characterization over Europe. Water Resour Manag 27(6):1675–1692

Bonaccorso B, Cancelliere A, Rossi G (2015) Probabilistic forecasting of drought class transitions in Sicily (Italy) using standardized precipitation index and North Atlantic oscillation index. J Hydrol 526:136–150

Botterill LC, Wilhite DA (eds) (2005) From disaster response to risk management, Australia's national drought policy. Springer, Dordrecht

Cancelliere A, Salas JD (2010) Drought probabilities and return period for annual streamflows series. J Hydrol 391:77–89

Cancelliere A, Ancarani A, Rossi G (1998) Susceptibility of water supply reservoirs to drought conditions. J Hydrol Eng 32(2):140–148

Cancelliere A, Nicolosi V, Rossi G (2009) Assessment of drought risk in water supply systems. In: Iglesias A et al (eds) Coping with drought risk in agriculture and water supply systems. Springer, Dordrecht, pp 93–109

CEC (2009) Commission of the European Communities. White paper. Adapting to climate change: towards a European framework for action, COM(2009) 147 Final

Chen ST, Yang TC, Kuo CM, Kuo CH, Yu PS (2013) Probabilistic drought forecasting in Southern Taiwan using El Niño-Southern Oscillation index. Terr Atmos Ocean Sci 24:911–924

Cittadino C, Landrini C (1994) Interventi di emergenza idrica del Dipartimento della Protezione Civile. In: Rossi G, Margaritora G (eds) La siccità in Italia 1988–90 [Water emergency intervention by the Civil Protection Department. The 1988–90 drought in Italy] (in Italian). Istituto Poligrafico e Zecca dello Stato, Rome, pp 185–193

Cutore P, Di Mauro G, Cancelliere A (2009) Forecasting Palmer index using neural networks and climatic indexes. J Hydrol Eng 14:588–595

De Vito L, Rusconi A (1994) La siccità poliennale 1988–90 in Italia [The 1988–90 drought in Italy]. In: Rossi G, Margaritora G (eds) La siccità in Italia 1988–90 [The 1988–90 drought in Italy] (in Italian). Istituto Poligrafico e Zecca dello Stato, Rome, pp 41–57

Duckstein L (1983) Trade-offs between various mitigation measures. In: Yevjevich V, Da Cunha L, Vlachos E (eds) Coping with droughts. Water Resources Publications, Littleton, pp 203–215

Dziegielewski B (2000) Drought preparedness and mitigation for public water supply. In: Wilhite DA (ed) Drought: a global assessment, vol 2. Routledge Publisher, London

EC (2007) Drought management plan report including agricultural, drought indicators and climate change aspects. Water scarcity and drought expert network technical report, 2008-023, Luxembourg

EC (2011) Water scarcity and droughts in the European Union, Report from Commission to the European Parliament and the Council, SEC (2011), 338

Eredia F (1922) Siccità del 1921 [Drought of 1921] (in Italian) Servizio Idrografico Centrale, Rome

Estrela T, Vargas E (2012) Drought management plans in Europe Union. The case of Spain. Water Resour Manag 26(6):1537–1553

Gudmundsson L, Seneviratne SI (2015) European drought trends. Proc Intl Assoc Hydrol Sci 369:75–79. https://doi.org/10.5194/piahs-369-75-2015

ISPRA (2018) Linee guida sugli indicatori di siccità e scarsità idrica da utilizzare nelle attività degli osservatori permanenti per gli utilizzi idrici [Guidelines on the indicators of drought and water scarcity to be used by the Observatories for water resources uses] (in Italian). Available at http://www.isprambiente.gov.it/files2018/notizie/LineeGuidaPubblicazioneFinaleL6WP1_concopertina.pdf. Last accessed May 2019

ISTAT (2018) Giornata mondiale dell'acqua. Le statistiche dell'ISTAT [World water day 2018. ISTAT statistics] Focus 22.3.2018. Available on-line: http://istat.it. Last accessed May 2019

Leone G (1994) Effetti della siccità sulle irrigazioni nel Mezzogiorno d'Italia [Drought impacts on irrigation in Southern of Italy]. In: Rossi G, Margaritora G (eds) La siccità in Italia 1988–90 [The 1988–90 drought in Italy] (in Italian). Istituto Poligrafico e Zecca dello Stato, Rome, pp 137–141

Marchetti G (1938) La eccezionale siccità dei primi quattro mesi del 1938 in Italia meridionale e centrale [The exceptional drought of first four months of 1938 in Southern and Central Italy] (in Italian). Annali dei Lavori Pubblici 7:237–246

McKee TB, Doesken NJ and Kleist J (1993) The relationship of drought frequency and duration to time scales. Paper presented at 8th conference on applied climatology. American Meteorological Society, Anaheinma, USA

MEDROPLAN (2007) Drought management guidelines and examples of application in Mediterranean countries. In: Iglesias A, Cancelliere A, Gabiña D, López-Francos A, Moneo AM, Rossi G (eds) EC MEDA water programme, Zaragoza. On line Available at www.iamz.ciheam.org/medroplan. Last accessed May 2019

Mishra AK, Singh VP (2010) A review of drought concepts. J Hydrol 391:202–216

Mishra AK, Singh VP (2011) Drought modelling: a review. J Hydrol 403:157–175

Munda G, Parrucini M, Rossi G (1998) Multi-criteria evaluation methods in renewable resources management: integrated water management under drought conditions. In: Beinat E, Nijkamp P (eds) Multicriteria analysis for land-use management. Kluwer, Dordrecht, pp 79–93

NDIC (2011) National Drought Information Center. U.S. Drought Monitor, University of Nebraska-Linvoln. Available at http://drought.unl.edu/dm. Last accessed May 2019

Nicolosi V, Caruso V, Rossi G, Cancelliere A, (2008) Gestione del rischio di siccità in un sistema idrico complesso: il caso del sistema Agri-Sinni. Atti IDRA 2008 [Management of drought risk in a complex water system: Agri-Sinni system. In: Proceedings of the IDRA2008], Perugia September 9–12, 2008, (in Italian), Morlacchi editore, Perugia, pp 1–10

Nicolosi V, Cancelliere A, Rossi G (2009) Reducing risk of shortage due drought in water supply systems using genetic algorithms. Irrig Drain 58:171–188

Pecora S (2019) Agenzia Prevenzione Ambiente Energia, Emilia Romagna, Servizio idrologia, Personal Communication

Rossi G (2000) Drought mitigation measures: a comprehensive framework. In: Vogt JV, Somma F (eds) Drought and drought mitigation in Europe. Kluwer, Dordrecht, pp 233–246

Rossi G (2009) European Union policy for improving drought preparedness and mitigation. Water Int 34(4):441–450

Rossi G (2017) Chapter 28: Policy framework of drought risk mitigation. In: Eslamian S, Eslamian F (eds) Handbook of drought and water scarcity, vol 3. CRC Press/Taylor and Francis, Boca Raton, pp 568–586

Rossi G, Cancelliere A (2002) Early warning of drought: development of a drought bulletin for Sicily. In: Proceedings of the international conference new trends in water and environmental engineering for safety and life, Capri, 24–28 June, pp 1–12

Rossi G, Cancelliere A (2013) Managing drought risk in water supply systems in Europe: a review. J Water Resour Dev 29(2):272–289

Rossi G, Margaritora G (eds) (1994) Siccità in Italia 1988–90 [Drought in Italy 1988–1990] (in Italian) Presidenza del Consiglio, Dipartimento Protezione Civile. Istituto Poligrafico dello Stato, Rome, p 193

Rossi G, Benedini M, Tsakiris G, Giakoumakis S (1992) On regional drought estimation and analysis. Water Resour Manag 6(4):249–277

Rossi G, Cancelliere A, Giuliano G (2005) Case study: multi-criteria assessment of drought mitigation measures. J Water Resour Plan Manag 131(6):449–457

Rossi G, Cancelliere A, Giuliano G (2006) Role of decision support system and multicriteria methods for the assessment of drought mitigation measures. In: Andreu J, Rossi G, Vagliasindi F, Vela A (eds) Drought management and planning for water resources. Taylor & Francis, Boca Raton, pp 204–240

Rossi G, Nicolosi V, Cancelliere A (2008) Recent methods and techniques for managing hydrological droughts, In: Lopez-Francos A (ed) Drought management scientific and technological innovations. options Méditerranéeennes. Zaragoza, serie A no. 80, pp 251–260

Rossi G, Nicolosi V, Cancelliere A (2011) Operational drought management via risk-based conjunctive use of water. In: Proceedings of XIVth IWRA congress, Porto de Galinhas, Recife, Brasil, September

Santos JF, Portela MM, Pulido-Calvo I (2014) Spring drought prediction based on winter NAO and global SST in Portugal. Hydrol Process 28(3):1009–1024

Spinoni J, Naumann G, Vogt JV, Barbosa P (2015) The biggest droughts in Europe from 1950 to 2012. J Hydrol Reg Stud 3:509–524

Tsakiris G, Pangalou D, Vangelis H (2007) Regional drought assessment based on the Reconnaissance Drought Index (RDI). Water Resour Manag 21(5):821–833

UN Secretariat General (1994) United nations convention to combat drought and desertification in countries experiencing serious droughts and/or desertification, particularly in Africa, Paris

UNISDR (2009) Drought risk reduction. Framework and practise contributing to the implementation f the Hyogo framework for action, UN Secretariat of the Internationa Strategy for Disaster reduction, Geneva, p 213. Available at www.unisdr.org/files/3608_droughtriskreduction.pdf

UNISDR (2014) Hyogo framework for action 2005–2015. Building the resilience of nations and communities to disasters. Available at https://www.unisdr.org/we/coordinate/hfa. Last accessed May 2019

UNISDR (2015) Sendai framework for disaster risk reduction 2015–2030 UNISDR. Available at https://www.unisdr.org/we/inform/publications/43291. Last accessed May 2019

Vicente-Serrano SM, Begueria S, Lopez-Moreno J (2010) A multiscalar drought index sensitive to global warning: the Standardized Precipitation Evapotranspiration Index. J Clim 23:1696–1718

Wilhite DA (1987) The role of government in planning for drought: where do we go from here? In: Wilhite DA, Easterling WE with Wood DA (eds) Planning for drought. Westview Press, Boulder, pp 425–444

Wilhite DA (ed) (2000) Drought: a global assessment, vols 1 and 2. Routledge Publisher, London

Wilhite DA, Botterill L, Monnik K (2005) National drought policy: lessons learned from Australia, South Africa and the United States. In: Wilhite DA (ed) Drought and water crises. Taylor& Francis, Boca Raton, pp 137–172

Yevjevich V (1967) An objective approach to definitions and investigations of continental droughts, Hydrological paper, vol 23. Colorado State University, Fort Collins

Yevjevich V, Hall WA, Salas JD (1983) Coping with droughts. Water Resources Publications, Littleton

Chapter 13
The Water-Food Nexus in Italy: A Virtual Water Perspective

Francesco Laio, Stefania Tamea, and Marta Tuninetti

Abstract Agriculture has a long-standing tradition over the Italian territory, and its geography is significantly heterogeneous across regions, due to different hydro-climatic conditions and local practices. This chapter examines the Italian water use in the agricultural sector by considering both the local water use and the reliance on external water resources occurring through the import of primary and derived commodities. The water assessment is carried out by means of the "water footprint" concept, which aims at quantifying the amount of water required for the production of a good, and the "virtual water trade", which tracks the exchange of water resources from producing countries, where water has been physically used, to consuming countries. In the first part of this chapter, the Italian virtual water balance is analysed considering the amount of water imported, exported and locally used for production. Overall, an increase of the virtual water import and a decline in the use of local water resources are observed: Italy relies on imported water resources for more about half of its food consumption. In the second part, the role of Italy in the international trade network is assessed. Results show that Italy is primarily a net importer of virtual water from other countries (e.g. France, Germany, Brazil, Indonesia), but, at the same time, it is also a net exporter towards the UK, the Mediterranean region, the USA and other minor countries. In the third part, the spatial variability of water use across Italy is explored, looking at the subregional spatial variability to highlight the production sites generating the largest water footprint.

13.1 Introduction

Food production is inextricably linked and reliant upon freshwater resources. In fact, the vast majority of global freshwater use (nearly 70% of the total withdrawal and around 90% of the total consumption) is devoted to the production of agricultural commodities, largely for human consumption (FAO 2011; Hoekstra and

F. Laio (✉) · S. Tamea · M. Tuninetti
Department of Environment, Land and Infrastructure Engineering,
Politecnico di Torino, Torino, Italy

© Springer Nature Switzerland AG 2020
G. Rossi, M. Benedini (eds.), *Water Resources of Italy*, World Water Resources 5,
https://doi.org/10.1007/978-3-030-36460-1_13

Mekonnen 2012). Water is thus a major factor controlling food availability. Rain-fed agriculture, sustained by precipitation-recharged soil moisture (the so-called green water), covers 80% of the cultivated land worldwide. On the remaining 20% of cultivated land, irrigated agriculture provides 42% of global food production (FAO 2011). At present, nearly 30% of green water resources and only 10% of maximum available blue water (i.e. water withdrawal from surface- and groundwa-ter bodies) are used (Oki and Kanae 2006). The total volume of blue water con-sumption from the agricultural, industrial and domestic sectors is 2000 $km^3 \cdot year^{-1}$ according to Wada and Bierkens (2014), and it is unevenly distributed worldwide, with the largest consumptions occurring in India, Pakistan, China, USA and Mexico.

In order to explore the nexus between food production and water consumption, Hoekstra and colleagues (e.g. Hoekstra et al. 2012) introduced the *water footprint* as an indicator of water use related to goods and services produced or consumed by an individual (or a country), separating green water from blue water. The notion of water footprint is tightly connected to that of virtual water content, which represents the amount of water that is conceptually embedded (though not physically present) in a traded good. The concept of virtual water content has been introduced by Allan (2003), who suggested that virtual water import, i.e. the water embedded in imported goods, was a mechanism that contributed to compensate for water shortage in the Middle East countries.

In light of the fact that agriculture is the major water-consuming sector, many studies have focused on water footprint of crops. In particular, they focused on the efficiency of water use, expressed by the crop water footprint (CWF) quantified as the volume of water evapotranspired during the growing season divided by the crop yield (Mekonnen and Hoekstra 2010; Hoekstra et al. 2012). In a framework where water resources suffer from increasing pressures from population growth, economic development and climate change, the international food trade is vital for food secu-rity (Hanjra and Qureshi 2010). Through the international trade of agricultural goods, water resources that are physically used in the country of production are "virtually" transferred to the country of consumption. Virtual water trade has often been recognized for its ability to improve physical and economic access to food commodities in water-scarce regions, allowing nations to save domestic water resources through the import of water-intensive products (Chapagain et al. 2006). Thus, food trade leads to a global redistribution of freshwater resources, although it is recognized that commodities are being traded, and not water.

This chapter focuses on the Italian consumption of water for agricultural produc-tion and on the prominent role of Italy in the global virtual water trade. The existing literature on these issues is very limited and includes the papers by Tamea et al. (2013), Antonelli and Greco (2014) and Santini and Rulli (2015). The chapter is orga-nized as follows: Sect. 13.2 describes the virtual water balance of Italy, quantifying at the country scale the total amount of water expended for food production, the amount embedded in food consumed locally, and the amount of water virtually moved through the international food trade. Section 13.3 further specifies the international dimension of the problem, analysing the role of Italy in the international virtual water trade. Section 13.4 provides information on the water footprint of agricultural goods at high spatial resolution. Finally, in Sect. 13.5, some conclusions are drawn.

13.2   The Virtual Water Balance of Italy

The analysis starts with considering the volumes of virtual water embedded in the production, consumption and trade of Italy, with volumes obtained by aggregating across a large number of agricultural products.

13.2.1   Virtual Water Volumes

The assessment of virtual water is based on the FaoStat dataset of the Food and Agriculture Organization (FAO 2018), which provides complete information about the national production and trade of agricultural goods in the period 1961–2013. The goods here considered include crops, processed crops, livestock primary and processed, and live animals, for a total of 208 produced goods and 272 traded goods. The number of produced goods is smaller because it refers only to primary products, as summing primary and derived goods would lead to double count the water needed to produce them. Care is used in avoiding the double counting of primary crops used as feed and of animal goods, produced from animals grown with the same crops. To do so, a feed ratio, f, i.e. the ratio between feed and total supply, of each category of goods is derived from the Food Balance Sheet of Italy (FAO 2018). After associating such ratio to each good in the category, the production of each good is reduced multiplying by $(1-f)$, thus neglecting the fraction of goods being used as feed.

Production and trade data are converted into equivalent virtual water volumes according to the virtual water content (VWC) per unit weight of the good. The virtual water content is computed starting from the public dataset WaterStat (Mekonnen and Hoekstra 2010) which provides the VWC of a large number of goods, differentiated by the country of production and averaged over the period 1996–2005. The dataset accounts for rain-fed and irrigated areas in each country and returns a weighted mean across the whole national production. A simple but robust method, the Fast-Track method (Tuninetti et al. 2017), is then applied to obtain the time-varying virtual water content from 1961 to 2013. The method assumes that the temporal variability of VWC is entirely defined by the variability of agricultural crop yield, and it has been verified in Tuninetti et al. (2017). The time-varying VWC is applied to all crop-based goods (about 75%), while the remaining goods are left with a VWC constant in time and equal to the national WaterStat data. The virtual water considered in the present analysis is green, i.e. soil water originating from precipitation, and blue, i.e. water withdrawn from surface or groundwater and provided to crops as irrigation. Unless differently specified, the sum of green and blue virtual water is considered. Where the blue virtual water is considered alone, the

VWC of each product is obtained from the total VWC multiplied by the ratio of blue to total VWC, as available in the WaterStat dataset (Mekonnen and Hoekstra 2010) and kept constant in time.

The volume of each produced and exported good is multiplied by the VWC of the good in Italy, year by year, assuming that the country mainly exports locally produced goods. The volume of each imported good is multiplied by the world average VWC of trade of such good, as the origin of imported goods could not be defined across the whole period. Indeed, the origin of imported goods can only be derived for the period 1986–2013 thanks to the availability of detailed trade matrixes from the FaoStat database (see Sect. 13.3). The world average VWC is computed year by year as a mean of all national time-varying VWC of the good weighted by the country exports. Finally, the virtual water volumes are summed across all goods to obtain the total volumes of virtual water embedded in the import, export and production of Italy.

13.2.2 Virtual Water Balance in Time (1961–2013)

A simple balance is applied to compute the virtual water embedded in the consumption of agricultural goods C, as:

$$C = P + I - E, \tag{13.1}$$

where I is import, E is export and P is production; stock variations are neglected, and supply is assumed to be entirely consumed, thus encompassing in this term also seed, waste and other uses. The temporal evolution of the virtual water balance of Italy, with reference to the sum of green and blue water, is shown in Fig. 13.1 with the positive (P, I) and negative (E, C) terms of the balance represented on the positive and negative axes, respectively.

The calculations show that in 2013, the last year analysed, Italy imported 71 km^3 of virtual water (31 km^3 in 1961) and exported 24 km^3 (4 km^3 in 1961), while it used 66 km^3 for agricultural production (70 km^3 in 1961) and 113 km^3 for consumption (97 km^3 in 1961). VW import increased markedly in the first decade, and then it reduced its pace with some oscillations and started a more regular increase after the mid-1980s, concluding with a decade of limited increase. The overall trend of Italian import is positive and in good agreement with the increase of global virtual water trade, and it appears to have only temporarily suffered from the global crisis in 2008. VW export has constantly increased across the whole time period, at a regular pace of 0.4 km^3/year and very large percentage increases, as also highlighted in previous studies about Italy (Tamea et al. 2013; Antonelli and Greco 2014).

The VW embedded in agricultural production of Italy had non-monotonic variations in the considered period, with a total increment of about 10 km^3 in the first two decades and a comparable decline started around the year 2000. Such decline follows the decrease of agricultural area in Italy, which dropped from 20.7 million hectares in 1961 down to 13.6 in 2013, with a harsh decline of arable land in particular, which lost almost 50% of its surface (FAO 2018). The marked increase of

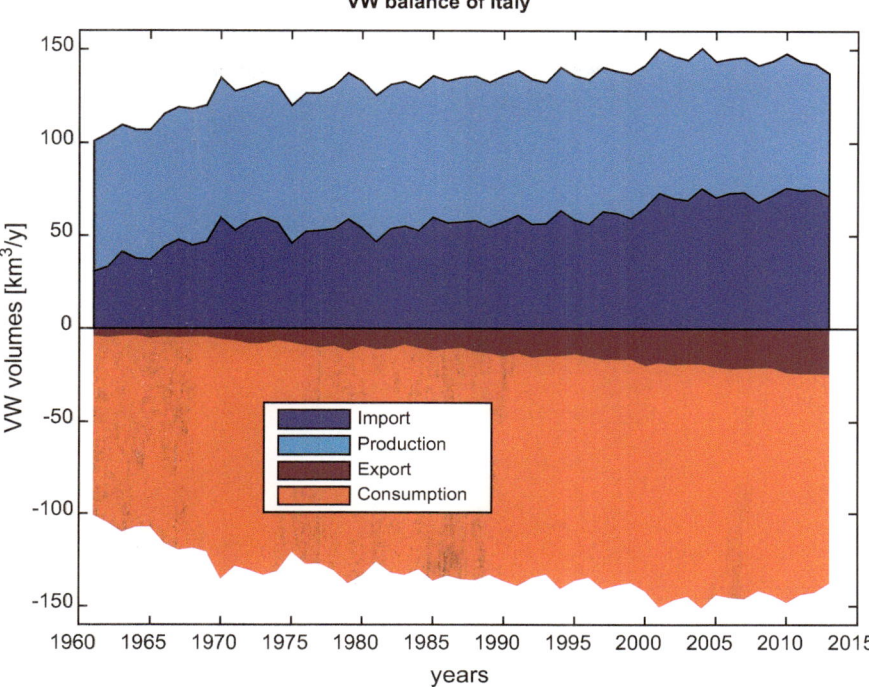

Fig. 13.1 The (green plus blue) virtual water balance of Italy with time-varying CWF

virtual water imports and the decline in the use of local water resources lead to a situation (in 2013) in which VW import overtook the VW of local production. This implies that Italy relies on imported (virtual) water for more than half of its water balance pertaining agricultural products, placing the country in a position of strong dependency on international trade and on external water resources. The VW dependency may reflect into a vulnerability of the country if a crisis would occur worldwide or in trade partner countries, as they could reduce the food (and embedded water) exported to Italy, reducing the country supply.

The virtual water embedded in the national consumption, as obtained from Eq. (13.1) and shown in Fig. 13.1, increased markedly in the first 15 years of analysis, nourished by the increase in both local production and import. A long phase of stagnation follows, with some fluctuations dictated by the variability of imports, likely compensated by the dynamics of national food stocks that are here embedded in consumption. It is worth relating the virtual water embedded in national consumption to the country population (data taken from the same data source: FAO 2018) and look at the per-capita virtual water balance.

Figure 13.2 shows that trends in all VW balance terms are confirmed, indicating their predominance over the population trend in Italy and highlighting that all increases occurred at a more-than-proportional rate with population. Per capita VW consumption has been relatively constant in time in the last decades, with an average

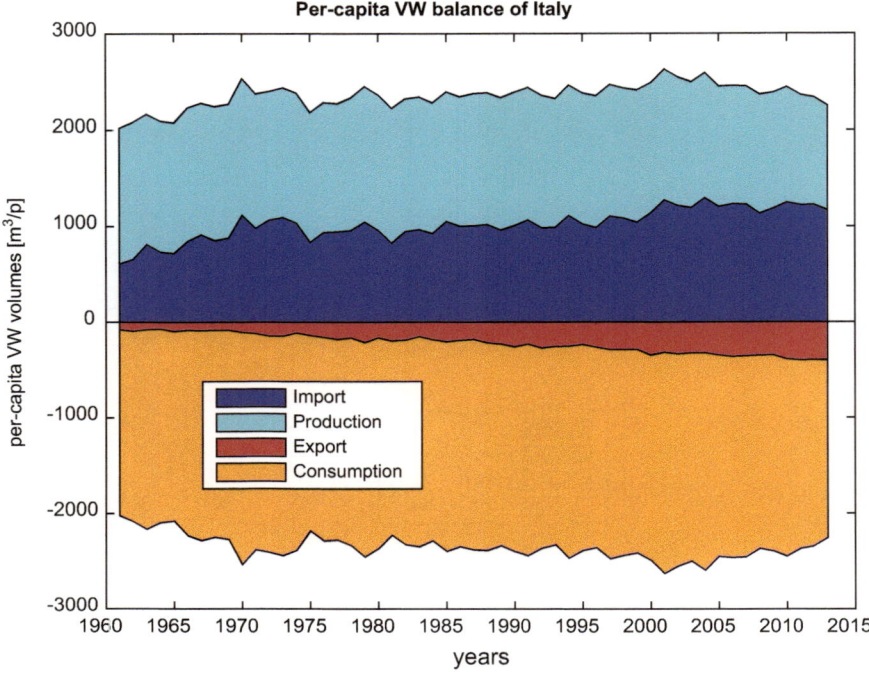

Fig. 13.2 The per-capita (green plus blue) virtual water balance of Italy

equal to 2100 m³/p/year (about 5800 litres/p/day), compared with a VW embedded in agricultural production of about 1300 m³/p/year (about 3600 litres/p/day). The difference between the two terms is provided by trade, in particular by the difference between per-capita VW import and VW export, whose averages equal about 1000 m³/p/year and 200 m³/p/year, respectively.

13.2.3   The Blue Virtual Water Balance

Previous results refer to the sum of green and blue virtual water. Here only blue virtual water is considered, including the water volumes provided as irrigation to satisfy the evaporative demand of crops plus the indirect use of blue water embedded in animal products plus the volumes for processing the agricultural goods. Volumes are obtained as mentioned above (Sect. 13.2.1), i.e. by multiplying the production and trade quantities of each good by the time-varying total (green plus blue) VWC and by the ratio of blue to total VWC, as given in the WaterStat dataset (Mekonnen and Hoekstra 2010). The blue VW of national consumption is obtained as in Eq. (13.1), and the four terms (positive P, I and negative E, C) are shown in Fig. 13.3.

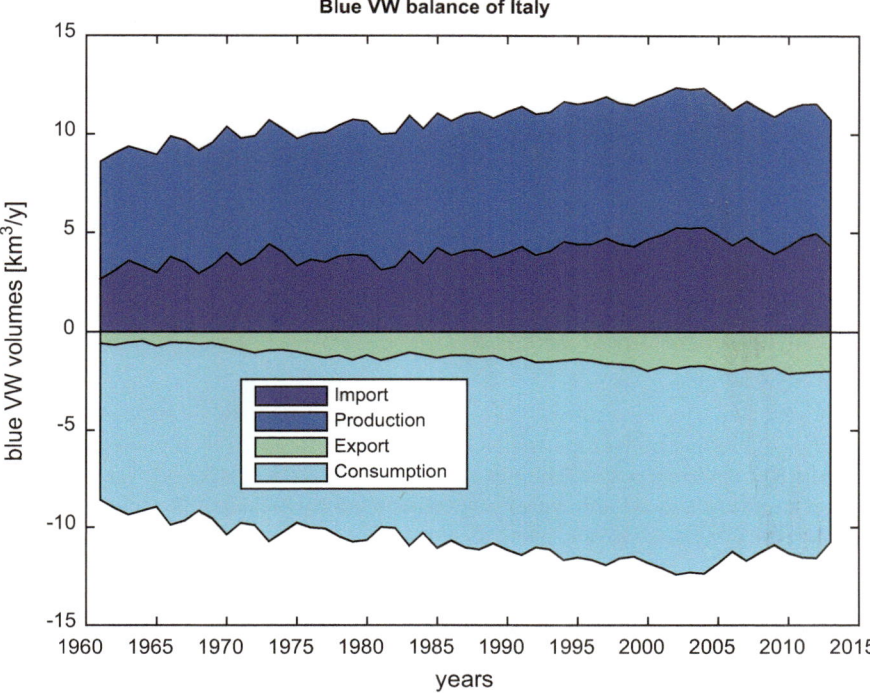

Fig. 13.3 The blue virtual water balance of Italy

In Italy in 2013, the blue VW embedded in trade was 4.3 km³ and 2.0 km³ for import and export, respectively, while the agricultural production and the national consumption required 6.4 km³ and 8.7 km³ of blue water. All terms have been increasing in time, at least until year 2000. The blue VW import increased markedly, with larger fluctuations in the first decade; the absence of a rate change after the 1970s (as seen for the total VW import) indicates that in such period Italy increased the share of blue water in its imports. The blue VW import shows oscillations around a constant value after year 2000, in line with the import of total VW. The blue VW export increased steadily throughout the whole period. The blue VW embedded in agricultural production has grown until the early 1990s and then stagnated and started decreasing in the last decade.

In order to correctly interpret the above data about the blue VW of national production and compare them with the results in Sect. 13.4, it must be noticed that the present analysis includes a large number of agricultural goods and animal products. In large part, the computed blue VWC refers to the optimal volume needed to satisfy an evaporative demand, and therefore it results in significantly lower than the agricultural water withdrawals reported in national and international reports. Withdrawals can be converted into effective irrigation by irrigation efficiencies. Accordingly, the values in Fig. 13.3 are compatible, for example, with the Sixth

Italian Census of Agriculture (ISTAT 2010) which indicates a volume of agricultural water withdrawals of 11.1 km³ (in 2010) or with the AquaStat database of FAO (FAO 2018), which indicates a volume of agricultural water withdrawals of Italy of 12.9 km³ (in 2007). Other values from the AquaStat database, such as 25.6 km³ in 1970, are probably justified by the much lower irrigation efficiency at that time, in particular in relation to rice cultivations.

13.3 Italy Within the Virtual Water Trade Network

13.3.1 Materials and Method

This section analyses the role of Italy in the international trade of agricultural commodities, including both crops and animal products, for 272 traded goods, under the perspective of virtual water. The virtual water trade (VWT) network for year 2013 (the most recent and reliable year) is reconstructed by means of the trade data provided by the FAO database and using the virtual water content detailed in Sect. 13.2.1.

Detailed trade data are available for each commodity, as metric tons exchanged per year, since 1986 (only cumulative export and import are reported in the Food Balance Sheets starting from 1961) as declared by the trading countries. Starting from these data, for each product a bilateral trade matrix, **F**, is constructed whose elements $F_{i,j}$ represent the trade flow from country i to country j. When divergent declarations exist from the exporter and importer countries, the flow is chosen from the country with higher reliability.

The trade matrix **F** is then converted into a VWT matrix (**VW**) by using the virtual water content of the commodity in the exporting country, assuming that countries export locally produced goods. When a country does export a given commodity, but it is not a producer itself of that commodity, this country is assigned the VWC of the closest producing country, with the assumption of similarities in the water use efficiency. Hence, the virtual water flow ($VW_{i,j}$) from country i to country j reads

$$VW_{i,j} = F_{i,j} \cdot VWC_i \qquad (13.2)$$

and it is expressed in cubic metres of water per year. Each country has a typical VWC for each crop and each year, as specified in Sect. 13.2. For each link (e.g. the import from Brazil to Italy), the virtual water volume is finally summed up over all traded food items, thus building up the network of all countries trading virtual water with Italy.

It is worth noting that using a detailed trade matrix to compute the virtual water flows allows one to obtain a more accurate estimates of the flows. In fact, in this case also the import flows are associated with the VWC of the origin country, rather than assigning them a world-average value as done in Sect. 13.2.2.

13.3.2   Results

Italy is one major global importer of virtual water, but it also has a relevant role in food (and thus, virtual water) export. As previously shown, Italy is a net virtual water importer, thus intensively relying on external water resources. In 2013, the VW volume imported (55 km³) exceeds of more than two times the VW volume exported (24 km³). The fact that the imported VW volume shown here is smaller than the value (71 km³) obtained in Sect. 13.2.2, using globally averaged VWC, suggests that Italy tends to import from countries that exhibit lower VWC (i.e. higher water use efficiency) than the global average. Over 60% of the Italian export is directed towards ten major countries, most of which localized in Europe (the USA is the only exception); nearly 60% of the Italian import comes from just ten major countries, four of which are outside the European territory (Fig. 13.4). Germany and France are the most important European partners: they import 6 km³ of virtual water from Italy (30% of the total Italian export) and export 11 km3 of virtual water towards the Italian territory (20% of the total Italian import). Rice, macaroni and cheese of cow milk are the most important products in terms of virtual water behind the Italy-France flow; rice, chocolate and cheese of cow milk are the most important products behind the Italy-Germany flow. Italy imports a consistent volume of virtual water from Germany (5.6 km³) mostly because of the import of pork meat, coffee and chocolate. Italy also relies on a similar amount of water coming from Indonesia

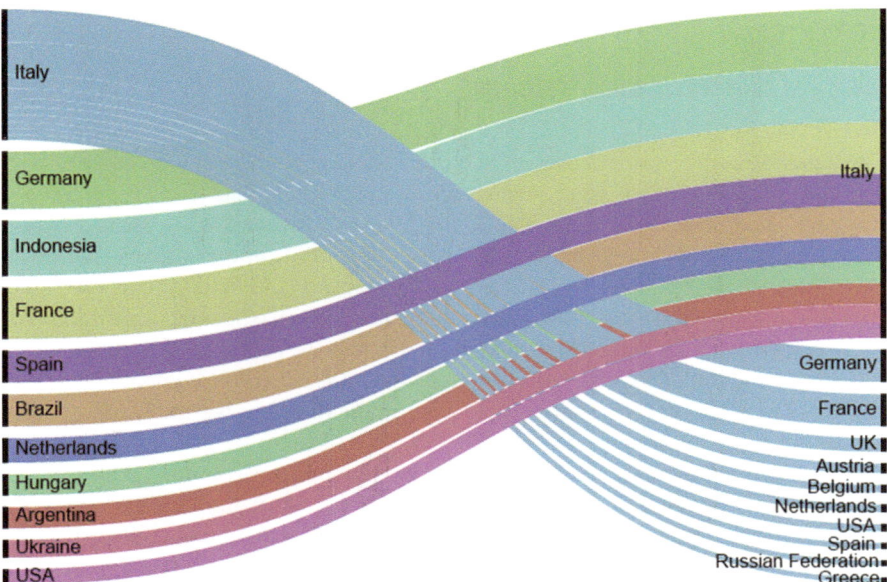

Fig. 13.4 Top-ten virtual water flows departing from and reaching Italy in 2013

through the import of palm oil and coffee. It imports other foreign water resources from Brazil (3.1 km^3), Argentina (2.1 km^3) and the USA (1.6 km^3).

13.4   High-Resolution Crop Water Footprint

All of the analyses performed in the previous sections pertained with aggregated data over the country. This section considers higher-resolution data to investigate the spatial variability of the water footprint of agricultural production within the Italian territory. Since the analysis focuses only on crops, the term "crop water footprint" is used here in place of the virtual water content per unit weight of crop.

13.4.1   Crops Under Consideration

Twenty different crop-based commodities are considered, which are largely cultivated across the Italian territory. The gross production of these products is 58 million tons in 2010 according to the statistics for ISTAT (2018), and it corresponds to more than 90% of the total Italian agricultural production. Moreover, these crops are central in the human diet covering 83% of the total food energy intake required by the typical Italian diet (FAO 2018, Food Balance Sheet) and also because some of these products are important components of the feed for livestock (e.g. maize).

13.4.2   Materials and Methods

The Italian water footprint of crop production is obtained with a detailed assessment that considers the following products: wheat, rye, barley, oats, rice, grain maize, sorghum, other cereals, tuberous plants, dried pulses, roots and bulbs, stalks, salads, fruits, oleifera seeds, sugar beet, fresh fruit, citrus fruit, grapes, olives. For each crop, the analysis evaluates (i) the crop water footprint (*CWF*) as a function of the crop actual yield and its evapotranspiration demand and (ii) the total water footprint (*WF*) generated by the crop production, which equals the product between the *CWF* and the production of a given year. Each assessment is accomplished both for green and for blue water resources, in order to explore the different dependencies of crop species on irrigation water.

The *CWF* value is evaluated for each crop at a spatial resolution of 5 by 5 arc minutes, corresponding to areas of about 7 by 7 km at the Italian latitude, although the input data present different spatial resolutions, which have required some preliminary elaborations. *CWF* estimates are referred to year 2010, which allows us to present the most updated assessment of the Italian WF given that studies are generally centred on the period 1996–2005 due to data availability at the global scale (e.g. Mekonnen and Hoekstra 2010; Tuninetti et al. 2015). *CWF* is defined in each pixel as the ratio between the water evapotranspired by the crop during the growing season, ET_a (mm), and the crop actual yield, Y_a (ton/ha), as

$$CWF = \frac{10 \cdot ET_a}{Y_a} \qquad (13.3)$$

where the factor 10 converts the evapotranspired water height expressed in mm into a water volume per land surface expressed in m³/ha.

The total water evapotranspired by the crop in a single growing season, ET_a, is obtained by summing up over the length of the growing period the daily actual evapotranspiration, $ET_{a,j}$ (j is the day). The length of the growing period is delimited by the planting and harvesting dates taken from the dataset MIRCA2000 provided by Portmann et al. (2010). This dataset distinguishes between rain-fed and irrigated production and provides the month in which the growing period starts and ends at 5 by 5 arc minute resolution, considering multi-cropping practices, for year 2000. The planting and harvesting dates are set in the middle of the month due to lack of more precise information. Despite some adjustments in the planting and harvesting dates of each crop may have occurred between 2000 and 2010, the growing seasons have been considered to remain the same of year 2000. Given that some products are missing in the MIRCA2000 dataset, associations are made arbitrarily between the missing products and the most similar classes available in the dataset (namely, maize for grain maize, other cereals; others annual for oats, roots and bulbs, stalks leaves and inflorescences, salad, fruits; potatoes for tuberous plants; soybeans for oleifera seeds; others perennial for fresh fruit, total olives).

Daily $ET_{a,j}$ is calculated following the approach proposed by Allen et al. (1998), namely,

$$ET_{a,j} = k_{c,j} ET_{0,j} k_{s,j} \qquad (13.4)$$

where $k_{c,j}$ is the daily crop coefficient, $ET_{0,j}$ is the daily reference evapotranspiration from a hypothetical well-watered grass surface with fixed crop height, albedo and canopy resistance and $k_{s,j}$ is the daily water stress coefficient depending on the available soil water content, with a value between 0 (maximum water stress) and 1 (no water stress).

The crop coefficient depends on crop characteristics and, to a limited extent, on climate. It is influenced by crop height, albedo, canopy resistance and evaporation from bare soil. During the growing period, $k_{c,j}$ varies with a characteristic shape divided into four growing stages (I, initial phase; II, development stage; III, mid-season; IV, late season). The crop coefficients and the proportional length of each growing stage have been derived from Chapagain et al. (2006) and are specific for 10 different climatic regions.

Monthly ET_0 data at 10 by 10 arc minute resolution are given by New et al. (2002) as a long-term average over 1961–1990. These data are converted to 5 by 5 arc minute data by subdividing each grid cell into four square elements and assigning them the correspondent 10 by 10 values. Daily $ET_{0,j}$ values are determined through a linear interpolation of the monthly ET_0 data and attributing the monthly value to the middle of the month.

The water stress coefficient typical of the cell varies along the growing period depending on the total available water content (*TAWC*) and the readily available water content in the root zone (*RAWC*). Specifically, *RAWC* is the portion of *TAWC* that the crop can actually use. These variables vary along the growing season depending on the rooting depth, which is different from crop to crop and is generally deeper under rain-fed conditions. The values of rooting depths and the fraction of *TAWC* that can be actually used by the plants have been derived from Allen et al. (1998). Coefficients $k_{s,j}$ are computed under rain-fed and irrigated conditions. In each cell, the presence of rain-fed and/or irrigated conditions is established based on the MIRCA2000 dataset, which gives the crop-specific area cultivated in year 2000 and the proportion of the total area that is cultivated under rain-fed or irrigated conditions. Hence, the provincial cropland available from the ISTAT dataset are firstly disaggregated at the 5'×5' cells according to the MIRCA2000 spatial distribution. Then, the total areas are split into rain-fed and irrigated areas using the mentioned proportions.

For irrigated production, the crop is assumed to transpire at full rate, i.e. irrigation is provided to compensate for the difference between potential (crop) and actual evapotranspiration. Hence, $k_{s,j}$ is equal to 1 throughout the growth period. For rain-fed production, $k_{s,j}$ depends on the daily soil water balance, and whenever water from precipitation is not sufficient for an optimal evapotranspiration, the crop goes stressed, and $k_{s,j}$ becomes lower than 1. The water balance is based on the 30 arc second maps of the available water content given by the Harmonized World Soil Database and the 10 arc minute maps of monthly precipitation given by New et al. (2002).

Since the daily $k_{s,j}$ is different in the two production types, as well as the $ET_{0,j}$ (i.e. the growing period can have different planting dates in rain-fed and irrigated conditions), the $ET_{a,j}$ (green and blue) is different in the two production types. In the case of rain-fed crops, the green component is equal to the total volume of evapotranspiration. In the case of irrigated crops, the blue component is equal to the volume of irrigation water provided to the crop, and the green component is the difference between the total evapotranspiration and the blue component. The overall evapotranspiration of green and blue water from the cell is obtained by the weighted mean of the green and blue evapotranspiration in rain-fed and irrigated conditions. Further details on the computation of the water stress coefficient can be found in Tuninetti et al. (2015).

Crop yields data for year 2010 have been calculated using the ISTAT data about production and cultivated area available at the province level for each study commodity: i.e. yield is the ratio between production and harvested area, and it is expressed as ton per hectare. In this study, yield is assumed as constant across each province due to the lack of more detailed data, thus assigning to each pixel the yield value of the relative province. Gridded production, which is required to evaluate the total WF, is obtained as the product between the gridded yield and the distributed area.

13.4.3   Results

Figure 13.5 shows the Italian water footprint associated with the production of 20 study products for year 2010. The overall Italian WF was 31 km^3 (green plus blue water), of which 11% (3.5 km^3) came from surface- and groundwater bodies to fill the irrigation demand along the growing season. The WF shown here is smaller than that shown in Fig. 13.1 is due to the different sets of products considered in the present analysis, which leaves out all animal-based and other minor plant-based products.

The Italian water footprint varies considerably across the territory due to differ-ent climatic conditions, soil properties and cultivated crops. The total WF (Fig. 13.5a) varies between 0.5 hm^3 and 40 hm^3 in a single cell. The cells with the largest WF values are found in Puglia where some gridded values exceed 50 hm^3. These large values, and the spatial heterogeneity in general, are determined not only by the annual potential evapotranspiration (which of course increases with decreasing lati-tude), but also by the extension of the cultivated area within the cell, and by the duration of the growing season for the cultivated crops, with permanent crops

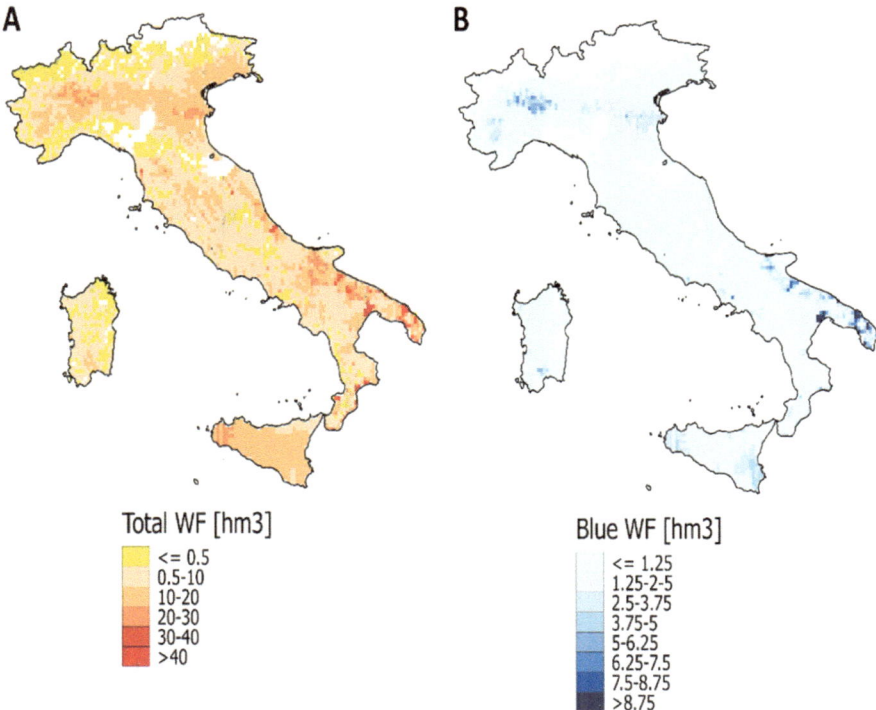

Fig. 13.5 The Italian water footprint at 5'×5' resolution referred to year 2010: blue plus green water footprint (**a**), blue water footprint (**b**)

Table 13.1 Gross irrigation volumes (or withdrawals) associated with the Italian production of 20 study crops. Average efficiencies are taken from Howell (2003); system prevalence is given by the ISTAT (2010, Table 2.39)

	Surface irrigation	Submerged irrigation	Irrigation sprinkler	Drip irrigation	Others	Total
Average efficiency [%]	40	14	75	90	53.5	
System prevalence [%]	27.2	34.8	26.8	9.6	1.5	
Irrigation volume [km³] (this study)	2.6	8.4	1.2	0.4	0.1	**12.7**

leading to larger footprints than temporary crops. The blue WF (Fig. 13.5b) also shows a significant variability with the largest WFs in Piemonte and Puglia, either in sites where rice is cultivated or in places where reliance on irrigation is stronger due to scarce precipitation and intensive agriculture.

It is quite interesting to notice that some areas, for example a large part of the Tuscany and Lazio regions, are characterized by a strong spatial heterogeneity in the water use, with high and low consumptions in nearby cells (Fig. 13.5a). This testifies of a landscape where agricultural activities were, in year 2010, interspersed with other economic activities, contrasting with other areas (e.g. Plain of Po river, Apulia) devoted to intensive agriculture practices.

The blue WF of crop production refers to the net irrigation water requirement; thus it refers to the amount of water that is actually evapotranspired by the crops. The net irrigation requirement rarely coincides with the total volume of water that needs to be withdrawn from surface- or groundwater bodies, namely, the gross irrigation requirement. This amount is generally larger due to the inefficiencies that characterize the water transportation and irrigation systems. In Table 13.1 the gross irrigation volumes required by the Italian agricultural production of the 20 study crops are summarized. These values have been obtained by allocating the total blue WF to the different irrigation systems according to the system prevalence given by the ISTAT database for year 2010. Accordingly, 27.2% of the Italian agriculture relies on surface irrigation, 34.8% relies on submerged irrigation (86% is used for rice production), 26.8% relies on irrigation sprinkler, and 9.6% relies on drip irrigation. The total blue WF is then divided by the characteristic efficiency of the irrigation systems, i.e. surface, submerged, sprinkler and drip irrigation, as provided by Howell (2003). The estimated gross irrigation volume is around 12.7 km³ of surface water plus groundwater, which is nearly four times the consumptive blue water use (i.e. blue WF). The estimated gross irrigation volume compares well with the ISTAT estimate, which is around 11.1 km³, and differs from it by only a 10%.

13.5 Concluding Remarks

Water used for food production represents a large share of the total amount of water consumed by human beings. In this chapter, the water footprint of over 200 agricultural goods in Italy has been quantified to measure the total water use for food production, as well as the total amount of virtual water imported to or exported from Italy. For 20 food items, the analysis has been extended to consider a higher spatial resolution (grid cells covering about 50 km^2). The obtained results depict a situation where most of the water footprint is in the form of green water, with the cultivation of only few food items (e.g. rice and maize) being dependent on relevant amounts of irrigation water. Also spatially, regions of Italy where the blue water footprint is high are limited to few agriculture-intensive areas. In general, the country relies markedly on foreign water resources to meet its water-for-food needs. The external water footprint of the Italian consumption is especially large both in nearby country as France and Germany and in faraway countries like Indonesia and Brazil. Globalization of water resources through the virtual water trade is thus already a reality for Italy.

References

Allan JA (2003) Virtual water – the water, food, and trade nexus. Useful concept or misleading metaphor? Water Int 28(1):106–113

Allen RG, Pereira LS, Raes D, Smith M (1998) Crop evapotranspiration-guidelines for computing crop water requirements, FAO Irrigation and Drainage Paper n.56. Food and Agriculture Organization of the United Nations, Rome

Antonelli M, Greco F (2014) L'impronta idrica dell'Italia, World Wide Fund (WWF) report, available online at https://www.wwf.it/?6781, (in Italian) accessed on May 2019

Chapagain AK, Hoekstra A, Savenije H (2006) Water saving through international trade of agricultural products. Hydrol Earth Syst Sci Discuss 10(3):455–468

FAO (2011) The state of the world, land and water resources for food and agriculture. Food and Agriculture Organization of the United Nations, Rome

FAO (2018) Food and Agriculture Organization of the United Nations, Statistics Division. www.fao.org/faostat/en/#data. Accessed on Apr 2019

Hanjra MA, Qureshi ME (2010) Global water crisis and future food security in an era of climate change. Food Policy 35(5):365–377

Hoekstra AY, Mekonnen MM (2012) The water footprint of humanity. Proc Natl Acad Sci 109(9):3232–3237

Hoekstra AY, Chapagain A, Aldaya A, Mekonnen AM (2012) The water footprint assessment manual: setting the global standard. Earthscan, London

Howell TA (2003) Irrigation efficiency. Encyclopedia of water science. Marcel Dekker, New York, pp 467–472

ISTAT (2010) Istituto Italiano di Statistica, Sixth Census of Agriculture. https://www.istat.it/it/archivio/66591. Accessed on 19 Nov 2018

ISTAT (2018) Istituto Italiano di Statistica. http://dati.istat.it/#. Accessed on Nov 2018

Mekonnen M, Hoekstra A (2010) The green, blue and grey water footprint of products, Value of water research report series, 47–48. UNESCO-IHE, Delft

New M, Lister D, Hulme M, Makin I (2002) A high-resolution data set of surface climate over global land areas. Clim Res 21:1): 1–1):25

Oki T, Kanae S (2006) Global hydrological cycles and world water resources. Science 313(5790):1068–1072

Portmann FT, Siebert S, Döll P (2010) MIRCA2000 – Global monthly irrigated and rainfed crop areas around the year 2000: a new high-resolution data set for agricultural and hydrological modeling. Glob Biogeochem Cycles 24(1)

Santini M, Rulli MC (2015) Water resources in Italy: the present situation and future trends. In: Antonelli M, Greco F (eds) The water we eat. Springer International Publishing, Switzerland. ISBN: 978-3-319-16392-5

Tamea S, Allamano P, Carr J, Claps P, Laio F, Ridolfi L (2013) Local and global perspectives on the virtual water trade. Hydrol Earth Syst Sci 17:1205–1215. https://doi.org/10.5194/hess–17–1205–2013

Tuninetti M, Tamea S, D'Odorico P, Laio F, Ridolfi L (2015) Global sensitivity of high–resolution estimates of crop water footprint. Water Resour Res 51(10):8257–8272

Tuninetti M, Tamea S, Laio F, Ridolfi L (2017) A Fast Track approach to deal with the temporal dimension of crop water footprint. Environ Res Lett 12(7):074010

Wada Y, Bierkens MF (2014) Sustainability of global water use: past reconstruction and future projections. Environ Res Lett 9(10):104003

Chapter 14
Adaptation Strategies to Climate Change for Water Resources Management

David J. Peres

Abstract The issue of climate change has received growing attention in recent years, from science to politics and mass media. It poses new challenges in water resources management, in all aspects: water scarcity, droughts, floods and landslides. Several scientific studies have been carried out to assess the potential future impacts of climate change on water resources in Italy. These studies indicate several negative impacts of global warming, with possible increases in the frequency and severity of almost all water-related hazards, though there are significant uncertainties in the climate projections on which the assessments are based. In recent years, Italy has provided itself of a series of documents that define the strategies for adaptation to climate change, culminating in the Italian Climate Change Adaptation Plan. This document provides detailed guidelines on how to face the new challenges posed by climate change, with important implications for water-related sectors. With reference to these sectors, the studies indicated, in general, the need of more investments on management strategies, nonstructural measures, blue-green infrastructures and awareness-raising. The chapter aims at reviewing, with respect to Italy, the main features of climate change, the related impacts on water resources and the main documents and guidelines that define the adaptation strategies and lines of action to contrast such impacts.

14.1 Introduction

The emission of greenhouse gases into the atmosphere, mainly due to industrial activities, is considered as one of the main factors that can induce changes in the climate. It follows that actions to contrast climate change (mitigation) and its impacts (adaptation) have a clear nexus with economic and social development policies. As a member of the European Union (EU), Italy has dedicated a high political and scientific attention to the issue of climate change. The EU has in fact promoted

D. J. Peres (✉)
Department of Civil Engineering and Architecture, University of Catania, Catania, Italy
e-mail: djperes@dica.unict.it

© Springer Nature Switzerland AG 2020
G. Rossi, M. Benedini (eds.), *Water Resources of Italy*, World Water Resources 5,
https://doi.org/10.1007/978-3-030-36460-1_14

programs to study and to face risks deriving from climate change, funding programs and issuing directives focused on the topic. Italy has complied with this European policy, acting at various administrative levels and actively engaging with studies and projects conducted by research institutions.

Attention to climate change in Italy and the possible impacts on society and individuals symbolically emerged from the end-of-year 2017 speech from the President of the Republic, Sergio Mattarella, of which we report an excerpt in the following:

We need to be prepared for the immediate future – interpret, understand new things. Innovations are so fast that they lead us into a new era that also poses unprecedented questions on the relationship between man, development and nature. Just think of climate change consequences such as droughts, limited water availability, devastating fires. In this regard, there is a growing awareness that has also received an impulse from the magisterium of Pope Francis [...]. Lifestyles, consumption, languages change; the jobs and the organization of production change, some professions disappear, others appear. Today the word future can also evoke uncertainty and worry. It has not always been this way. Scientific discoveries and the technical evolutions in history have accompanied a positive idea of progress. However, changes need to be controlled to prevent them from producing injustice and marginalization. The real mission of politics consists exactly in the ability challenging these innovations by guiding the processes of change, so that the new season becomes more fair and sustainable.

As can be seen, the speech is permeated by the concept of uncertainty, that, in the case of climate change, may refer to both psychological and scientific aspects. The former are associated with concepts from concern to fear, often stressed by politics and media; the latter are more commonly recalled by the scientific literature, especially when it comes to applying the climate models to predict the future climate and to assess the potential impacts of related changes. In fact, the various studies, guidelines and plans prepared for Italy always take into account, to a certain degree, this kind of uncertainty. Uncertainties in predicting the future climate are due to both difficulties in climate modeling and in predicting the success of greenhouse reduction policies. Main difficulties in climate modeling are determined by simplifications in the representation of the physical processes taking place in the atmosphere, as well as in the representation of boundary conditions. With reference to climate projections in Italy, for example, a modeling difficulty is related to the fact that the Mediterranean Sea is connected with the oceans only through the Strait of Gibraltar, which is not resolved by modeling grids. Hence, the accurate estimation of flows as inlet from the oceans and the main river basins (Po, Nile, Black Sea, etc.) is important and constitutes a specific source of uncertainty affecting climate modeling in the Mediterranean area (Castellari et al. 2014). On the side of greenhouse gas emission prediction, the Fifth Coupled Model Intercomparison Project (CMIP5) has proposed four possible future scenarios, better known as Representative Concentration Pathway (RCP), each related to a degree of success of climate change mitigation policies; yet, it remains difficult to predict which scenario is the most plausible.

Without neglecting the scientific limitations in the assessment of climate change impacts, the Italian government, through its Ministry for Environment and the Protection of the Land and the Sea (MATTM), has financed several studies, with operational support from various research institutions. The outcomes of these studies provide a comprehensive picture of the possible impacts of climate change in Italy and how to cope with them. This chapter aims at providing an overview of the main features of climate change and its impacts and guidelines for adaptation policies in Italy, based on the following documents:

- The Adaptation Strategy issued by the MATTM (2014)
- The report on future climate in Italy, produced by the National Institute for Environmental Protection and Research (ISPRA 2015)
- The Adaptation Plan produced for the MATTM by the Euro-Mediterranean Center on Climate Change (CMCC 2017)
- The reports on climate indicators in Italy, in particular the last one released (ISPRA 2018)

Though sometimes they are misleadingly used interchangeably, there is a specific difference between the words "strategy" and "plan", while the former provides a strategic vision for adaptation at the national level, the latter defines modalities by which the strategy is pursued. Thus, they are based on different elements (see Fig. 14.1).

14.2   Climate Change Assessments

Several studies have been conducted to assess possible future climate changes in Italy, which refer to two distinct approaches: (a) analysis of trends in meteorological or hydrological series of multi-decadal length and (b) assessment of future anomalies based on projections resulting from the application of climate models. The former studies have the advantage of being based on observed data, and thus they are not subject to climate modeling uncertainty. However, in this case, the assessment of future impacts can be based only on the extrapolation of observed trends, which obviously may not remain the same as in the past, also in view of several possible scenarios for climate change mitigation policies. In contrast, the projections based on climate models allow taking into account these scenarios. Regarding the first type of approach, since year 2006, ISPRA publishes a report on climate indicators, in which some historical trend analyses are carried out, based on observed data. Figure 14.2 shows some trend graphs taken from the report updated to year 2017 (ISPRA 2018): while on the one hand the temperature trends are quite evident, there does not seem to be any clear trend in annual precipitation at the national scale. The report indicates similar results for precipitation indicators that represent seasonal characteristics of rainfall and the frequency or magnitude of intense rain events.

An interesting analysis based on regional climate models' (RCMs) future projections can be found in the study carried out by ISPRA (ISPRA 2015). The study is based on the analysis of the data from four regional climate models, available as part of the Med-CORDEX initiative (Ruti et al. 2016). In particular, the most significant elements of knowledge and uncertainty on the evolution of the future climate in Italy are evidenced by analysing the variations of a set of internationally recognized indicators, each one indicating a specific characteristic of climate (see Table 14.1)

Fig. 14.1 Elements for the National Adaptation Strategy and the National Adaptation Plan according to MATTM (2014)

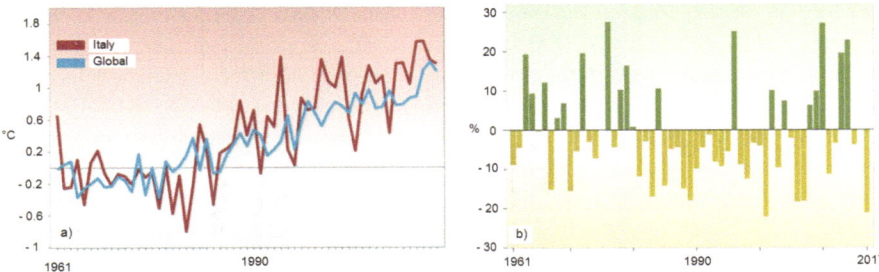

Fig. 14.2 Anomalies of (**a**) global and Italian mean temperature and (**b**) mean total annual precipitation, with respect to the baseline period 1961–1990 (ISPRA 2018)

Table 14.1 Main indices proposed by the expert team on climate change detection and indices (ETCCDI), commonly used to analyse possible future impacts of climate change

Category	Name	Description
Temperature indices	Number of tropical nights (TR20)	Annual count of days when TN (daily minimum temperature) > 20 °C
	Number of frost days (FD0)	Annual count of days when TN (daily minimum temperature) < 0 °C
	Number of summer days (SU25)	Annual count of days when TX (daily maximum temperature) > 25 °C
	Warm spell duration index (WSDI)	Annual count of days with at least 6 consecutive days when TX > 90th percentile
	Cold nights (TN10P)	Percentage of days in which the minimum daily temperature is less than the 10th percentile of the respective normal climatological distribution
	Cold days (TX10P)	Percentage of days with maximum daily temperature less than the 10th percentile of the respective normal climatological distribution
	Hot nights (TN90P)	Percentage of days in which the minimum daily temperature is less than the 90th percentile of the respective normal climatological distribution
	Hot days (TX90P)	Percentage of days with maximum daily temperature less than the 90th percentile of the respective normal climatological distribution
Precipitation indices	Simple precipitation intensity index (SDII)	Annual precipitation divided the count of rainy days in a year (rainfall ≥ 1 mm)
	Precipitation in most rainy days (R95P)	Annual total precipitation when daily precipitation > 95th percentile
	Maximum daily precipitation (RX1day)	Maximum value of precipitation in one day
	Maximum length of dry spell (CDD)	Maximum number of consecutive days with daily precipitation < 1 mm

More in detail, three time horizons are considered, representing periods of 30 years (2021–2050, 2041–2060 and 2061–2090), for which both the average and extreme climate indices of temperature and precipitation are calculated, relatively to intermediate- and high-level emission scenarios (respectively, RCP4.5 and RCP8.5). The analysis is focused on climate extremes and their variations, as they are considered particularly relevant for envisioning adaptation strategies, due to the impacts they produce on the environment and on society in general. Future climatic variations are estimated as the difference between the values in a future period and in the baseline one (1971–2000). The reference values obviously vary across models and are produced by applying models in hindcast mode, i.e. by performing simulations referring to past periods. The main results of the analysis are shown in Table 14.2. As can be seen, it emerges an overall increase in temperatures and heat waves, and a decrease in annual precipitation totals, with some tendency towards a progressive reduction of rainy days, which in turn may become more intense.

However, as also mentioned in the report itself, the assessment presents the drawback of analysing the projections regardless of their reliability, which could have been measured indirectly through the evaluation of their ability to reproduce the past and present climate – an aspect addressed for some regions of Italy (Peres et al. 2017; Mascaro et al. 2018).

A more recent assessment of the impacts of climate change was carried out by the CMCC and presented within the National Adaptation Plan (see Sect. 14.4). The CMCC analysis is, in some aspects, more detailed than that of ISPRA, since climate change is determined by homogeneous climatic areas determined by cluster analysis applied to a wide set of climate indices. Another innovative aspect is the analysis of climate change also for the sea. However, the study presents the limit of using the projections of a single regional climate model (namely, the COSMO-CLM RCM nested on the CMCC-CM GCM), and not of an ensemble of several models, which would have allowed to take into account model uncertainty. In spite of these significant methodological differences with the ISPRA study, it can however be said that, at national level, the results of the two are quite similar.

14.3   Italian Adaptation Strategy

The National Strategy for Adaptation to Climate Change (MATTM 2014) has the main objective of developing a nationwide vision on how to deal with the impacts of climate change, and to identify a set of actions and guidelines to cope with it, so to minimize the related risks. The document represents the reference point for the implementation in Italy of adaptation measures coordinated by the competent institutional authorities at national, regional and local administrative levels. In the Strategy, the set of actions and guidelines are identified with reference to the socio-economic and environmental sectors most vulnerable to climate change (see

Table 14.2 Assessment of climate change impacts in Italy, according to the report by ISPRA (2015), based on the analysis of four RCMs from the MED-CORDEX initiative

Temperature	Precipitation
Constant increase of temperatures over time. In RCP8.5 the increase is twice compared to the RCP4.5 Average temperature in Italy will increase across the twenty-first century between 1.8 and 3.1 °C (ensemble mean 2.5 °C) in RCP4.5 and between 3.5 and 5.4 °C in RCP8.5 (ensemble mean 4.4 °C). These variations are fairly uniform over the country The most pronounced increase in average temperature is expected in the summer season, with variations in a century between 2.5 and 3.6 °C in RCP4.5 and between 4.2 and 7.0 °C in RCP8.5. In spring the least significant increase is expected, with variations in a century between 1.3 and 2.7 °C in RCP4.5 and between 2.8 and 4.8 °C in RCP8.5. The sign of variations is positive (temperature increase) in all seasons and consistent for all models, but the magnitude of variations often differs significantly from one climate model to another Temperature extremes show equally important and significant variations. All the models agree in indicating a reduction in frost days and an increase in tropical nights, summer days and heat waves, though with some significant differences in their magnitude.	Rainfall change projections are much more uncertain than for temperature, and the two scenarios have not so clearly distinguishable impacts Models prevalently indicate a slight decrease of national average of the cumulative annual precipitation Overall, the changes forecast for 2061–2090 are between in the range -8% to +5%; the ensemble mean indicates a reduction of about 1.5%. In the RCP8.5 scenario, this interval widens (resulting between -15% and + 2%), and the ensemble mean moves in the direction of a reduction in precipitation The spatial distribution of the expected future planned variations is very different from one climate model to another. The models seem to indicate that northeastern regions are less likely affected by reduction in precipitation The national average values slightly decrease in spring, summer and autumn and weakly increase in winter. Localized significant variations of precipitation are however present, which can reach a decrease up to 150–200 mm in spring or summer and an increase up to 100–150 mm in winter Some representative indices of the frequency and intensity of precipitation extremes indicate a progressive reduction of rainy days, causing more intense and less frequent rainfall events. However, the magnitude of these variations is very uncertain and mostly weak

Table 14.3). As can be seen, a wide range of sectors is treated, which include agriculture, health and energy. Yet, many sectors are water-related: water resources (quantity and quality); desertification, land degradation and droughts; and hydrogeological risks (floods and landslides). Given the scope of this chapter, only these four sectors will be considered here, for which some key messages useful to face climate change can be delivered.

With reference to the sector of water resources (quantity and quality), it may be argued that Italy has learnt to deal with scarcity problems in the past. Thus, it can be considered mostly ready to face climate change challenges. Nevertheless, in several locations in Italy, chronic deficiencies in infrastructures, management and governance affect the efficiency of exploitation of available resources. One main factor of vulnerability is the temporal and spatial heterogeneity of resource availability, which makes management more critical. In many cases, inappropriate water

Table 14.3 Sectors and subsectors considered for the analysis of climate change impacts

Sector	Subsector
Water resources (quantity and quality)	
Desertification, land degradation and drought	
Hydrogeological risk (floods and landslides)	
Biodiversity and ecosystems	Terrestrial ecosystems
	Marine ecosystems
	Internal and transition water ecosystems
Forests	
Agriculture, aquaculture and fishing	Agriculture and food production
	Sea fishing
	Aquaculture
Coastal areas	
Tourism	
Health (risks and impacts of climate change, environmental and meteo-climatic determinants of health)	
Urban settlements	
Critical infrastructures	Cultural heritage
	Transportation and infrastructures
	Hazardous industries
Energy (consumption and production)	
Special cases	Alpine and Apennines area
	Hydrographic district of the Po river

Adapted from MATTM (2014)

management is also related to ecological quality, as overexploitation causes widespread and profound alteration of the natural hydrological regimes (Peres and Cancelliere 2016). The assessment of potential future impacts of climate change on water resources involves a high degree of uncertainty because of the limitation in the skill of the global and regional climate models to simulate many aspects of greatest interest for water resources. Hence, trend analysis, based on observed data, has to be given more consideration, respect to other sectors, with the restrictions already mentioned in Sect. 14.2. Given these features of climate change, and the current knowledge, it is important that adaptation measures are identified from time to time on the basis of local conditions. A holistic approach should be adopted for the effectiveness and efficiency of adaptation strategies. Due to the complexity of the problems, participatory approaches should be encouraged, as they are the only that can ensure success in water adaptation measures.

In facing the challenges relatively to the sector of desertification, land degradation and droughts, it has to be considered that about one third of the Italian territory is subject to these phenomena (MATTM 2014). Climate change will potentially increase the action of erosion and induce higher salination and loss of organic fraction in soils. Drought phenomena may increase the risk of fires and water stress with effects in both wetlands and dry areas, sometimes in combination with water scarcity events. Especially in the southern and insular regions of Italy, poverty and land degradation can increase their effects as a result of climate change. The management of natural resources (water, soil, subsoil and vegetation) will require a full implementation of current national, European and global policies, encouraging the use of the best available technologies and knowledge, taking into account local specificities.

With respect to hydrogeological risks, it must be pointed out that effective adaptation strategies cannot disregard monitoring and detailed analysis of available data. This is fundamental for identifying accurately the areas at highest risk, so that priorities can be assigned to the most urgent structural interventions, also considering that these are costly, and thus are not economically sustainable to be implemented for all prone areas. Many large infrastructures that have impacts on hydrogeological risks have been built in the 1970s and may need restoration, so it is also important to conduct an in-depth analysis on infrastructures such as dams, by taking into account the risks of events such as dam-break caused by earthquake, landslides, floods and combined or sequential events. It is necessary to evaluate the indirect impacts on the dams and in general on water storage infrastructures occurring in the present and which will likely increase because of climate change. The design criteria of these infrastructures must be revised by updating past hydrological-hydraulic criteria. About this point, the guidelines to plan and design the activities to contrast hydrogeological risk produced by *Italia Sicura* (Safe Italy) (Menduni et al. 2017), the government's mission structure against hydrogeological risk and development of hydraulic infrastructures (active from May 2014, but dismissed on June 2018), provided some specific indications on how to design hydraulic infrastructures in a climate change context. In particular, it suggests to overcome the traditional design usually based on a single reference scenario, typically characterized by a given design return period, since climate change could induce variations of the event corresponding to that return period within the life cycle of a hydraulic work. Hence the response of the work for a range of return periods, around the reference one, should be assessed. For example, an infrastructure designed for the 100-year flood will have to be evaluated in its behaviour even for the 50- and 200-year flood, and, where available, technical solutions that still guarantee acceptable results of effectiveness should be preferred.

From an administrative level, greater coordination between the actions and policies implemented by the various territorial governance bodies, at different geographical and temporal scales, is also essential to cope with hydrogeological risks in a changing climate. The adaptation actions in the hydrogeological sector must carefully balance between structural and nonstructural countermeasures, taking into account the complexity and fragility of the Italian territory. Actions are needed to extend and increase the awareness of population, i.e. to improve their perception of hydrogeological risks. It is essential to identify adaptation measures in updating flood risk management plans required by the EU Water Framework Directive.

The Adaptation Strategy document is accompanied by an appendix that provides a detailed overview of proposals for action, divided into various types:

- Nonstructural or "soft" actions: such as the definition, implementation, improvement and enhancement of early warning systems, insurance policies against damages resulting from climate change
- Actions based on an eco-systemic or "green" approach: construction of infrastructures that respect the principle of hydrological invariance, through permeable materials
- Structural and technological or "grey" actions: i.e. mainly infrastructures that offer an storage capacity allowing to limit the flow rates but not the downstream volumes, or debris flow retaining infrastructures
- Short- and long-term actions: the first to be implemented within year 2020, the second even after
- Transversal actions between sectors (soft, green or grey): for instance, awareness-raising campaigns in areas affected by hydrological cycle modifications that involve the participation of citizens and associations, are transversal for sectors of water resources and hydrogeological risk

This classification of measures is conserved within the Adaptation Plan, where they are described more in-depth and their implementation priorities are set.

14.4   Italian Adaptation Plan to Climate Change

The National Plan for Climate Change Adaptation, whose non-definitive version for public consultation has been completed on July 2017 (CMCC 2017), has the following three main objectives: (a) to describe the most urgent adaptation actions identified in the national strategy (cf. Sect. 14.3), detailing the timeline and who is responsible for their implementation; (b) to provide indications for improving the uptake of possible opportunities; and (c) to promote the coordination of actions at different territorial and administrative levels. The plan includes three parts: (1) context analysis, climate scenarios and climate vulnerability, (2) adaptation actions and (3) participatory, monitoring and evaluation tools (MRA) for the plan. Figure 14.3 summarizes the main elements of each part, which are more in detail described hereafter.

The first part of the plan is dedicated to the assessment of climate change impacts and vulnerabilities. It starts with the identification of the terrestrial and marine boundaries where climate change has similar effects and thus where the possible implications on extreme events and on the amount and seasonal distribution of water resources are expected to occur (other details on this analysis are shown on previous Chap. 5, at Sect. 5.4). In particular, firstly the "homogenous climatic macro-regions" (HCMRs) for terrestrial and marine areas are identified. Precisely, these are macro-regions having similar climatic conditions during a past reference

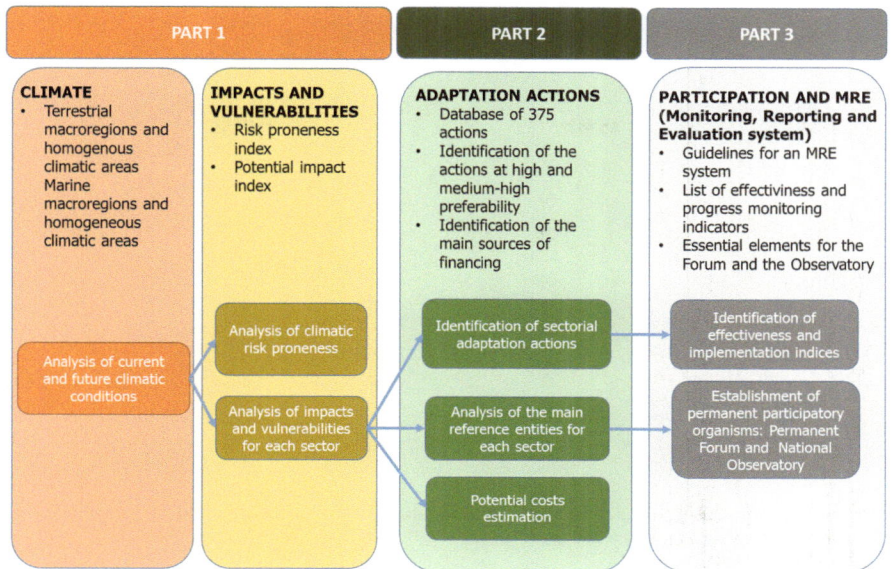

Fig. 14.3 Structure and essential elements of the Italian Climate Change Adaptation Plan. Adapted from CMCC (2017)

period. Then, areas that will have to cope in the future with the same climatic anomalies are delimited within the HCMRs, thus defining the "homogeneous climatic areas" (HCAs). The HCMR delimitation is based on a set of climatic indicators similar to those on Table 14.1, which have been assessed by the use of the E-OBS gridded observational dataset (Haylock et al. 2008), for the reference period 1981–2010. To the maps of these indices, cluster analysis has been applied. This has resulted in HCMRs, where subsequently, future climate projections have been analysed. To this aim the regional climatic model COSMO-CLM forced by the CMCC-CM global model (Scoccimarro et al. 2011), driven by the two scenarios RCP4.5 and RCP 8.5, has been used. The climatic projections of the selected RCM have been used to compute the anomalies of the same indicators used in HCMR delineation.

By a further application of cluster analysis, the areas with similar anomalies for the future have been grouped so to form the HCAs. These areas are used as a support for the subsequent sectorial analysis (see sectors in Table 14.3). Finally, the HCAs have been assumed as the geographic reference areas for the identification of adaptation measures and to foster the cooperation across adjacent territories in plan implementation, so to optimize the use of resources devoted to the application of synergic adaptation options. The analysis of current and future climatic conditions has been integrated with those relative to the assessment of the characteristics of the territory in terms of exposition, sensitivity and adaptation capacity, with the aim at assessing the risk, so to define the most appropriate adaptation strategies to be

Fig. 14.4 Methodology used within the Italian Adaptation Plan for the assessment of climate change risk of a given territory. Adapted from CMCC (2017)

implemented. Figure 14.4 briefly illustrates the methodology used within the plan for risk propensity assessment, which derives from a combined analysis of hazard, exposure, sensitivity and adaptation capacity.

These analyses are finally summarized into eight infographical sheets (six terrestrial and two marine), reporting a synthesis of current climatic zonation, prevalent climate anomalies, climate risk and potential impact and main "threats" and "opportunities" related to climate change. Overall, the results indicate a significantly higher number of threats than opportunities.

The second part of the Plan is dedicated to the analysis of the adaptation actions, the socio-political roles for their implementation, the necessary financial resources and the identification of funding sources. Based on the analyses of the first part of the Plan, and considering the rules, the policies and the pre-existing plans, the most important actions to face future climate changes are illustrated for each sector. For each HCA, the actions proposed in the Plan are associated with the impacts and objectives to pursue. This has led to a very detailed database of 375 adaptation actions. Table 14.4 shows the five macro-categories by which and the actions are grouped: information, governance, participatory and organizational processes. Each macro-category groups a series of categories to which specific measures are associated. As can be seen from the table, the Plan recommends a wide range of actions which surely need considerable economic resources and time. Some actions may however have lower costs than others and produce significant benefits. For instance, forecasting and early warning systems, if well designed and combined with participatory and risk-awareness campaigns, may allow an important decrease of the casualties related to hydrogeological risks, with way lower costs than grey infrastructures for flood or landslide mitigation. This is an important issue in Italy, since many climate-related causalities have been partially caused by a low awareness of

Table 14.4 Specific adaptation actions recommended by the Italian Adaptation Plan with an indication of main reference bodies for each macro-category; many of the indicated actions are transversal across several sectors, even not water-related

Macro-category	Main reference bodies	Category	Some specific actions
Information	MATTM, Ministry of Education, University and Research (MIUR), Civil Protection Department, Regional Agencies for Environmental Protection (ARPA), ISPRA, Basin District Authorities, research bodies, universities	Research and assessment (R&A)	R&A on climate change risk, resilience, vulnerability and relative components
			R&A on climate change adaptation impacts and solutions
			Climate model projections
		Monitoring, data collection and model development	Climate-related data harmonization and standardization
			Climate-related databases and informative portals
			Climate-related DSS and integrated information and communication systems
			Forecasting and early warning systems for climate-related hazards
			Climate, physical, chemical and biological indicators
		Dissemination, perception, awareness and training	Climate change information initiatives
Governance	Government, MATTM, Ministry of Health, Ministry of Labour, Higher Institute of Health (ISS), regional and local administration bodies, research institutions and associations, orders of professionals	Plans and strategies	Climate change plans and strategies
		Guidance	Good practices for climate change management
		Law and regulation adjustments	Law and regulation adjustments related to climate change
		Economic and financial instruments	Insurance and other risk-transferring instruments/compensations for adverse climate change effects
			Environmental certifications
			Economic and financial incentives for climate change adaptation measures
			Investment plans for climate change adaptation

(continued)

Table 14.4 (continued)

Macro-category	Main reference bodies	Category	Some specific actions
Participatory and organizational processes	Government, Basin District Authorities, regional administrative bodies, land reclamation syndicates, integrated water service managing bodies	Organization and management	Organization of civil protection at the local level
			Diversification of business strategies to better face climate change
		Partnership and participation	Cross-sector coordination, tables, committees and networks to better face climate change
Adaptation and improvement of plants and infrastructures	Government, MATTM, Ministry of Economy and Finance (MEF), Ministry of Economic Development (MiSE), Regional and local administrative bodies, integrated water service managing bodies land reclamation syndicates	Defense systems, networks, storage and transmission	Conversion of irrigation systems in more resilient systems
			Maintenance, improvement and interconnection of networks/flexible transmission systems for electricity production
			Maintenance, improvement and interconnection of water distribution networks
			Flood control dams
		Plants, materials and technologies	Vehicles and machines: increase of energy efficiency
			Improvement of domestic and industrial cooling systems
			Increase of water resources storage capacity
			Low environmental impact structures
Solutions based on ecosystem services	MATTM, Basin District Authorities, regional and local administrative bodies, land reclamation syndicates	Forest and agroforest ecosystems	Silvicultural management for biodiversity protection and conservation
			Silvicultural management for risk prevention
		River, coastal and marine ecosystems	River requalification and river bank maintenance/conservation, reconstruction and re-naturalization of coastal areas
			Restoration and management of wetlands
			Conservation, reconstruction and re-naturalization of coastal areas
			Protection and management of marine habitats
		Redevelopment of the built environment	Residential construction/urban green
			Improved sustainability of road drainage systems
			Urban green
		Integrated solutions	Increase of territorial ecological connectivity (green infrastructure)

Adapted from CMCC (2017)

potential risks and thus wrong behaviour. Table 14.4 is a brief synthesis of the actions database. Within the database, a detailed scheme describing the main information for the implementation of each action is provided. This information includes description, implementation time, sectors and involved climatic macro-regions, reference institutions, effectiveness indicators and monitoring. The database offers an overview of actions, allowing for their selection and their classification following hierarchic criteria.

The third part of the Plan is focused on participation, monitoring and evaluation tools and is divided into three main items related to (i) development of a participatory approach for drawing up and implementing the Plan, (ii) criteria for defining indicators of effectiveness of adaptation actions and (iii) methods for monitoring and evaluating the effects of adaptation actions. In fact, the participation of the population is a fundamental element for successful implementation of actions. Relatively to the Plan, a public consultation phase was envisaged in the process of drafting its text, when a questionnaire was filled by more than 700 persons, allowing to identify which actions are considered to be the most urgent by the population. Surely, this provides a basis for the Plan to be widely accepted by the population and thus less related barriers to its implementation.

The Plan also discusses the establishment of permanent participation bodies: the Permanent Forum and the National Observatory on Climate Change. The Permanent Forum has the objective to promote information, education and decision capacity of citizens and stakeholders, relatively to climate change adaptation and the actions envisioned by the Plan. The Forum has two specific objectives: (1) awareness-raising and information on the meaning of "climate change adaptation" and (2) knowledge diffusion and education on solutions and practices for adaptation. The Observatory instead is an organism formed by regional and local (municipality) representatives for the identification of territorial and sectorial priorities and for the subsequent monitoring of the effectiveness of adaptation actions. One critical issue in the observatory structure is that in many cases, regional and local entities seldom have the necessary technical and scientific competences to provide a reliable feedback in the monitoring tasks.

The final section of the Plan goes in-depth on the implementation, monitoring and evaluation system (MRE). The MRE is based on the identification of a series of specific indicators of effectiveness and implementation of the adaptation options. Two types of MRE indicators are defined: progress and effectiveness indicators. Some examples of the indicators are shown in Table 14.5. As can be seen, the assessment of the progress and the effectiveness of Plan actions is a quite complex task. Of course the MRE system requires by itself considerable investments. Moreover, these indicators may lack from a holistic view, i.e. do not allow to take into account possible interactions (and correlations) across the single aspects and actions.

Overall, the Plan is a quite ambitious; due to limited financial resources that hamper its complete implementation, the intervention priorities seem not sufficiently defined, also considering that different regions and local communities have significantly higher climate change vulnerability.

Table 14.5 Examples of effectiveness and progress indicators to assess implementation of the Italian Plan for Climate Change Adaptation (CMCC 2017)

Macro-category	Effectiveness indicators	Progress indicators
Information	Improvement of basic knowledge (data, information and available knowledge) – number of publications (peer reviewed and not) Estimation of the environmental and economic costs and benefits (Euro) of the various adaptation solutions or inaction Increase in the number of cognitive tools (decision support tools (DST), other tools (technologies, methodologies, etc.) to support adaptation Increase in the number of companies/businesses that assess the risks and opportunities deriving from climate change Improvement of risk mapping tools for land and urban planning Performance of predictive models on real case studies Reduction in the number of casualties related to extreme climate events Increase of the geographical extent of monitoring networks Increase of public availability of information and data Number of people involved in awareness-raising campaigns	Investments in research projects about the impacts of climate change and adaptation (Euro) (also considering different fund types) Number of climate change studies and projects funded (national and international) Number of studies assessing climate change vulnerability and risk (by sector and by region) Maps of risk and vulnerability developed for specific sectors and geographical areas Production of geographic information systems, smartphone applications and web applications for information collection Number of early warning systems updated to take account of climate change and adaptation Number of monitoring systems implemented Number of communication and dissemination actions/events per year Dissemination material about climate change Number of public administrators who have received adaptation training
Governance	Reduction of damages (economic and environmental) related to drought events Decreased transportation delays (frequency, timing) due to extreme weather conditions Increase in the number of buildings protected from river and marine inundation Reduction in the number of hospitalizations caused by extreme temperature and weather events Reduction of the frequency of hydrogeological events Reduction in the number of new structures built in vulnerable areas Water resources savings Reduction of inhabited areas in the defenseless coastal zone Reduction of areas prone to hydrogeological risk Reduction of infrastructures damage (Euro) thanks to preventative works	Number of sectorial programs and plans that take into account climate change adaptation Number of cities with a mobility and urban traffic plan that considers climate change impacts and adaptation Number of cities with sustainable urban drainage solutions (SUDS) Built area in the proximity of marine inundation-prone zones Number of properties damaged by river or marine inundation Status of implementation of regulations for new buildings in terms of climate proofing National/international funds for adaptation to climate change Expenditure in terms of economic incentives to support adaptation actions (Euro) Investments for the purpose of public and private adaptation (Euro) Investments in planning and emergencies management (Euro)

Organizational and participatory processes	Continuity of crop productivity Increase in the number of actors/organizations involved in intentional support networks relevant for adaptation Increase of budget for regional and national coverage of inspections Number of involved actors, covered sectors and represented areas	Number of technical reports, publications and scientific communications relevant to civil protection organization Number of typical products recognized by labelling by official production consortia in response to climate change Number of networks developed to support the decision-making and political process Sectors represented in the networks Number of meetings held (for each network) Number of guidelines produced
Adaptation and improvement of plants and infrastructures	Improvement of the ecological status of water Reduction of leakage from water transmission and distribution networks Electrical production of thermoelectric plants equipped new-generation cooling systems Increase of water resource availability	Conversion of irrigation systems (ha) Number of interruptions in supply systems of (a) water or (b) energy due to extreme weather events Extension of urban expansion areas taking into account risk maps Number of innovative machines for the development of a sustainable and efficient management of shared agriculture Net increase in reservoir capacity
Solutions based on ecosystem services	Reduction of damages (economic or environmental) due to hydrogeological instability Improvement of the ecological status of water Coastline stability Reduction of damage (economic or environmental) due to flooding Reduction (absolute (m) and relative (%)) of coastal erosion Stability of riparian structures after flood events Increase (absolute (km) and relative (%)) of road drainage systems Increase (absolute (m²) and relative (%)) of the public green area	Mapping of flood prediction and early warning systems Enlargement (ha) of reconverted coastal wetlands Conservation status of coastal habitats and species Length (km) of roads with new drainage systems Surface (ha) redeveloped as urban green Number of interventions in the field of green infrastructure

The Plan also seems having a weak ambition in introducing climate change adaptation within river basin planning: for instance, removal of "buried streams" existing in many Italian cities is not sufficiently stressed. The Plan provides only few indirect suggestions on how to introduce climate change in the design and construction of new infrastructures. Another issue is related to the fact that there is no attempt for coordinating the different territorial levels of climate change adaptation planning, in a hierarchical way. In other words, nothing ensures the consistency of regional and local adaptation Plans with the National one.

In spite of these limitations, the Plan traces a comprehensive view of climate change risk in Italy, providing a detailed set of measures to face the related challenges, and, at the end, it can be considered one of the most advanced National adaptation plans in Europe.

14.5   Conclusions

Mainly based on guidelines and directives from the European Community, Italy has produced several documents that assess in detail present and future impacts of climate change and propose adaptation actions to contrast them. The most important document, the National Plan of Adaptation to Climate Change, offers a comprehensive vision of the issue and provides a series of guidelines for correct water resources management in the presence of climate change. From an impact point of view, the assessments based on simulations from regional climate models mostly agree in indicating a future increase in annual average temperatures and a decrease in total annual rainfall, with some tendency towards an exacerbation of extreme phenomena such as heat waves and short and intense rain events. Apart from the new challenges posed by these climate modifications, the Italian peninsula has been subject to droughts, floods and landslides, since ancient dates, mainly for its natural proneness due to geological, geomorphological and climatic characteristics. This proneness has been aggravated by poorly controlled urban development in the years of economic boom. When they will be implemented, the indications and actions provided by the Plan will solve many water resources management issues. In most cases, these issues are just past problems that will be potentially exacerbated by future impacts of climate change. Nevertheless, the path for implementation of the actions is still long, also due to the limited financial resources that the government addresses to adaptation. It is therefore desirable that the challenges posed by climate change will give a decisive impulse for investments in defense from hydraulic and hydrogeological risks, which have always been insufficient in Italy: it is then that the "threats" of climate change may turn into "opportunities".

References

Castellari S, Venturini S, Ballarin Denti A, Bigano A, Bindi M, Bosello F, Carrera L, Chiriacò MV, Danovaro R, Desiato F, Filpa A, Gatto M, Gaudioso D, Giovanardi O, Giupponi C, Gualdi S, Guzzetti F, Lapi M, Luise A, Marino G, Mysiak J, Montanari A, Ricchiuti A, Rudari R, Sabbioni C, Sciortino M, Sinisi L, Valentini, Viaroli P, Vurro M, Zavatarelli M (eds) (2014) Rapporto sullo stato delle conoscenze scientifiche su impatti, vulnerabilità ed adattamento ai cambiamenti climatici in Italia [Report on the status of scientific knowledge of climate change impacts, vulnerability and adaptation in Italy], Ministry of the Environment and the Protection of the Land and the Sea](in Italian), Rome

CMCC (2017) Piano nazionale di adattamento ai cambiamenti climatici [National plan for climate change adaptation], Ministry of the Environment and the Protection of the Land and the Sea, Rome. http://www.minambiente.it/sites/default/files/archivio_immagini/adattamenti_climatici/documento_pnacc_luglio_2017.pdf (in Italian)

Haylock MR, Hofstra H, Klein Tank AMG, Klok EJ, Jones PD, New M (2008) A European daily high-resolution gridded dataset of surface temperature and precipitation. J Geophys Res (Atmospheres) 113:D20119. https://doi.org/10.1029/2008JD10201

ISPRA (2015) Il clima futuro in Italia: analisi delle proiezioni dei modelli climatici regionali [The future climate in Italy: analysis of the regional climate model projections], Environmental Status Pub.n.58/2015, ISBN 978-88-448-0723-8. http://www.isprambiente.gov.it/it/pubblicazioni/stato-dellambiente/il-clima-futuro-in-italia-analisi-delle-proiezioni-dei-modelli-regionali (in Italian)

ISPRA (2018) Gli indicatori del clima in Italia nel 2017 – Anno XIII [Climate indicators in Italy for the 2017 – Year XIII], Environment Status Pub. n. 80/2018, ISBN: 978-88-448-0904-1.   http://www.isprambiente.gov.it/it/pubblicazioni/stato-dellambiente/gli-indicatori-del-clima-in-italia-nel-2017

Mascaro G, Viola F, Deidda R (2018) Evaluation of precipitation from EURO-CORDEX regional climate simulations in a smallscale Mediterranean site. J Geophys Res Atmos 123(3):1604–1625

MATTM (2014) Strategia nazionale per l'adattamento ai cambiamenti climatici [National strategy for climate change adaptation], Ministry of the Environment and the Protection of the Land and the Sea, Rome. http://www.minambiente.it/sites/default/files/archivio/allegati/clima/documento_SNAC.pdf (in Italian)

Menduni G, Brath A, Iannarelli E, Zarra C (2017) Linee guida per le attività di programmazionee progettazione degli interventi per il contrasto del rischio idrogeologico [Guidelines to plan and design the measures to contrast hydro-geological risk], ItaliaSicura – Presidency of the Council of Ministers, Italian Hydro-technical Association, ISBN 9788894874006 (in Italian)

Peres DJ, Cancelliere A (2016) Environmental flow assessment based on different metrics of hydrological alteration. Water Resour Manag 30:5799. https://doi.org/10.1007/s11269-016-1394-7

Peres DJ, Caruso MF, Cancelliere A (2017) Assessment of climate-change impacts on precipitation based on selected RCM projections. Eur Water 59:9–15

Ruti PM et al (2016) MED-CORDEX initiative for Mediterranean climate studies. Bull Am Meteorol Soc 97(7):1187–1208

Scoccimarro E, Gualdi S, Bellucci A, Sanna A, Fogli PG, Manzini E, Vichi M, Oddo P, Navarra A (2011) Effects of tropical cyclones on Ocean heat transport in a high resolution coupled general circulation model. J Clim 24:4368–4384

Part V
Conclusions

Chapter 15
The Future of Water Management in Italy

Marcello Benedini and Giuseppe Rossi

Abstract Different paradigms have characterized the evolution of water resources management in Italy. The paradigm changes are described, focusing on the main strengths and weaknesses of the current status of water infrastructure, legislation framework and governance of water services. Technical and management priorities are discussed, as key issues of an agenda for the future, relevant to protection and use of water resources, as well as to increase of resilience to hydrologic extremes. Priorities for the future of water policy take into account a large spectrum of actions, relevant to technical and institutional innovations, aiming at improving the expected living conditions of the country. The technological innovations are connected to the development of social and political issues in an improved ethical perspective which requests proper information and qualified education in coping with water resources problems.

15.1 Paradigm Evolution

Italy has a wide tradition in water uses, pollution control and flood risk mitigation. The governments after the country unification (1861) played an important role in the design and construction of water infrastructures necessary to provide municipal water supply, sewer systems and irrigation, essential drivers to modernize the country. The improvement of urban centres, the reclamation and the enhancement of agriculture, as well as the increase of hydropower plants (the "white coal") were essential elements of the economic development during the 1921–1940 period.

After the Second World War, the development of new hydropower schemes was one of the main objectives of the national policy aimed at assuring the energy necessary for recovering and improving the economic level of the country. At the same

M. Benedini (✉)
Italian Water Research Institute (Retired), Rome, Italy
e-mail: benedini.m@iol.it

G. Rossi
Department of Civil Engineering and Architecture, University of Catania, Catania, Italy

© Springer Nature Switzerland AG 2020   357
G. Rossi, M. Benedini (eds.), *Water Resources of Italy*, World Water Resources 5,
https://doi.org/10.1007/978-3-030-36460-1_15

time, the establishment of the Southern Italy Development Fund allowed the construction of large multipurpose water systems for improving the conditions of the southern regions, including the islands of Sicily and Sardinia.

This long stage of the water commitment, named "hydraulic mission" by Allan (2003), coincides with the "industrial modernity" and is characterized by the adoption of an engineering and economic approach to water resources exploitation (as discussed in Chap. 2). With the support of universities and research institutions, renowned designers and advisors such as engineers, geologists and experts in soil mechanics and hygiene contributed to the outstanding activities of many firms, which worked also in several foreign countries, especially for the construction of dams and for the development of hydropower schemes.

The paradigm changed when new needs arose in order to reduce pollution in rivers, lakes, aquifers and coastal water and to face water-related disasters (flooding and landslides) as well as severe droughts and beginning of desertification. The increasing awareness for environment, supported by a sensitive and strong public opinion, founded the basic guidelines in the directives about the quality of water bodies issued by the European Union (EU 2000). The occurrence of frequent dramatic flooding, such as that of Florence in 1966, boosted politic and scientific institutions to pay growing attention to the identification and mitigation of hydraulic and geological risks.

More recently (EC 2015), a comprehensive approach, aimed at achieving an integrated, sustainable and equitable management, is becoming the new paradigm for the water and soil management. Last orientations to apply successfully this comprehensive approach claim not only the triple consolidated dimension of economic, environmental and social sustainability but also the key role of the governance dimension. Today, this is perceived as an essential basic condition to maximize welfare in a socially equitable way, without compromising the natural systems.

15.2   Strengths of Italian Water Development

As the previous chapters discuss, the water resources management in Italy shows several strengths. They are recalled in this paragraph.

(a) *Almost all water resources are inland resources.*

A very propitious geographic feature of Italy is that almost all water resources are entirely located in the national territory, while the limited number of catchments shared with neighbouring countries has been the object of agreements for common utilization, mostly in the case of hydropower schemes or conjunctive management of pre-alpine lakes.

(b) *A large set of water infrastructures exists in all the Italian regions.*

A rich heritage is in the municipal aqueducts, in some large water supply schemes, as well in the networks of channels and pipelines to meet irrigations demand. Several dams in rivers can increase the surface water availability for various purposes, and innumerable wells assure the exploitation of groundwater.

(c) *The modernization process of municipal water services is in progress.*

The establishment of municipal services proposed by the Galli Law (36/1994), aiming at unifying water supply, sewerage and wastewater treatment, as well as at enlarging the size of the territorial unit to include several municipalities, has produced significant improvement in the quality of urban services throughout the country, with only a few exceptions in some regions.

(d) *The water supply to agricultural and industrial users is fair.*

A large part of the irrigation systems operate in "consortium", an administrative institution that contributes to achieve minor water consumption and reduced costs. Likewise, the aggregated complexes of industrial plants have developed several common opportunities for suitable water supply and disposal of wastewater and sludge.

(e) *The innovations in electric power generation plants are successful.*

Thanks to the worldwide acknowledged pioneer expertise, most Italian hydropower schemes are transformed in order to play a new role after the growth of the renewable energy sources, as well as to comply with the constrains imposed by environment preservation. Significant improvement concerns the adoption of pumped storage generation plants, besides more efficient machinery. Several mini- and microgeneration plants contribute to meet the local energy demand. The largest thermoelectric schemes have been located preferably in coastal areas, where they can use seawater for cooling, thus saving precious freshwater for other important purposes.

(f) *An improved cooperation between productive and research sectors is showing promising results.*

The increased ties between the productive sectors and the scientific community contribute to tackle and solve the most complex water problems in an efficient way.

(g) *The European Directives have a positive impact on water planning.*

The European Directives 2000/60/EC and 2007/60/EC have fostered the planning process in the whole country for improving the water quality in natural bodies and mitigating the flood risk. In several cases, the organizational delays in the start-up of the new interventions are disappearing, and the mandatory deadlines to comply with the European Directives are fulfilled.

(h) *The growing role of national organisms improves the performance of subnational bodies.*

Although many commitments in water field have been transferred to the regions, in accordance to a general reform of Italian administration, some new organisms established at central level help to give homogeneity to the working rules of several subnational authorities, regional agencies and local companies for service management. A significant example is the Institute for Environmental Protection and Research (ISPRA), which coordinates the monitoring of hydrometeorological data and defines the criteria for surveying the status of rivers, lakes and aquifers, now committed to the District Authorities. Similarly, the Authority for Regulation of Energy, Networks and Environment (ARERA) provides homogeneous guidelines to the Integrated Water Service for computation of tariffs and for monitoring the quality of service for customer satisfaction.

(i) *Emphasis on sustainable criteria and social goals.*

Notwithstanding some difficulties, the sustainability paradigm begins to pass from a mere principle declaration to the practical application. A growing number of water management companies are now publishing the sustainability report as an annex to their economic budget, in accordance to the European Directive 95/2014. Ethical principles are explicitly declared in Law 221/2015, which draws the access to drinking water up to customers in uneasy economic conditions. The adoption of ethical principles by several utility companies demonstrates the growing awareness on these requirements.

15.3   Weaknesses of the Italian Experience

Along with the strengths that have been mentioned above, several weaknesses characterize the water sector in Italy. The main cases are discussed below.

a') *Sectorial approach to the water issues.*

The sectorial approach still affects the development of some water infrastructures and the implementation of the legislation and institutional commitments for water system management. A coordinated planning of water resources and soil defense is delayed, in spite of the comprehensive approach declared in many acts. In many cases, the competition among various ministries at central level and different authorities at regional level contributes to increase delays and ineffectiveness of the actions for achieving the prospected reforms.

b') *Controversies between the different levels of governance.*

Some persisting controversies among the institutions at national, regional and local levels have an impact on the new administration lines on water governance. Very often, this is the case of municipal water services, the protection environment and the defense from water-related disasters. In spite of the reform of municipal water service, based on a territory larger than the municipal area, several municipalities continue to operate water supply, sewerage and wastewater treatment in an autonomous way.

c') *Shortcomings in the maintenance of hydraulic facilities.*

A persistent lack of maintenance hinders the efficient running of several existing plants. Even though leakages and wastes exist in all water uses, municipal supply mains and distribution networks show big losses in the relevant water systems. For the whole country, a loss of 2.7 km^3/year is estimated, equivalent to the 40% of the total withdrawal in 2015. A consequence of this is also an annual energy waste for the induced increase in running of power plants.

d') *Slow advances in the achievement of the objectives of the water-related planning tool.*

In several cases, the achievement of the water quality status recommended by the 2000/60/EC Directive is delayed, as well as that of flood risk reduction recommended by the 2007/60/EC Directive. Particularly inadequate is the control of groundwater quality. Several aquifers are overexploited, due to the lack of regular licenses and abstraction monitoring. Many urban centres, transportation structures and cultivated lands are still subject to a high risk of flooding.

e') *Emergency actions continue to prevail on prevention measures for coping with the hydrological extremes.*

Despite the priority declared for prevention approach in coping with floods and droughts, political and technical actions seem to favour an intervention during and after the disasters for people security. Therefore, interventions are mostly oriented to recovering the damaged structures, very often with unexpected costs. Limited financial resources contribute to an inadequate approach for increasing the resilience to flood risk and water shortage impact.

f') *Scarce preparation to the foreseeable impact of climate changes.*

Several proposals included in the National Plan for Climate Change Adaptation (2017), prepared by Ministry of Environment in accordance to European Guidelines, and describing a very detailed list of actions in several sectors (agriculture, health, energy, etc.), have a very limited implementation in the daily activity of many sectors of water resources management.

15.4   Key Issues of an Agenda for the Future

An attempt to define some essential points of an agenda for the future of Italian water resources requires a critical analysis of some strategies recently proposed at international level. In order to solve global and local water problems, the most recent guidelines address the challenge of a sustainable and equitable water management derived from the "2030 Agenda Sustainable Development Goals" of the United Nations (UN 2015; Giovannini 2018). The large spectrum of suggested actions should be implemented in order to achieve technical, institutional and social innovations, necessary for water management. Similar actions can also increase the cooperation between the institutions for water governance and improve the participation of responsible people.

From a technical and management point of view, an agenda for the future of Italian water resources should consider the following priorities, many of which have been claimed already two decades ago (IRSA 1997, 1999):

- Better knowledge of water availability, water demand and ecosystem requirements, to achieve through improved monitoring systems and more efficient surveys on water services performance
- Adoption of suitable pricing systems and adequate economic incentives for the users, necessary to enhance effective water saving
- Improved and up-to-date tools for the measurement of the meteorological, hydrologic and hydraulic data required for the management and protection of water resources
- Suitable criteria and tools for the evaluation of sediment transport in water bodies, including the siltation of reservoirs
- Efficient and reliable models for the hydrological events, the hydraulic aspects in the water body and the rational use of water, adopting the advanced tools of the information technology
- Improved procedures for seawater and brackish water desalination, in order to increase the resource for domestic and industrial use in the coastal areas
- Advanced technologies for treating and reusing wastewater and sludge disposal
- New technologies for the performance of water services, to reduce losses and waste in the withdrawal, conveyance and distribution systems
- Availability of drinking water 24 h a day and 7 days a week for all citizens, also in order to reduce the use of expensive bottled water
- Planned effort for a continuous maintenance of the water infrastructures, especially for the efficiency of the various uses and wastewater treatment
- Adaptation of the storage facilities to the increased variability of runoff and occurrence of drought events due to climate change
- More effective approach for reducing the flooding risk, through actions aimed at prevention, by imposing hydraulic invariance and respect of hydraulic risk constraints in the general land planning
- Water quality indicators in watercourses, able to better interpret the ecosystem health, taking into account watershed stress, physical habitat, chemical exposure and biological response
- Ecologically based classifications of physical habitat of rivers and the definition of their occurrence, for a better and more efficient river flow evaluation
- Groundwater protection from pollutant discharges and saline intrusion in coastal areas
- Wise application of the sustainability paradigm, avoiding the use of surface water above the rate of natural renewability and avoiding the overexploitation of groundwater
- Development of advanced information systems able to provide the amount of data necessary for the solution of the complex water problems
- Adoption of robust ecological, economic and social indicators, able to improve decision-making process when a compromise between human and environmental requirements is necessary

An agenda for the future in water resources requests also a scenario based on some priorities in water policy. In the last decades, the unlimited growth, the capitalism as the only model for economy and society and the human being as consumer have been predominant (Soderbaum 2009). An expected scenario will deal also with ecology, human security needs and welfare status.

A scenario based on "business as usual" will include more compulsory concepts, like the increasing population, urbanization, economic growth and energy generation, under the new risk of climate change and the constraint of an improved environment protection. A scenario based on "social-ethical change" will take into account the limits of an uncontrolled growth, the shift from one-dimensional monetary thinking to multidimensional impacts and the evolution of environmental and social attitudes in the political concern.

Whichever will be the future scenario, an improved management of Italian water resources should include at least the following list of policy priorities:

- An updated legislation to simplify the water-related plans
- Better rules for regulating withdrawal from surface water bodies, which may include revising permits in order to assure ecological flow, and to avoid overexploitation of groundwater
- A reform of the responsible institutions, necessary for improving the coordination among the different bodies of the public administration, at central and regional level
- A better governance of municipal water services
- Transparent information to the water services customers, also for activating a responsible public participation
- Implementation of the principle of hydraulic and hydrological invariance in the urban and land planning, in order to avoid the increase of peak flow and flood volume due to the growth of impervious areas
- A better coordination in the warning and alert/alarm procedures of flood risk, particularly between municipalities and civil protection systems
- A plan for mitigating the risk of drought and water shortage at each water supply level
- A primary effort for maintaining the existing hydraulic structures and for realizing the new works needed for the future management of water resources

The preliminary conditions for the specific measures to adopt include two conceptual aspects. The first is a multidisciplinary approach to face the water issues complexity that is expected to increase in the future. The second aspect concerns the utmost importance of recognizing an ethical responsibility for water (Selborne 2000; Falkenmark and Folke 2002; Rossi Paradiso 2011).

Unfortunately, today the most common approaches to water problems are far from any ethical perspective. Often a "cynic view" is prevalent, based on the belief that the true driving forces behind water governance and management are always an economic and political power, with the conviction that any attempt at change could not be successful. Another perspective focuses the lack of equity in water allocation as a pretext to fight all the economic systems. Finally, a persistent "utopian idealism", stressing the importance of the principle declarations and "right charts", and considering them as a "panacea", is not sufficient for guiding the hard and long process of achieving better water resources protection and safety from water-related disasters (Rossi 2015).

15.5   Concluding Remarks

The analysis of paradigm change in water resources exploitation and the attempt to identify strengths and weaknesses of current situation confirm that the increasing complexity of water resources management requires enlarged perspectives in the definition of an agenda for the future. In terms of water problems to face, an integrated water management, including issues of water supply, water quality and water-related risk, has to take into account also the links with food security and energy development. Besides the consolidated approach, based on engineering, economic and environmental sciences, a multidisciplinary approach is required, including social, political and ethical dimensions.

Integrated, sustainable and equitable water management requests a planning activity formulated on thorough evaluations of several assessments connected to the water, which is conceived a natural resource to be better recognized, protected and used. Such an activity requires collecting reliable data and their use by means of the best available expertise.

A comprehensive view of the problems of water quantity, water quality and water excess at river basin scale is the basic premise of planning tools. A wise balance between human needs and ecosystem protection is fundamental for the design of the water infrastructures, while an adequate regulation of institutional aspects is essential for the coordination of different levels of government with an improved public participation in decision-making process. The utmost goal is an efficient and effective operation of water systems with satisfaction of users and environment protection.

A great ethical challenge is expected for all the persons involved in water concerns, at any level of responsibility, in scientific and practical sectors.

This encourages looking to the future with the hope that the water resources will play always their fundamental role for the development of Italy and its people, with the opportunity of transferring technological, management and cultural innovations to other countries in order to contribute to the economic, social and vital development.

It is worthwhile to remember how water has been the main reason of life for centuries, starting from the ancient Italian civilization that gave it divine attribute. The conclusion of this book can be therefore a sentence borrowed from the "Canticle of the Creatures", historically one of the first documents written in Italian language, due to Saint Francis of Assisi. At the beginning of the thirteenth century, expressing his gratitude to God, he wrote a sentence that in modern English sounds: "Praised be you, my Lord, for Sister Water, so very useful, and humble and precious, and chaste".

References

Allan JA (2003) Virtual water: the water, food and trade nexus. Water Int 28(1):4–11

EC (2015) European Commission. The role of science, technology and innovation policies to foster the implementation of the sustainable development goals, Report of the Expert Group "Follow-up to Rio+20, notably the SDGs". DG RTD

EU (2000) European Union. Towards a sustainable strategic management of water resources: evaluation of present policies and orientation for the future. European Commission, Study 31, ISBN 92, Luxembourg

Falkenmark M, Folke C (2002) The ethics of socio-ecohydrological catchment management towards hydrosolidarity. Hydrol Earth Syst Sci 6(1):1–9

Giovannini E (2018) L'utopia sostenibile [Sustainable utopia] (in Italian). Laterza, Bari-Roma

IRSA (1997) Water Research Institute. Long range study on water supply and demand in Europe. Level A: studies at country level. Italy. ICWS, Final Report

IRSA (1999) Water Research Institute, Un futuro per l'acqua in Italia. [A Future for Water in Italy] (in Italian). Quaderni IRSA n.109, Rome, Italy p 235

Rossi G (2015) Achieving ethical responsibilities in water management: a challenge. Agric Water Manag 147:96–102

Rossi Paradiso G (2011) Responsabilità etica nella gestione delle acque [Ethical responsibility in water management] (in Italian). L'Acqua 3:60–78

Selborne L (2000) The ethics of freshwater use: a survey. World Commis. on the ethics of scientific knowledge and technology. UNESCO, Paris

Soderbaum P (2009) Science, ideology and sustainable development: an actor oriented approach. In: Biswas A, Tortajada C, Izquierdo R (eds) Water Management in 2020 and Beyond. Springer, Berlin, pp 125–136

UN (2015) (United Nations) Transforming our world: the 2030 agenda for sustainable development. http://www.org/ga/search/view_doc.asp?symbol=A/RES/70/1&Lang=E